# PREFACE

This report is a chronological compilation of narrative summaries of news reports and government documents highlighting significant events and developments in United States and foreign aeronautics and astronautics. It covers the years 1996 through 2000. These summaries provide a day-by-day recounting of major activities, such as administrative developments, awards, launches, scientific discoveries, corporate and government research results, and other events in countries with aeronautics and astronautics programs. Researchers used the archives and files housed in the NASA History Division, as well as reports and databases on the NASA Web site.

## TABLE OF CONTENTS

PREFACE ............................................................................................................................... i

JANUARY 1996 ................................................................................................................... 1
FEBRUARY 1996 ................................................................................................................. 5
MARCH 1996 ........................................................................................................................ 9
APRIL 1996 ......................................................................................................................... 13
MAY 1996 ........................................................................................................................... 17
JUNE 1996 ........................................................................................................................... 22
JULY 1996 ........................................................................................................................... 25
AUGUST 1996 .................................................................................................................... 29
SEPTEMBER 1996 .............................................................................................................. 33
OCTOBER 1996 .................................................................................................................. 37
NOVEMBER 1996 .............................................................................................................. 41
DECEMBER 1996 ............................................................................................................... 45

JANUARY 1997 .................................................................................................................. 49
FEBRUARY 1997 ............................................................................................................... 54
MARCH 1997 ...................................................................................................................... 59
APRIL 1997 ......................................................................................................................... 64
MAY 1997 ........................................................................................................................... 69
JUNE 1997 ........................................................................................................................... 74
JULY 1997 ........................................................................................................................... 79
AUGUST 1997 .................................................................................................................... 84
SEPTEMBER 1997 .............................................................................................................. 89
OCTOBER 1997 .................................................................................................................. 93
NOVEMBER 1997 .............................................................................................................. 99
DECEMBER 1997 ............................................................................................................. 103

JANUARY 1998 ................................................................................................................ 107
FEBRUARY 1998 ............................................................................................................. 116
MARCH 1998 .................................................................................................................... 121
APRIL 1998 ....................................................................................................................... 127
MAY 1998 ......................................................................................................................... 133
JUNE 1998 ......................................................................................................................... 140
JULY 1998 ......................................................................................................................... 149
AUGUST 1998 .................................................................................................................. 157
SEPTEMBER 1998 ............................................................................................................ 161
OCTOBER 1998 ................................................................................................................ 166
NOVEMBER 1998 ............................................................................................................ 171
DECEMBER 1998 ............................................................................................................. 174

JANUARY 1999 ................................................................................................................ 179
FEBRUARY 1999 ............................................................................................................. 183

| | |
|---|---|
| MARCH 1999 | 187 |
| APRIL 1999 | 191 |
| MAY 1999 | 196 |
| JUNE 1999 | 201 |
| JULY 1999 | 207 |
| AUGUST 1999 | 213 |
| SEPTEMBER 1999 | 219 |
| OCTOBER 1999 | 224 |
| NOVEMBER 1999 | 232 |
| DECEMBER 1999 | 239 |
| JANUARY 2000 | 246 |
| FEBRUARY 2000 | 250 |
| MARCH 2000 | 254 |
| APRIL 2000 | 257 |
| MAY 2000 | 261 |
| JUNE 2000 | 266 |
| JULY 2000 | 271 |
| AUGUST 2000 | 275 |
| SEPTEMBER 2000 | 282 |
| OCTOBER 2000 | 288 |
| NOVEMBER 2000 | 297 |
| DECEMBER 2000 | 304 |
| APPENDIX A: TABLE OF ABBREVIATIONS | 311 |
| APPENDIX B: BIBLIOGRAPHY | 317 |

# JANUARY 1996

*8 January*

Furloughed NASA employees returned to work at the end of a federal government shutdown of 14 to 21 days. NASA had allowed some critical employees to return before others. The government shutdown, resulting from a budget stalemate between the Republican Congress and the Democratic White House, forced NASA to cease operations and to clear thousands of its employees from field centers across the country. Upon resuming its normal proceedings, NASA faced a backlog of work because of the disruption. Workers returning to Houston's Johnson Space Center (JSC) were especially anxious to catch up, because the Center had only three days to prepare for a scheduled Shuttle launch. A snowstorm delayed for a few more hours the return of workers to Marshall Space Flight Center (MSFC) in Huntsville, Alabama.[1]

*9 January*

A delegation from the U. S. Congress traveled to Russia to meet with Russian Space Agency officials. Representatives Jerry Lewis (R-CA) and F. James Sensenbrenner Jr. (R-WI), members of the subcommittee in charge of NASA's funding—the Subcommittee on Veterans Affairs, Housing and Urban Development, and Independent Agencies—led the contingent. Their goal was to keep intact the coalition of countries dedicated to making the International Space Station (ISS) a reality. Faced with Russian insistence that the planned space station connect with the Russian space station *Mir*, the U.S. lawmakers warned Russian officials not to pursue radical changes in the plans for the space station. Sensenbrenner noted that obstinacy on the part of the Russians might result in the coalition's building the ISS without Russian involvement.[2]

*12 January*

G. Porter Bridwell retired after nearly 40 years of service at MSFC in Huntsville, Alabama. Bridwell had served as Director of MSFC from 1994 until his retirement. He left the Center with an annual operating budget of US$2.5 billion and more than 3,000 employees. Bridwell had spent most of his career at MSFC, serving briefly as Acting Director of Stennis Space Center in 1987, and at NASA Headquarters from 1993 to 1994. Bridwell had also been a member of the space station redesign team. During his tenure, Bridwell had received NASA's Exceptional Service Medal, the Outstanding Leadership Medal, and the Exceptional Achievement Medal, attaining the rank of Meritorious Executive.[3] NASA announced that J. Wayne Littles would take over as the new Director of MSFC.[4]

Republican presidential candidate Patrick J. Buchanan announced that he would cancel a controversial political advertisement using images of the Space Shuttle *Challenger* accident. Responding to harsh criticism from New Hampshire residents, Buchanan removed the footage of *Challenger* from the television spot. Buchanan stated that the purpose of the advertisement, his first in the crucial primary state of New Hampshire, was to demonstrate his service to President Ronald Reagan during difficult times. However, in the home state of teacher-astronaut Christa McAuliffe, who had died in the accident, many residents felt that the campaign's use of the

---

[1] NASA, "NASA Headquarters Regroups After Longest Federal Shutdown," *HQ Bulletin*, 5 February 1996; John Makeio, "Federal Employees Face Backlog Here," *Houston Chronicle*, 9 January 1996.
[2] Ben Lannotta, "Congress to Russia: Drop Mir–Station Link Idea," *Space News*, 15–21 January 1996, 3, 20.
[3] Martin Burkey, "Bridwell Bids MSFC Farewell with Optimism," *Huntsville Times* (AL), 12 January 1996.
[4] NASA, "Littles Named Director of Marshall Space Flight Center," news release 96-3, 16 January 1996.

images exploited the event for political gain. The political controversy brought to the foreground the lingering public dismay over the lost *Challenger*, even as the 10-year anniversary of the tragedy neared.[5]

Shuttle *Endeavour* lifted off from Kennedy Space Center (KSC) in Cape Canaveral, Florida, beginning a mission focused primarily on plucking the Japanese Space Flying Unit from orbit for analysis. Commander Brian Duffy led a six-person crew, which included Japanese astronaut Koichi Wakata. *Endeavour* also carried a small satellite to deploy. KSC officials delayed *Endeavour*'s launch for 23 minutes because of a series of communications problems between the Shuttle crew and the flight-control team. *Endeavour*'s flight was the first of eight Shuttle missions planned by NASA for 1996.[6]

*16 January*
Lockheed Martin announced its intention to cut 200 jobs at its facility for the production of external fuel tanks for NASA's Space Shuttle. NASA's budget reductions had caused the layoffs, according to Lockheed. Lockheed Martin's facility in New Orleans employed more than 2,500 workers in 1996.[7]

NASA released time-lapse photographs taken by the Hubble Space Telescope (HST), making the images immediately available on the Internet. Based on these images, lauded as the deepest images of the universe ever taken, scientists raised their estimate of the number of galaxies in the universe from 10 to 50 billion; "Suddenly, Universe Gains 40 Billion More Galaxies," the *New York Times* reported. Using a narrow "keyhole" view and focusing the HST on a portion of the sky only the width of a dime, telescope operators probed deep into the universe. The images revealed the process of a star's death, as well as new information about how galaxies evolve. HST astronomer Howard E. Bond estimated that the Sun would die out in about 5 billion years.[8]

Space Shuttle *Endeavour*'s crew used the spacecraft's 50-foot (15-meter) robotic arm to snare NASA's Office of Aeronautics and Space Technology Flyer (OAST-Flyer) satellite. The OAST-Flyer, valued at US$10 million, weighed 2,600 pounds (1,200 kilograms).[9]

*18 January*
Astronauts aboard Space Shuttle *Endeavour* tested new NASA spacesuits in extreme cold. In preparation for work on the ISS, U.S. astronaut Winston E. Scott stood nearly motionless in space for 35 minutes, in temperatures nearing 100° below 0°F (-73.33°C), to test the suit's insulating capacity. In completing the exercise, Scott became the second African American man to walk in space. The newly improved suits, produced at a cost of US$10 million each, proved an

---

[5] Howard Kurtz, "Television Ad Backfires on Buchanan," *Washington Post*, 12 January 1996; Associated Press, "Buchanan Chastised for Using Challenger Image in TV Ad," 12 January 1996.
[6] Mark Carreau, "Shuttle Blasts Off in Pursuit of Satellite After Slight Delay," *Houston Chronicle*, 12 January 1996.
[7] Associated Press, "Lockheed Martin Cuts Shuttle Program Jobs," 16 January 1996.
[8] John Noble Wilford, "Suddenly Universe Gains 40 Billion More Galaxies," *New York Times*, 16 January 1996; Kathy Sawyer, "NASA Takes Portrait of Universe," *Washington Post*, 16 January 1996; NASA, "New Hubble Images Available," news release N96-2, 17 January 1996.
[9] Associated Press, "Shuttle's Astronauts Grab a 4-Ton Japanese Science Satellite from Its Orbit," 17 January 1996; Mark Carreau, "Shuttle Dodges Military Satellite on Way To Retrieve Japanese Craft," *Houston Chronicle*, 13 January 1996; Michael Cabbage, "Satellite Stowed," *Florida Today* (Brevard, FL), 14 January 1996.

effective barrier against the cold. The gloves and the boots of the suits contained heating elements. NASA made the thermal improvements to the spacesuits after the astronauts became cold during the February 1995 spacewalk, causing its early termination.[10]

At a conference of the American Astronomical Society in San Antonio, Texas, a team of U.S. astronomers announced the discovery of two new stars with the characteristics to sustain life within their solar systems. The astronomers identified the bodies, both visible to the naked eye, as star 70 Virginis in the constellation Virgo and a star orbiting 47 Ursae Majoris in the Big Dipper. Geoffrey Marcy and Paul Butler, researchers affiliated with the University of California at Berkeley and San Francisco State University, spent nine years monitoring 120 Sun-like stars in order to discover the two stars.[11]

*21 January*
Space Shuttle *Endeavour* landed safely after a successful nine-day mission. The Shuttle crew accomplished its primary task of capturing a Japanese satellite. The early-morning landing, in which the Shuttle returned to Earth under a shroud of darkness, was a rare occurrence.[12]

*22 January*
NASA released its analysis of the data from the Galileo probe mission to Jupiter, information that had caused scientists to rethink their theories about Jupiter's formation. The data suggested that Jupiter does not have the three-tiered cloud structure scientists had anticipated, and that the amount of helium present on Jupiter is half of scientists' projections. NASA Associate Administrator for Space Science Wesley T. Huntress Jr. touted the data collected as beyond NASA's highest expectations. The Galileo probe had been one of NASA's most challenging missions undertaken to obtain scientific data. According to NASA, the probe had made the most difficult planetary atmospheric entry ever attempted, surviving entry speeds of more than 100,000 miles (160,000 kilometers) per hour and temperatures twice as hot as the surface of the Sun.[13]

*23 January*
George W. S. Abbey became the seventh Director of JSC in Houston, Texas. Abbey had been Acting Director since Carolyn Hunter had stepped down from the post on 4 August 1995. According to NASA Administrator Daniel S. Goldin, Abbey's distinguished NASA career made him the right man for the job. Administrator Goldin announced, "George Abbey is uniquely qualified to lead the Johnson team into the future. Over the course of his eminent career with NASA, he has distinguished himself as an innovator and pioneer at all levels of Agency management." Abbey had begun working at NASA in 1967 after serving as a pilot in the U.S. Air Force. He had worked on the Apollo Spacecraft Program before becoming Director of Flight Operations. NASA had recognized Abbey's superior service by awarding him its Exceptional Service Medal and two Distinguished Service Medals.[14]

---

[10] Michael Cabbage, "Staying Warm in Cosmic Cold," *Florida Today* (Brevard, FL), 18 January 1996; *Jet Magazine*, "Navy Captain Winston Scott Becomes Second Black To Walk in Space," 5 February 1996, 24.
[11] Kathy Sawyer, "Two Planets Discovered that Might Sustain Life," *Washington Post*, 18 January 1996.
[12] Associated Press, "Shuttle Crew Lands After Success with 2 Satellites, 2 Space Walks," 21 January 1996.
[13] NASA, "Galileo Probe Suggests Planetary Science Reappraisal," news release 96-10, 22 January 1996.
[14] NASA, "Abbey Named Director of Johnson Space Center," news release 96-11, 23 January 1996.

*28 January*
NASA commemorated the 10-year anniversary of the *Challenger* tragedy. In advance of the anniversary, NASA Administrator Daniel S. Goldin reaffirmed NASA's dual commitments to exploring the frontiers of the universe and to maintaining the highest safety standard possible. According to Goldin, "the best way to honor the memories of the crew of *Challenger*, and of all the men and women, who have given their lives to explore the frontiers of air and space, is to continue their bold tradition of exploration and innovation." In the decade following the accident, NASA had instituted multiple safety changes in the Shuttle program, including the nine changes in the program's structure recommended by the Presidential Commission on the Space Shuttle Challenger Accident (Rogers Commission). By the end of January 1996, NASA had launched 49 Shuttle missions since the *Challenger* accident and had successfully deployed the HST. Memorial observances at KSC and JSC took place at 11:39 a.m. (EST), the exact time that *Challenger* exploded.[15]

*30 January*
NASA and the Russian Space Agency announced that the United States and Russia had reached an agreement to extend the Shuttle–*Mir* collaboration until 1998, with both countries reaffirming their commitments to building the ISS. At the time of the agreement, Russia was continuing to negotiate with the international space community to maintain the Russian station *Mir* as a long-term portal to the ISS. Through additional Shuttle flights to the *Mir* outpost, the United States planned to deliver thousands of pounds of materials to space, which otherwise would have required Russian launches. The United States offered Russia this significant assistance in the hope that the cash-strapped country would be able to keep its commitments to the ISS project. Vice President Albert A. Gore Jr. and Russian Prime Minister Viktor S. Chernomyrdin announced concurrently that U.S. astronaut William M. Shepherd and Russian cosmonaut Sergei K. Krikalev would be the first crew members to live aboard the ISS.[16]

NASA announced completion of the exterior of the U.S. modules constructed to house the astronauts aboard the ISS, as well as more than 80,000 pounds (36,000 kilograms) of flight hardware for the station. With the completion of the modules, the United States successfully attained the first benchmark in the long and expensive process to complete its part of the ISS. Boeing workers had completed the project at MSFC. NASA planned to launch the first node of the ISS in December 1997.[17]

*31 January*
NASA announced that, in collaboration with industry leaders and university scholars, it had developed an instrument that generates the world's most intense source of commercial x-rays, more than 100 times stronger than conventional x-rays. "This new optical instrument provides

---

[15] NASA, "Space Shuttle Program Changes and Accomplishments Since 1986," news release, January 1996; NASA, "Administrator Goldin Issues Statement on Challenger Observance," news release 96-2, 16 January 1996.
[16] NASA, "NASA and RSA Agree To Extend Shuttle–*Mir* Activities," news release 96-18, 30 January 1996; Warren E. Leary, "U.S. To Help Defray Russia's Expenses for the Space Station," *New York Times*, 1 February 1996.
[17] NASA, "Exterior of US Space Station Modules Completed; Flight Hardware on Track for Launch in 1997," news release 96-17, 30 January 1996; NASA, "United States' Space Station Modules Are Right on Track," *HQ Bulletin*, 20 February 1996, 3; The Boeing Company, "Exterior of U.S. Space Station Modules Completed Flight Hardware on Track; First Launch in 1997," news release, 5 February 1996.

something never before possible; a capability to control the direction for x-ray beams," said Walter M. Gibson, professor at State University of New York at Albany. Using capillary optics, the instrument could focus more precisely on a target in a research or medical situation. Daniel C. Carter, Director of MSFC, postulated that the new instrument would have numerous medical and commercial applications, including improvements in drug research, medical imaging, and forensic science.[18]

## FEBRUARY 1996

*1 February*
Astronomers at California Institute of Technology announced the discovery of what they believed to be the most distant galaxy ever glimpsed from the Earth. The team of scientists, including Thomas A. Barlow, Limin Liu, Wallace L. W. Sargent, and Donna S. Womble, had stumbled across the unnamed galaxy while studying a quasar silhouetted by the galaxy's light. The team used the W. M. Keck Observatory atop the Mauna Kea volcano in Hawaii. In comparison to other documented galaxies, their find was a relatively young galaxy—probably formed less than 1 billion years after the Big Bang. The galaxy resides near the constellation Virgo. Ironically, since the galaxy lies 14 billion light-years away from Earth, the scientists had no way of knowing whether their discovery still actually existed. The scientists hoped that their discovery would offer new understanding of the period when stars began to congregate into galaxies.[19]

*2 February*
Space Shuttle Program Director Bryan D. O'Connor announced his resignation from NASA, effective at the end of February 1996. O'Connor stated that the transition in Shuttle management had presented him with an opportunity to leave NASA without causing undue disruption. In the months before his resignation, O'Connor had voiced some concerns about NASA's planned transfer of responsibility for Shuttle flight operations to United Space Alliance, a private contractor formed by Lockheed Martin and Rockwell International. In announcing his departure, O'Connor praised NASA workers and astronauts for their exemplary service. After the 1986 *Challenger* accident, O'Connor had played an important role in restoring the Shuttle program and restructuring NASA. O'Connor, a former U.S. Marine Corps pilot and astronaut, had directed a major overhaul of the space station program, taking over as Director of the Space Shuttle Program in 1994.[20]

*6 February*
NASA announced its plan to shift authority over the Shuttle program from NASA Headquarters in Washington, DC, to Johnson Space Center (JSC) in Houston, Texas. JSC also would take control over construction of the International Space Station. The shift was one of NASA Administrator Daniel S. Goldin's efforts to restructure NASA and remove layers of bureaucracy.

---

[18] NASA, "NASA Helps Invent Revolutionary X-ray Instrument," news release 96-19, 31 January 1996; NASA, "NASA Helps Invent Revolutionary X-ray Device," *HQ Bulletin*, 4 March 1996, 3.
[19] California Institute of Technology, "Astronomers Discover the Most Distant Galaxy," news release, 31 January 1996; Reuters, "Scientists Discover Farthest Galaxy Yet," 1 February 1996; John Noble Wilford, "New Galaxy May Shed Light on Universe," *New York Times*, 1 February 1996.
[20] NASA, "O'Connor To Leave NASA," news release 96-23, 2 February 1996; Kathy Sawyer, "NASA Space Shuttle Director Resigns," *Washington Post*, 3 February 1996.

Goldin's mission centered on controlling costs and improving efficiency. With the change in organization, key managers at JSC and other regional centers such as Kennedy Space Center (KSC) gained new authority to act without consulting Goldin or other officials at NASA Headquarters in Washington, DC. The change caused some observers to comment that NASA was shifting back to the less strictly hierarchical model of operation that had characterized it before the *Challenger* accident. In response to the *Challenger* tragedy and the ensuing *Report of the Presidential Commission on the Space Shuttle Challenger Accident* (Rogers Commission Report), NASA had centralized some of its operations, to avoid the poor communication between regional centers that had contributed to the tragedy.[21]

*12 February*
Japan's National Space Development Agency (NASDA) experienced a setback when one of its prototypes for a small, robotic spacecraft landed in the ocean and sank. The Japanese officials had anticipated the ocean landing immediately after the launch of the Hyflex shuttle, the *Hope-X*, from an island in southern Japan. In spite of the loss of its shuttle, NASDA did not consider the exercise an entire failure. The *Hope-X* had launched successfully, separated from its rocket at a height of 70 miles (113 kilometers), and then returned to Earth 19 minutes later, as planned. The problem occurred when the craft splashed into the ocean, and a rope connecting the 1-ton (900-kilogram or 0.9-tonne) shuttle to its flotation device broke, causing it to sink. NASDA reported that it had intended to collect data during the flight to test the shuttle's fitness for reentry, but had been unable to procure much of the information needed. Japan had manufactured the US$37 million shuttle domestically, planning the exercise as a part of its effort to bolster its fledgling space program.[22]

NASA announced the selection of eight proposals for its newly inaugurated Advanced Concepts Research Projects (ACRP) program. NASA had received more than 100 proposals for the program, founded in September 1995 to identify and support new ideas and technologies that might eventually improve the U.S. space program. The ACRP program, allowing up to US$250,000 in support for each selected proposal, included proposals covering a wide spectrum of technologies and fields, such as: "fusion-based space propulsion, optical computing, robotics, interplanetary navigation, materials and structure, ultra-lightweight large aperture optics, and innovative modular spacecraft architectural concepts."[23]

*13 February*
The television series *Home Improvement* aired an episode showing members of Shuttle *Columbia*'s crew in scenes shot in space. The Shuttle's crew had captured the film during Mission STS-73 in October 1995. The airing marked the first time that NASA had allowed the filming of operations in space, specifically for a television series. The NASA crew included Kenneth D. Bowersox, Catherine G. Coleman, Frederick W. Leslie, Kathryn C. Thornton, and

---

[21] NASA, "NASA Headquarters Begins Migrating Functions to the Centers," *HQ Bulletin*, 20 February 1996, 1; Associated Press, "NASA Shifts Shuttle, Station Authority from Washington to Houston," 7 February 1996; Larry Wheeler, "Goldin's Plan To Decentralize NASA Raises Challenger Concerns," *Florida Today* (Brevard, FL), 7 February 1996.
[22] Associated Press, "Japanese Shuttle Sinks," 12 February 1996.
[23] NASA, "Innovative Space Concepts Selected for Negotiations," news release 96-28, 12 February 1996.

Albert Sacco Jr. Bowersox, who had appeared previously on *Home Improvement*, praised the event as a public relations achievement for NASA.[24]

*17 February*
NASA launched a space probe aimed at a faraway asteroid only about twice the size of Manhattan Island. Scientists intended the probe, called Near Earth Asteroid Rendezvous (NEAR), to reach the asteroid in three years. If the mission succeeded, NASA planned for the probe to orbit the asteroid for nearly one year before eventually crashing into the rock. Astronomers hoped that the data gathered during the exercise would provide new information about the formation of the universe. With a budget of US$120 million, the probe mission was a relatively cheap foray into the gathering of space data. The probe's restricted budget was necessary because the new Discovery program had cost US$150 million, and NASA had a three-year cap on the development of new spacecraft. Johns Hopkins University successfully engineered the craft within the stringent guidelines.[25]

*20 February*
Receiving congratulations from the rest of the international space community, Russia celebrated the 10-year anniversary of the Russian space station *Mir*. The pathbreaking spacecraft had hosted a steady stream of astronauts for long-term stays in space throughout the late 1980s and 1990s, transcending many of the changes in the political structure of its home country. The Soviet space program had launched the first *Mir* module on 20 February 1986 from the desert of Kazakhstan, but because of Cold War tensions, the Soviets had not released specific information about the station as it entered orbit module by module. Rockets had transported the individual modules into space, where cosmonauts assembled them. The fully assembled space station weighed 130 tons (120,000 kilograms or 120 tonnes). As tensions between the United States and the former Soviet Union lessened, Shuttle missions to *Mir* had become frequent. During the 1990s, many countries, including the United States, had paid the fledgling Soviet space administration, which became the Russian Space Agency, hundreds of millions of dollars to send their astronauts to *Mir*. Thus, *Mir* had provided the USSR with a needed source of revenue. However, since the fall of the Soviet Union, the Russian space program had received only one-fifth of the government funding provided to the Soviet space program. In 1996, even with *Mir* in an advanced state of decline, Russian officials remained committed to preserving the aging craft's lifespan. Marcia S. Smith, an aerospace expert with the Congressional Research Service, described *Mir*'s state as akin to "a car that has 200,000 miles [320,000 kilometers] on it." Nevertheless, the space station had provided scientists with the only spacecraft in space consistently occupied by humans for a decade, a significant contribution that deserved commemoration.[26]

*21 February*

---

[24] Ann Hodges, "'Home Improvement' Project Takes On a New Kind of Space," *Houston Chronicle*, 12 February 1996.
[25] NASA, "NEAR Sent To Study Eros," *HQ Bulletin*, 20 February 1996, 3; Associated Press, "Asteroid Visit," 17 February 1996.
[26] European Space Agency, news release, 15 February 1996; Associated Press, "Russia-Space," 20 February 1996; Associated Press, "Russian Space Program Drifting Toward Disaster?" 19 February 1996.

Russia launched the *Soyuz* spacecraft carrying two cosmonauts. The Russian Space Agency intended the mission to *Mir* to bring *Mir* residents Yuri P. Gidzenko, Sergei V. Avdeyev, and Thomas Reiter back to Earth after their lengthy stay aboard the space station.[27]

*22 February*
Mission STS-75, the 19$^{th}$ mission of Shuttle *Columbia*, launched from KSC in Cape Canaveral, Florida. A seven-person crew embarked on the mission, scheduled to last 13 days. Scott J. Horowitz, the first Shuttle pilot with a PhD, captained STS-75. NASA astronauts Andrew M. Allen, Jeffrey A. Hoffman, and Franklin R. Chang-Diaz; Swiss astronaut Claude Nicollier; and two Italian astronauts, Umberto Guidoni and Maurizio Cheli, made up the crew. One of the mission's priorities was to place in flight the Italian tethered satellite system (TSS), which scientists had designed to harness new sources of power for the Space Shuttle. An attempt in 1992 to deploy the TSS had failed when a bolt in the system's reel mechanism caused the tether to jam. The U.S.- Italian satellite had cost US$443 million to produce. The satellite, extended on a 12-mile (19-kilometer) tether, had the potential to use its orbit to create electrical power for other spacecraft. NASA planned to fly the satellite for 22 hours. To retrieve the satellite, the crew intended to retract the tether until the satellite was about 2 miles (3 kilometers) from the Shuttle, allowing it to hover at this distance while the Shuttle maneuvered into position to pick it up.[28]

*25 February*
NASA announced that the tethered satellite experiment had failed. The 12-mile (19-kilometer) tether, attaching the 0.5-ton (450-kilogram or 0.45-tonne) satellite to Shuttle *Columbia*, broke unexpectedly just as the tether had nearly reached its full extension. Before the tether broke, the experiment to use an orbiting satellite to produce electricity had been proceeding smoothly. The astronauts had been in the process of extending the satellite from the Space Shuttle for more than 5 hours, when they noted a decrease in tension and realized that the tether and satellite had broken away. NASA Commander Andrew M. Allen responded to the break immediately, directing the Shuttle away from the satellite to prevent a collision. However, the satellite never came close to the Shuttle; it was more than 18 miles (29 kilometers) away only a few minutes after breaking free. NASA officials could not immediately identify the cause of the break. The astronauts aboard the Shuttle managed to record images of the satellite trailing away from the Shuttle. This footage, a black and white rendering of the tether disappearing into space, reached Flight Command Center in Houston, Texas, shortly after the event. "Those are some tether dynamics we did not want to see," stated one Mission Control operative. NASA dismissed the idea of attempting to retrieve the satellite, explaining that, although it was technically possible, such a mission would be too dangerous and too costly in consumption of fuel.[29]

---

[27] NASA, *Aeronautics and Space Report of the President: Fiscal Year 1996 Activities* (Washington, DC, 1996), p. 102.
[28] NASA, "Space Shuttle Mission STS-75—Press Kit," news release 96-27, February 1996; William J. Broad, "Shuttle Blasts Off for High-Wire Act To Unfurl 12.5 Miles of Electrical Cable in Space," *New York Times*, 23 February 1996; William Harwood, "Shuttle Begins Mission To Deploy Space Power Ball on 12.5-Mile Tether," *Washington Post*, 23 February 1996.
[29] NASA, "STS-75 Mission Control Status Report No. 10," 26 February 1996; Steve Marshall, "Satellite Tether Snaps; Shuttle Safe," *USA Today*, 26 February 1996; William Harwood, "Science Satellite Is Lost as Shuttle Tether Breaks," *Washington Post*, 26 February 1996.

*26 February*
NASA formed an independent panel to review the failure of the tethered satellite experiment and the loss of the satellite. NASA selected Kenneth J. Szalai, Director of Dryden Flight Research Center, to chair the panel. The formation of such an investigative body in the wake of a costly loss conformed to standard practice under NASA's Space Flight Operations Contingency Plan. In a NASA press release announcing the panel's formation, Szalai stated, "given the public investment in the tethered satellite, it is important that we find out what went wrong."[30]

*27 February*
Continuing its efforts to cut costs by using private contractors, NASA awarded McDonnell Douglas Aerospace a US$500 million, eight-year contract to provide "fixed-price medium-light (Med-Lite) class expendable launch vehicle services."[31] NASA's Goddard Space Flight Center managed the program, and the Orbital Sciences Corporation served as the major subcontractor. At the time of the agreement, NASA had planned three Med-Lite missions launching from both the East Coast and the West Coast.[32]

**MARCH 1996**

*3 March*
NASA scientist Benjamin F. Chao explained to the *New York Times* the results of his research on Earth's orbit. A geophysicist at Goddard Space Flight Center, Chao attributed the Earth's slightly changed orbital pattern to the increased construction of dams. As the planet's population has grown, people had attempted to maximize use of Earth's limited water resources by building reservoirs. Using geophysics, international data, and theoretical calculations, Chao postulated that this increase in impounded water on the Earth's surface had actually helped speed up the Earth's orbit. Chao first made public his findings in the *Geophysical Research Papers.* Some members of the scientific community challenged his conclusions.[33]

*Columbia* astronauts set a series of small fires inside the Shuttle to help NASA scientists improve their design of smoke detectors and fire extinguishers for use in a weightless environment. The fires spread quickly within the enclosed vehicle. Astronaut Jeffrey A. Hoffman marveled at the behavior of the flames, which was completely different in a weightless environment from that typically observed on Earth. The fires scattered quickly and burned much more intensely than on Earth. NASA had never before tested the fire-detection equipment in a space environment. The fire equipment worked properly, and the crew put out all fires without incident.[34]

*4 March*

---

[30] NASA, "NASA To Form Independent Review Panel," news release 96-39, 26 February 1996.
[31] Less than two years into the agreement, McDonnell Douglas Aerospace merged with the Boeing Company.
[32] NASA, "McDonnell Douglas Aerospace Awarded Contract To Provide Med-Lite ELV Service," news release 96-40, 27 February 1996; The Boeing Company, "Boeing Completes McDonnell Douglas Merger," news release, 31 July 1997.
[33] Malcolm W. Browne, "Dams for Water Supply Are Altering Earth's Orbit, Expert Says," *New York Times*, 3 March 1996.
[34] Associated Press, "Shuttle Astronauts Set Fires To Check Safety Equipment," 3 March 1996.

The Communist Party of China acknowledged for the first time that a failed satellite launch occurring on 15 February 1996 had killed six people and injured 57 others. The Long March 3B rocket had exploded, destroying the *15 Intelsat* satellite and damaging 80 homes. The secretive government of the People's Republic of China had denied the major accident, announcing simply that a setback had occurred. However, foreign news sources had captured video footage of the launch rocket as it careened out of control and exploded in a highly populated area. The Chinese government ended the news blackout, which had prevented its own citizens from learning of the disaster, stating China's intention to delay future launch plans.[35]

Henry McDonald became Director of Ames Research Center (ARC) in Mountain View, California. NASA had selected McDonald for the position in January 1996 after the selection of former Director Ken K. Munechika as NASA's first Director of Moffett Federal Airfield. A former professor of mechanical engineering at Pennsylvania State University, McDonald brought to NASA both industrial and academic experience.[36]

*5 March*
For the first time in its history, NASA allowed a researcher to direct an experiment from his home university, rather than from Marshall Space Flight Center (MSFC). Martin Glickman led a group of scientists from Rensselaer Polytechnic Institute, taking control of a crystal experiment aboard Shuttle *Columbia*. Both NASA and the scientific community praised the experiment as a crucial step in the relationship between astronauts and academic scientists. Because they did not have to travel to NASA's MSFC in Huntsville, Alabama, the scientists were able to use their own instruments and data systems. Using remote control, they repeatedly cooled and heated the crystal samples on the Shuttle. Scientists hoped that allowing universities to control experiments from their own institutions would facilitate greater student participation in experiments conducted in space.[37]

*7 March*
NASA released Hubble Space Telescope (HST) images revealing much of the surface of Pluto for the first time. The Pluto imaging team declared the quality of the pictures better than they had expected. "Hubble has brought Pluto from a fuzzy, distant dot of light, to a world which we can begin to map, and watch for surface changes," remarked Marc W. Buie, a member of the team. The HST took snapshots of nearly the entire surface of the planet during 6.4 days of its rotation. The pictures showed Pluto's complexity, revealing nearly a dozen distinct provinces, and confirming the existence of a polar ice cap, which scientists had suspected but had been unable to verify. NASA officials were optimistic that the images would pave the way for a proposed Pluto flyby mission in the coming years. Pluto remained the only planet not yet visited by a spacecraft.[38]

*8 March*
NASA agreed to restructure and extend its contract with USBI Company, which had previously assembled and refurbished the Shuttle's solid rocket boosters. The new contract extended the

---

[35] Reuters, "China: Fiery Launch Killed 6, Injured 57," 4 March 1996.
[36] NASA, "McDonald Named Director of Ames Research Center," news release 96-8, 19 January 1996.
[37] Associated Press, "Campus Takes Control of Shuttle Experiment," 5 March 1996.
[38] NASA, "Hubble Reveals Surface of Pluto for First Time," news release 96-43, 7 March 1996.

partnership for 45 months, paying US$500 million to USBI to support the Shuttle program. Based on the terms of the agreement and on USBI's production schedule, NASA estimated that the partnership would cover approximately seven Shuttle missions annually for the duration of the contract.[39]

*9 March*
Space Shuttle *Columbia* touched down at Kennedy Space Center (KSC) in Cape Canaveral, Florida, after a 16-day mission clouded by the disappointment of losing a tethered satellite. NASA delayed the landing for one day because of poor weather on the coast of Florida.[40]

*12 March*
Having received 550 proposals, NASA's Office of Life and Microgravity Science and Applications awarded US$17 million to 168 scientists from 32 different states, to conduct microgravity research. NASA and the scientific community hypothesized that further investigation into the effects of low gravity on physical processes would lead to major advances in fluid physics and material sciences. The scientists hoped that if they could control the process of the formation of materials, they could develop and improve a wide range of materials and products, including metal alloys, semiconductors, ceramics, glasses, and polymers.[41]

*13 March*
NASA announced that it had developed a prototype for new software to teach pilots how to fly commercial aircraft. The improved technological capabilities of computers had allowed Steve Casner, a scientist at NASA's ARC, to develop a program imitating the flight management system of an automated "glass" cockpit and allowing pilots to use their laptops to simulate flying. Pilots could load the program onto their own computers, supplementing the training they had received through classroom instruction and flight simulators. The program featured five windows: "a control/display unit, mode control panel, two maps showing the aircraft's lateral track and its vertical track, and a flight mode enunciator showing which flight systems [were] currently controlling the airplane."[42]

NASA selected astronaut Wendy B. Lawrence to replace Charles J. Precourt as Director of Operations, Russia, making Lawrence the primary contact between NASA and Russian Space Agency officials. As Director of Operations, Russia, Lawrence became responsible for overseeing the training and preparation of U.S. astronauts at Gagarin Cosmonaut Training Center in Star City, Russia, outside of Moscow. Lawrence was the sixth astronaut to serve in this position.[43]

*14 March*
NASA released its blueprint for the future, the "NASA Strategic Plan 1996." The plan laid out NASA's vision for the United States' space exploration; noted its history; and discussed possible

---

[39] NASA, "NASA Restructures/Extends USBI Contract for Solid Rocket Booster Elements," news release C96-c, 8 March 1996.
[40] Associated Press, "Shuttle Lands Safely After Trying Mission," 10 March 1996.
[41] NASA, "Scientists Receive Microgravity Research Grants," news release 96-44, 12 March 1996.
[42] NASA, "NASA Develops New Laptop Pilot Training Tool," news release 96-47, 13 March 1996.
[43] NASA, "Lawrence To Replace Precourt as NASA Manager in Russia," news release 96-49, 13 March 1996.

changes in strategy, committing NASA to a threefold mission: scientific research, space exploration, and development and transfer of technology. Furthermore, the plan outlined NASA's quest to address science's fundamental questions, such as, how did the universe, galaxies, stars, and planets form and evolve? and, does life in any form, however simple or complex, carbon-based or other, exist elsewhere than on planet Earth? NASA intended the document as the starting point for future decisions regarding the allocation of its resources.[44]

*20 March*
At the Lunar and Planetary Science Conference, Adriana C. Ocampo, a geologist at NASA's Jet Propulsion Laboratory, announced the discovery of a chain of impact craters in Chad. According to a team of scientists, the presence of the craters suggested that a sizable comet or asteroid might have struck ancient Earth. Scientists discovered the craters using the Spaceborne Imaging Radar-C (SIR-C), which flew on Shuttle *Endeavour*'s Mission STS-59 in 1994. The impact craters were only the second chain of large craters ever discovered on Earth and the first impact craters found using the SIR-C technology. Ocampo warned that scientists still needed to excavate the ground extensively before they could be sure that the craters were indeed impact structures.[45]

*22 March*
Shuttle *Atlantis* launched from KSC in Cape Canaveral, Florida, setting out for the *Mir* space station. Space Shuttle Mission STS-76 was the third Space Shuttle–*Mir* docking mission. The Shuttle carried six astronauts, including Shannon W. Lucid, destined for a five-month stay on the Russian space outpost. Complications arose when flight controllers at Johnson Space Center in Houston, Texas, detected a potential problem during the launch. A minor leak had occurred in one of the Shuttle's three hydraulic systems used to move the wing flaps. After analyzing the extent of the problem, however, NASA officials determined that the mission could go on as planned. NASA planned the mission to last nine days, connecting with *Mir* on the third day.[46]

Former NASA astronaut Robert F. Overmyer died while test piloting the experimental Cirrus VK-30 aircraft near Duluth, Minnesota. Overmyer was performing a series of stalls at an altitude of more than 8,000 feet (2,400 meters) when the aircraft "departed from controlled flight." Overmyer had served in the U.S. Marine Corps from 1959 to 1986, becoming a NASA astronaut in 1969, and piloting Shuttle Mission STS-5 in 1982. After leaving NASA and the Marine Corps in 1986, Overmyer worked for McDonnell Douglas Aerospace.[47]

*24 March*
NASA astronaut Shannon W. Lucid became the first female resident of the *Mir* space station. Lucid traveled aboard Shuttle *Atlantis* to reach *Mir*, planning to remain on the Russian station for five months. She joined Russian cosmonauts Yuri I. Onufrienko and Yury V. Usachev for the long-term stay. *Atlantis*'s crew unloaded more than 5,000 pounds (2,300 kilograms) of supplies

---

[44] NASA, "NASA Strategic Plan 1998," NASA policy directive (NPD)-1000.1, *http://www.hq.nasa.gov/office/nsp/NSPTOC.html* (accessed 5 February 2008); NASA, "NASA Releases Strategy for the Future," news release, 14 March 1996.
[45] NASA, "Chain of Impact Craters Suggested by Spaceborne Radar Images," news release 96-55, 20 March 1996.
[46] NASA, "Space Shuttle Mission STS-76—Press Kit," news release 96-46, March 1996; William Harwood, "Shuttle Takes Off for Russian Mir Linkup," *Washington Post*, March 23, 1996.
[47] Edward H. Phillips, "Overmyer Dies in Crash," *Aviation Week and Space Technology* 144, no. 14 (1 April 1996): 29.

and 1,000 pounds (450 kilograms) of experiment samples and equipment during the Shuttle's docking. Lucid replaced U.S. astronaut Norman E. Thagard, who had spent 110 days with the Russian crew. Lucid's arrival generated some controversy. In attempting to convey Russia's openness to having a woman aboard *Mir*, cosmonaut Yuri N. Glazkov commented that the crew would appreciate Lucid's presence, because "we know women love to clean." Not surprisingly, some people resented his remark.[48]

*27 March*
NASA astronauts Michael R. "Rich" Clifford and Linda M. Godwin made an historic spacewalk outside the orbiting *Mir* space station. For the first time, astronauts exited *Mir* with the Shuttle docked at the station. Clifford and Godwin, wearing jet packs as a precaution, scaled the passageway connecting Shuttle *Atlantis* with the *Mir* complex, installing several experiment boxes on the outside of the space station. Some of these boxes were "cosmic dust catchers," meant to gather samples of the materials that might smash into orbiting space stations. The spacewalk and associated experiments were part of ongoing research in preparation for the construction of the new International Space Station. NASA regarded the spacewalk as more dangerous than usual, because the Shuttle was docked with *Mir* and, thus, unable to give chase quickly if one of the astronauts broke away from the spacecraft.[49]

Preparing to take over the administration of the Shuttle program from NASA, the Rockwell-Lockheed Martin joint venture, United Space Alliance (USA), named James C. Adamson as its chief operating officer. Adamson came to USA from Lockheed Martin, where he had served since 1994 as a group vice president and general manager of Lockheed's Engineering and Science Services. Adamson had served as an astronaut at NASA from 1984 to 1992, spending more than 334 hours in space as a part of Missions STS-28 and STS-43.[50]

**APRIL 1996**

*1 April*
Shuttle *Atlantis* landed at Edwards Air Force Base in California's Mojave Desert. The nine-day mission nearly came to an abrupt halt when *Atlantis*'s crew had difficulty opening the Shuttle's cargo bay doors during descent. After considering an emergency landing, NASA controllers directed *Atlantis*'s crew to open the doors manually. The crew successfully performed this necessary maneuver only to confront dangerous weather that prevented their landing at Kennedy Space Center.[51]

*4 April*
An international team of researchers announced that, while tracking Comet Hyakutake in March 1996, they had discovered x-rays emitting from the comet. German scientists working at Max Planck Institute for Extraterrestrial Physics in Garching, Germany, had collaborated on the

---

[48] William Harwood, "U.S. Woman To Start Stay on Russian Space Station," *Washington Post*, 24 March 1996; Associated Press, "Female Astronaut Settles In on Mir," 25 March 1996.
[49] Associated Press, "Spacewalking Astronauts Venture Outside Atlantis-Mir Complex," 27 March 1996; Todd Halverson, "Walkers Prepare for Historic Trek Outside Mir Station," *Florida Today* (Brevard, FL), 27 March 1996.
[50] *Houston Chronicle*, "United Space Alliance Taps New Top Officers," 27 March 1996.
[51] Associated Press, "Shuttle Back Safely After Mir Mission," 1 April 1996.

project with NASA scientists based at Goddard Space Flight Center in Greenbelt, Maryland. Konrad Dennryl, who had directed the use of Germany's ROSAT (Röntgensatellit) for the project, called the discovery of the x-rays "a thrilling moment." The comet emitted far stronger radiation signals than the scientists had expected, about 100 times brighter than anticipated. The scientists could not explain why the comet gave off such intense x-rays, but offered two preliminary hypotheses: 1) the gaseous molecules surrounding the comet absorbed x-rays from the Sun before reemitting the same x-rays; or 2) violent collisions between the comet material and the supersonic wind and particles from the Sun produced the extraordinarily strong x-rays. The scientists vowed to ascertain conclusively why the comet x-rays behaved as they did.[52]

*9 April*
Rockwell International Corporation pled guilty to charges of violating federal law in disposing of hazardous waste. Federal prosecutors had charged the company with violating the law after a 1994 explosion at a Rockwell rocket-testing facility in Southern California killed scientists Otto K. Heiney and Larry A. Pugh. Rockwell's Chief Executive Officer Donald R. Beall called the incident "a corporate failure for which we accept responsibility." By pleading guilty, Rockwell acknowledged liability and agreed to pay a US$6.5 million fine. According to the U.S. Department of Justice, the case was only one aspect of a long-term investigation into Rockwell's compliance with environmental regulations, an investigation that Rockwell's guilty plea did not end. Federal auditors were also looking into the possibility that Rockwell had overcharged NASA and the U.S. Department of Defense for disposal costs.[53]

NASA made its high-altitude research plane available to the disaster-recovery team responsible for containing hazardous leakage at the California Gulch Superfund site in Leadville, Colorado. The ER-2 plane, a U-2 spy plane adapted for civilian use, captured thousands of measurements per second, and its spectrometer mapped hundreds of square miles during each reconnaissance flight. Scientists used NASA's plane to map minerals and other substances, to help them detect the location of acid leaks from mines and heavy-metal contamination. Officials estimated that the ER-2 plane's identification of contaminated areas had saved cleanup agencies about one year and more than US$500,000.[54]

*11 April*
The U.S. Space Foundation inducted into its Space Technology Hall of Fame three technologies that NASA had originally developed for its astronauts: antishock trousers, flame-retardant seat materials, and the radiation barrier. The Space Foundation recognized the technologies as pathbreaking discoveries that had enhanced the quality of life on Earth. Scientists and engineers at Ames Research Center had developed the first two technologies, and members of the Apollo program at Johnson Space Center (JSC) had discovered the radiation barrier. NASA employees had often referred to the dual use of its technologies as spin-offs, recognizing that such inventions served to validate the expenditure of taxpayers' funds on NASA's research projects. According to a NASA spokesperson, the technologies "are not merely secondary applications.

---

[52] NASA, "First X-rays from a Comet Discovered," new release 96-66, 4 April 1996; John Noble Wilford, "Scientists Seek Explanation for X-rays Seen from Comet," *New York Times*, 5 April 1996.
[53] Andy Pasztor, "Rockwell Probe To Continue After Unit Pleads Guilty to Waste-Disposal Charge," *Wall Street Journal*, 9 April 1996; Mark Reed, "Firm Agrees To Pay Fine in Fatal Blast," *Los Angeles Times*, 9 April 1996.
[54] *New York Times*, "NASA Research Plane Aids in Waste Cleanup," 9 April 1996.

They provide direct, quantifiable, and invaluable benefits to the American taxpayer and the domestic economy." The U.S. Space Foundation made the inductions at the National Space Symposium, in Colorado Springs, Colorado. The Foundation had inducted 25 technologies into the Hall of Fame since the establishment of the awards in 1988.[55]

*12 April*
NASA and the scientific community celebrated the 15th anniversary of the Space Shuttle program. NASA had adopted the Shuttle as a means of controlling the costs of exploring space. However, NASA conceded that, since its deployment, the Shuttle had flown only nine times per year, compared with the 24 to 60 Shuttle flights per year anticipated. Furthermore, the cost of flying the Shuttle had reached almost US$500,000 more per mission than expected. During its first 15 years, the Shuttles had flown 76 missions, including 51 following the *Challenger* accident. The Shuttles had transported 545 tons (490,000 kilograms or 490 tonnes) of payloads to space, bringing back to Earth 18.5 tons (17,000 kilograms or 17 tonnes) of payloads. During the same period, 440 astronauts from 12 countries flew aboard the Shuttles. NASA experts predicted that the Shuttle would remain the primary vehicle of space exploration well into the 21st century.[56]

NASA completed the first step in its transition to using a single contractor to conduct all Shuttle operations, signing two novation agreements with the newly formed United Space Alliance (USA). NASA had announced in 1995 that it would consolidate its Shuttle operations under a single contract with USA, on a noncompetitive basis. When Rockwell International Corporation and Lockheed Martin Corporation developed USA as a joint venture, the two companies already held nearly 70 percent of the dollar value of all Shuttle-related contracts. NASA officials expected that the move to a single contract would reduce costs by eliminating duplicative work and streamlining management.[57]

*15 April*
NASA disclosed the startling discoveries of Rice University astronomer C. Robert O'Dell and graduate student Kerry P. Handron, who had used images captured by the Hubble Space Telescope to pinpoint thousands of gigantic, tadpole-shaped objects surrounding a dying star, the first time that scientists had observed the forms in such abundance. The scientists hypothesized that the final outbursts of the dying star probably caused the gaseous knots, each several billion miles across. O'Dell concluded that if a dying star caused the gaseous knots, then trillions of these forms might litter the universe.[58]

With funding from NASA's Office of Life and Microgravity Sciences and Applications and the Utah Agricultural Experiment Station, researchers at Utah State University successfully developed a strain of wheat suitable for growth in space. The space wheat, called USU-Apogee, produced a yield equivalent to 600 bushels of grain per acre (21,000 liters of grain per 4,100 square meters), a rate three times more than the rate achieved in previous experiments.

---

[55] NASA, "NASA Life-Saving Technologies Enter Space Hall of Fame," news release 96-69, 10 April 1996.
[56] Martin Burkey, "A Lean, Mean Teen: Shuttle Turns 15," *Huntsville Times* (AL), 13 April 1996; Seth Borenstein, "Costly History Dogs Shuttle on Anniversary," *Orlando Sentinel* (FL), 12 April 1996.
[57] NASA, "Shuttle Contracts Take First Step Toward Consolidation," news release 96-73, 12 April 1996.
[58] NASA, "Hubble Finds Thousands of Gaseous Fragments Around Star," news release 96-74, 15 April 1996.

Researchers had worked for more than a decade developing the wheat. During the tests, USU-Apogee had thrived under the difficult conditions of artificial sunlight and high levels of carbon dioxide. In conditions approximating the environment in space, the wheat grew on short stalks, produced an unusually large number of seeds, and maintained green leaf tips (reflecting proper calcium levels). Scientists hoped that astronauts would be able to grow the wheat in the future International Space Station, as a more economical means of providing food for long-term space residents than frequent Shuttle supply missions.[59]

*17 April*
NASA Administrator Daniel S. Goldin informed employees that NASA Headquarters in Washington, DC, would reduce its staff by more than 50 percent, citing "increasing budget pressures." Under the reduction plan, staff at NASA Headquarters would decrease from 1,430 to between 650 and 700, in less than one year's time. The Clinton administration had continued to push for government-wide budget cutbacks throughout the 1990s. Don Fuqua, President of the Aerospace Industries Association, described the staff reduction as a case of the federal government following the example of the private sector. According to Fuqua, before NASA decided to reorganize, most commercial aerospace business had already reorganized for greater efficiency.[60]

*18 April*
Chinese scientists attending a conference under the direction of the World Monuments Fund announced that space radar images had allowed them to locate and study two ancient, previously undetectable sections of the Great Wall of China. Spaceborne Imaging Radar C/X-Band Synthetic Aperture Radar, aboard Shuttle *Endeavour*'s Mission STS-59, produced the images in 1994. Diane L. Evans, a scientist at NASA's Jet Propulsion Laboratory, called the discovery an unexpected benefit of NASA's research: "Archaeology wasn't one of our original science objectives, but the imaging radar data has been found to be very useful for this type of research." Because the radar images penetrated vegetation and loose sand, archaeologists were able to use the images to explore the site without excavation. The radar images revealed sections of the wall built in two different periods—one during the Ming Dynasty and the other during the Sui Dynasty.[61]

*23 April*
Russia successfully launched the 21-ton (19,000-kilogram or 19-tonne) *Priroda* module, the last major component of the aging *Mir* space station, aboard a robotic rocket from Baikonur Cosmodrome in Kazakhstan. Russian scientists designed the *Priroda* module primarily as a laboratory for observing Earth. The module also contained equipment for NASA astronaut Shannon W. Lucid's stay aboard the station. NASA officials at JSC monitored the launch. *Priroda* carried a magnetically levitated experiment platform, a spectrometer, and a "glove box,"

---

[59] NASA, "Utah State University Develops Space Crop," news release 96-75, 15 April 1996.
[60] NASA, "Additional Personnel Reductions Planned for Agency Headquarters," news release 96-76, 17 April 1996; Kathy Sawyer, "NASA Speeds Up Downsizing," *Washington Post*, 18 April 1996; Mike Causey, "NASA To Lighten Pay Load," *Washington Post*, 18 April 1996; Warren Ferster and Jennifer Heronema, "NASA Cuts Headquarters Jobs Deeper, Faster," *Space News*, 22–28 April 1996, 4–5.
[61] NASA, "Space Radar Reveals Ancient Segments of China's Great Wall," news release 96-77, 18 April 1996; Associated Press, "Shuttle Found Lost Sections of the Great Wall in '94," 22 April 1996.

allowing astronauts to handle potentially hazardous material. The module was the sixth major portion of the *Mir* complex launched by Russia since the Soviet Union began the project in 1986. Budget shortfalls in the Russian space program had caused some concern that the final module might not reach *Mir* as planned.[62]

*24 April*
Without holding any subcommittee hearings on the matter, the U.S. House Committee on Science and Technology voted to trim NASA's budget for 1997 by US$450 million. NASA operated under a US$13.8 billion budget in 1996. Most of the cuts were from the budget of the initiative Mission to Planet Earth.[63]

*30 April*
Harry C. Holloway resigned from his post as NASA Associate Administrator for the Office of Life and Microgravity Sciences and Applications. Holloway was the first person to hold the appointment after Administrator Daniel S. Goldin created the position in March 1993. Holloway planned to return to the School of Medicine at Uniformed Services University. During his tenure, Holloway had established programs to conduct experimental studies in orbit and instituted the strategic plan for the Human Exploration and Development of Space. Holloway had been instrumental in strengthening ties between NASA and the National Institutes of Health, overseeing the signing of 10 agreements between the two agencies.[64]

At the High Energy Astrophysics Division of the American Astronomical Society, astronomers working with NASA's Rossi X-ray Timing Explorer (RXTE) spacecraft announced the discovery of the fastest vibrations ever recorded in a star system. The vibrations occur when massive neutron stars begin to die out, expelling the outer layers of their solid crusts and creating powerful x-rays. The astronomers working with the RXTE recorded x-rays flickering at up to 1,130 times per second. Scientists were unsure why this process occurred. According to NASA, "the possibility that the RXTE has detected actual waves in neutron stars or a very fast rotation period of one such star is of great scientific interest."[65]

## MAY 1996

*1 May*
NASA announced its 1996 astronaut candidate class, which included the first set of identical twins ever selected—Mark E. Kelly and Scott J. Kelly. The sibling U.S. Navy pilots joined 33 other astronauts in the class. NASA selected 10 pilots and 25 mission specialists from an applicant pool of more than 2,400 candidates, including members of the U.S. Army, Air Force, Coast Guard, Marine Corps, and Navy. Six members of the class held PhDs in their respective fields, and one was an MD. NASA selected the astronaut class, the largest in nearly 20 years,

---

[62] Mark Carreau, "Russia Launches Final, Major Mir Component," *Houston Chronicle*, 24 April 1996.
[63] Brett Davis, "Panel Sets $450 Million NASA Budget Cut," *Huntsville Times* (AL), 25 April 1996.
[64] NASA, "Science Head To Leave NASA," news release 96-80, 30 April 1996.
[65] NASA, "NASA Spacecraft Discovers Fastest Stellar Vibrations Yet," news release 96-81, 30 April 1996; Associated Press, "NASA Instrument Detects Fastest Vibrations Yet Seen in Universe," 30 April 1996.

from 16 different states; in addition, one astronaut candidate had been born in Argentina and another in the United Kingdom.[66]

*2 May*
Joseph B. Gurman of NASA's Goddard Space Flight Center announced the pathbreaking discoveries of a team of scientists monitoring the Solar and Heliospheric Observatory (SOHO) spacecraft. According to Gurman, SOHO had revealed unexpected activity on the Sun, as well as the best images yet of the plume-like structures, extending from the Sun. SOHO's scientists were excited and surprised to find that "movies made from SOHO ultraviolet data show that there is continuous motion and action everywhere on the Sun." The scientists observed that the Sun's plumes extend more than 13 million miles (21 million kilometers) into space. NASA and the European Space Agency (ESA) collaborated on the SOHO project.[67]

Former U.S. Representative Norman Y. Mineta received NASA's Distinguished Service Medal, the highest civilian honor awarded by NASA. Mineta had served 21 years in the U.S. Congress, representing California's Silicon Valley in the House of Representatives. As chairperson of the House Committee on Public Works and Transportation, he had played a vital role in bolstering the nation's transportation infrastructure and fostering technological research. Mineta remarked that serving over the long term as the senior member of the House Committee on Science and Technology had given him the opportunity to see firsthand NASA's significant achievements. When he accepted the award at a private ceremony, Mineta said, "I'm deeply honored to be recognized by NASA for any contributions I made furthering the mission of our nation's space operations."[68]

*3 May*
NASA's Galileo spacecraft discovered a giant iron core taking up half of the diameter of Jupiter's moon Io, as well as a large hole in Jupiter's magnetic field. Galileo Project Scientist Torrence V. Johnson of NASA's Jet Propulsion Laboratory (JPL) identified Io as the most "geographically active body in the [s]olar [s]ystem." Galileo scientists also measured the intense heat Jupiter generates, noting that the planet has a massive gravity field. Galileo managed to come within 559 miles (900 kilometers) of Jupiter during its 1995 flyby, allowing scientists to locate the iron core and the hole in the magnetic field. Although they were still grappling with the significance of the "completely unexpected finds," scientists reported their early analysis of the data in *Science* magazine.[69]

*10 May*
At Marshall Space Flight Center (MSFC), NASA announced that it had taken a major step forward in preparations for the International Space Station (ISS): the station's air purification system had passed a month-long test. NASA scientists tested the ISS's air pressure and the ability of its air system to regulate the mixture of carbon dioxide and oxygen that the astronauts

---

[66] NASA, "NASA Selects Astronaut Class of 1996," news release 96-84, 1 May 1996; Marcia Dunn, "In NASA First Identical Twins Selected for Astronaut Corps," *Huntsville Times* (AL), 2 May 1996.
[67] NASA, "Unexpectedly Active Sun and New Insights into Solar Plumes Found," news release 96-87, 2 May 1996.
[68] U.S. Newswire, "Norman Y. Mineta Receives NASA's Distinguished Service Medal," 2 May 1996.
[69] J. D. Anderson, W. L. Sjogren, and G. Schubert, "Galileo Gravity Results and the Internal Structure of Lo," *Science* 272, no. 5262 (3 May 1996): 709–712; NASA, "Galileo Finds Giant Iron Core in Jupiter's Moon Io," news release 96-89, 3 May 1996.

would breathe. In the simulation, scientists used a 6,200-cubic-foot (176-cubic-meter) module from which they had removed the oxygen. The air system not only proved it could maintain the proper levels of both oxygen and carbon dioxide, but also performed satisfactorily at a reduced power level. Engineers hoped to reduce the power level of the ISS during its nighttime orbit. The system, called the Atmosphere Revitalization Subsystem, demonstrated that it could provide healthy working and living conditions for the astronauts on board the ISS.[70]

*15 May*
NASA and the ESA made available images taken of Comet Hyakutake as it approached the Sun. The U.S. Naval Research Laboratory had captured the pictures during the appearance of the comet in late April 1996. The Naval Research Laboratory had built a special instrument—the Large Angle Spectrometric Coronagraph (LASCO)—to record the images. Operating aboard SOHO, LASCO suppressed the glare of the Sun, exposing the comet and its tails. In addition, the instrument produced the equivalent of a time-lapse movie of Hyakutake, thereby documenting the comet's tails as they changed direction in reaction to the Sun's forces. LASCO also allowed scientists to observe coronal mass ejections—the expulsion of hot gases. Researchers used the data to learn more about how comets interact with the Sun and about the behavior of comet tails.[71]

*17 May*
Peter H. Diamandis, President of Angel Technologies Corporation, an aerospace firm located in St. Louis, Missouri, offered a US$10 million prize to the first private aviators to complete a suborbital flight into space. Diamandis stated that the object of the prize was to motivate private companies to invest in space exploration, someday making spaceflights open to the public. "We're looking to spark a constructive competition," Diamandis explained. To claim the prize, the privately financed and built aircraft would have to ascend to 62 miles (100 kilometers) high, twice within a period of two weeks, delivering three passengers safely to space and back. Diamandis claimed that creating an entirely reusable Space Shuttle would considerably decrease the cost of going to space, opening the space frontier to people who had not trained as professional astronauts. Diamandis intended to announce the winner of the "X Prize" on 18 May, beneath the St. Louis Gateway Arch. Creator of the Voyager and top contender for the prize Burt Rutan, as well as NASA Administrator Daniel S. Goldin, planned to attend the award ceremony.[72]

*19 May*
Shuttle *Endeavour* lifted off from Kennedy Space Center (KSC) in Cape Canaveral, Florida, on Mission STS-77, scheduled to last 10 days. John H. Casper, making his fourth spaceflight, commanded the crew, and Curtis L. Brown Jr. piloted the Shuttle. Four mission specialists, Daniel W. Bursch, Marc Garneau, Mario Runco Jr., and Andrew S. W. Thomas, completed the flight crew. NASA planned various science experiments for the mission, including deployment of a retrievable satellite to test the feasibility of using satellites with inflatable parts. NASA's Office of Space Access and Technology sponsored most of the payloads aboard *Endeavour*. The Shuttle also carried a privately developed SPACEHAB module, holding nearly 3,000 pounds

---

[70] NASA, "Space Station Air Purification System Completes Major Test," news release, 10 May 1996.
[71] NASA, "SOHO Images Comet Hyakutake's Close Encounter with the Sun," news release, 15 May 1996.
[72] Paul Hoversten, "A Space Race with a Lofty Goal: Cash," *USA Today*, 17 May 1996.

(1,400 kilograms) of science experiments. NASA had contracted with SPACEHAB Inc. in 1990 to construct research laboratories. Canadian astronaut Marc Garneau was making his second trip into space.[73]

*20 May*
*Endeavour*'s crew successfully tested a giant inflatable antenna, deploying the structure and gathering data during its orbit. Project Manager Steven Bard, a NASA scientist at JPL, called the successful test a giant step toward developing lightweight, cost-effective equipment for space research. The silver, inflatable antenna, about the size of a tennis court, cost US$14 million to develop. The Shuttle astronauts used the Shuttle's robotic arm to launch the small satellite carrying the antenna. The crew situated *Endeavour* about 400 feet (122 meters) immediately above the antenna, to observe its behavior after deployment, but when the antenna made some unexpected tumbling rotations, the astronauts moved the Shuttle to a safer distance. After completing the test, the astronauts set off a series of small explosions to cut loose the antenna. Before the mission began, NASA officials had determined that deflating and recapturing the antenna would be too risky and costly to attempt.[74]

*21 May*
The Boeing Company, Honeywell Space Systems, and the S. P. Korolev Rocket and Space Corporation Energia (RSC Energia), based in Kaliningrad, Russia, announced a new venture to reduce significantly the cost and time involved in transporting payloads to the *Mir* space station. Boeing estimated that the service would save one year and would cut the cost of transporting a standard payload to space by at least US$7 million. The three companies split duties to maximize efficiency. Boeing assumed control of the overall management and payload-system integration; Honeywell accepted responsibility for providing payload-interface units and payload-integration support; and RSC Energia signed on to provide launch services and to oversee installation on *Mir*. Decreasing the expenses involved in sustaining *Mir* had become urgent in the 1990s, when both Russian and American space agencies had to respond to their governments' respective budget cuts.[75]

Scientists revealed that their analysis of the data from NASA's Galileo probe had offered new insights into Jupiter's climate. The probe found winds of 400 miles (640 kilometers) per hour, extending about 10,000 miles (16,000 kilometers) from the planet's core to its cloud top and producing storms that could last for centuries. The findings indicated that heat at Jupiter's core fueled the winds. Galileo released a probe, which parachuted 400 miles (640 kilometers) through Jupiter's atmosphere to gather its data, enduring temperatures ranging from -171°F to 305°F (-113°C to 152°C). The probe exploded after gathering and transmitting the data.[76]

---

[73] Warren E. Leary, "Shuttle Begins 10-Day Technology Flight," *New York Times*, 20 May 1996; NASA, "Space Commercialization and Technology Demonstrations Highlight Shuttle Mission STS-77," news release 96-83, May 1996.
[74] William Harwood, "Inflatable Antenna Passes Test," *Washington Post*, 21 May 1996; Associated Press, "Space Shuttle Deploys Giant Antenna," *Washington Times*, 21 May 1996.
[75] The Boeing Company, "Boeing, Honeywell, and Energia Join in Offering New Commercial Payload Service to Russian Space Station," news release, 21 May 1996.
[76] Paul Hoversten, "Jupiter's Winds: 10,000-Mile-Deep 'Giant Flywheel'," *USA Today*, 22 May 1996. NASA, "Galileo Probe Data Spurs New Concepts for Jupiter's Circulation and Formation," news release 96-103, 21 May 1996.

*22 May*
The trend of commercializing space exploration continued, as Pepsi-Cola Company announced the company's plan to use Russian cosmonauts aboard the *Mir* space station to record the first TV advertisement filmed in space. PepsiCo planned to record images of a giant replica of a Pepsi can in space. PepsiCo's Chief Marketing Officer Massimo F. d'Amore called space "the ultimate frontier of global marketing." The Russian Space Agency expected to receive nearly US$5 million for the advertising stunt, providing a welcome infusion of cash for the struggling, postcommunist space program. In response to its competitor's advertisement plans, the Coca-Cola Company furnished *Endeavour*'s Mission STS-77 with a soda dispenser equipped for use in space. The machine cost Coca-Cola Company US$1.5 million to develop.[77]

*23 May*
NASA revealed that the joint U.S.-Italian team investigating the failure of the Italian tethered satellite system (TSS) had determined that the tether experiment conducted aboard Mission STS-75 had revealed numerous fundamental flaws in space physics and plasma theories. The TSS experiment had ended with the tether breaking before reaching its full 13-mile (21-kilometer) length. The scientists concluded that a number of theoretical models, widely accepted for as long as 30 years, were incorrect and in need of revision. According to Noble Stone, a NASA scientist at MSFC, the team's most significant finding was that tether current flows between the satellite and the orbiter were three times greater than theoretical models had predicted. Because the amount of power generated was directly proportionate to the current, the harnessed energy had the potential to furnish thrust for rebooting a space station, satellite, or shuttle in a decaying orbit. Stone suggested that tether systems might some day supplement solar arrays as sources of power for long-term space platforms.[78]

*29 May*
Shuttle *Endeavour* made a smooth sunrise landing at KSC in Cape Canaveral, Florida, bringing to its close a fruitful 10-day research mission. The Shuttle crew celebrated their record achievement of 21 hours flying in close proximity to the satellites they had released. NASA planned to move *Endeavour* to Palmdale, California, for refurbishing.[79]

*31 May*
A team of NASA and university researchers announced the discovery of large quantities of ethane and methane gases in Comet Hyakutake, the first time scientists had detected such gases in a comet. The discovery led researchers to postulate that at least two basic types of comets exist. If validated by further research, the hypothesis would have significant ramifications for scientific theories regarding the formation of the Sun and planets. Michael J. Mumma, the lead researcher on the project, called the discovery of ethane in the comet "a blinding surprise." The

---

[77] Melanie Well, "Pepsi, Coke Go into Orbit," *USA Today*, 22 May 1996.
[78] NASA, "Early Findings from Tethered Satellite Mission Point to Revamping of Space Physics Theories," news release 96-106, 23 May 1996; NASA Marshall Space Flight Center "Fall Science Meeting Highlights Tether Satellite Results," *http://spacescience.spaceref.com/newhome/headlines/ast15oct96_1.htm* (accessed 27 October 2008).
[79] Stefanie Asin, "Welcome Back, Endeavour, Shuttle Lands at Kennedy Space Center After 10-Day Mission," *Houston Chronicle*, 30 May 1996.

team of researchers included Karen Magee-Sauer of Rowan College, New Jersey, and NASA scientists Neil Dello Russo, David X. Xie, and Charles Kaminski.[80]

## JUNE 1996

*4 June*

NASA and the Italian Space Agency released the investigative report on the failed tethered satellite system (TSS) experiment conducted during Shuttle *Columbia*'s Mission STS-75 in February 1996. The 358-page report, produced by a board established to investigate the incident, concluded that a breach in the layer of insulation surrounding the tether conductor, possibly the result of contact with debris, led to the failure. According to the report, the breach in insulation allowed an electrical current to jump from a copper wire in the tether to an electrical ground, causing the breakage: "the tether failed as a result of arcing and burning of the tether, leading to a tensile failure after a significant portion of the tether had burned away." The board noted that NASA scientists had already announced findings regarding tether current and voltage measurements, concluding that the unexpected breakage of Columbia's tethered satellite did not indicate any fundamental problem with the concept of the TSS. Furthermore, the board recommended that during future experiments, scientists should take greater precautions to ensure that neither debris nor high-voltage arcing compromised the tether's strength.[81]

Arianespace, the commercial arm of the European Space Agency, suffered a major setback in its bid to maintain control of the satellite-launching market in the face of competition from U.S. companies. Only seconds after the launch of Ariane 5, a huge rocket intended to launch satellites, Arianespace had to blow up the rocket when it veered severely off course. Arianespace had spent 10 years and more than US$7 billion to develop the 674-foot (205-meter), 821-ton (750,000-kilogram or 750-tonne) vehicle. Also lost in the explosion were satellites valued at more than US$500 million. Arianespace had projected a launch reliability of 98.5 percent for the rocket, considering failure so unlikely that the company had guaranteed a free launch to customers if the launch failed. A spokesperson for Arianespace acknowledged the disappointment of the failed test, but promised that the company would continue as the world's leader in launching satellites. Aerospace industry analysts predicted that U.S. companies such as Lockheed Martin would likely gain an increased foothold in the market because of the European company's misfortune.[82]

*7 June*

The experimental DC-XA flight vehicle completed a successful test flight, flying nearly 2,000 feet (610 meters) above the desert of the White Sands Missile Range in New Mexico and landing safely. NASA researchers and administrators hoped that the rocket would one day replace the Space Shuttle as the primary means of space exploration. The reusable, unpiloted rocket, taking off and landing from a vertical position, would cut significantly the costs of delivering payloads into orbit. NASA announced that the vehicle had been renamed Clipper Graham, in honor of

---

[80] NASA, "Chemical Measurements of Comet Hyakutake Suggest a New Class of Comets," news release 96-108, 31 May 1996.
[81] NASA, "Tethered Satellite Investigation Report Released," news release 96-112, 4 June 1996; Reuters, "Damaged Insulation Ruined Test in Space," 5 June 1996; Associated Press, "Study Finds Metal Debris Punctured Tether Pulled by Satellite," 5 June 1996.
[82] Kathleen Day, "Losing a Rocket and a Satellite Edge?" *Washington Post*, 5 June 1996; Craig Whitney, "Costly Failure: Space Launch Is Aborted," *New York Times*, 5 June 1996.

Lieutenant General Daniel O. Graham. Graham, who died on 31 December 1995, had served as an advisor to President Ronald Reagan and had been an early proponent for the development of reusable rockets. Administrator Daniel S. Goldin expressed optimism that NASA would be able to replace the Shuttle with the DC-XA flight vehicle: "We are going to fly the [S]huttle safely until we can replace it. But, by God, we are going to replace the [S]huttle."[83]

*10 June*
A team of astronomers announced at the American Astronomical Society convention that they had discovered ordinary vinegar in a stellar cloud 25,000 light-years away from Earth. The presence of an organic molecule, which may have played a role in the formation of life, excited scientists. "Acetic acid could have been one of the first steps toward the chemicals of life," explained Lewis E. Snyder, a member of the research team that made the discovery. Scientists had discovered complex molecules in space before, but formic acid, discovered in 1975, was the only other basic molecule detected before the discovery of the vinegar.[84]

*12 June*
Kennedy Space Center (KSC) scientists began the most comprehensive testing to date on how plants produce food and oxygen in space. The experiment, part of NASA's development of a Controlled Ecological Life Support System, called for the analysis of 128 potato plants and 6,500 wheat seeds. NASA scientists planned for the study to last from one to three years. NASA agricultural engineer John Sager, praising NASA's earlier success in producing crop species for space, explained the experiment's significance: "If we plan to live in space . . . we must determine if this system will be as successful over longer periods of time." Scientists were using KSC's Biomass Production Chamber to test the crops in a space-like environment. NASA had begun in 1987 to test strains of crops and plants for their ability to produce oxygen and food in space.[85]

*13 June*
The Smithsonian Institution named former U.S. Navy pilot and aviation executive Donald D. Engen as Director of the National Air and Space Museum. Engen succeeded Martin O. Harwit, who had resigned amidst criticism of an exhibition about the Enola Gay atomic bomb mission. Some World War II veterans had deemed the exhibition too sympathetic to Japan, and some members of Congress had faulted the National Air and Space Museum for deviating from its mission of exhibiting the history of the United States' aeronautical achievements. Smithsonian employees praised Engen's appointment, suggesting that he would restore the museum's credibility.[86]

*19 June*
The U.S. Space Walk of Fame, based in Titusville, Florida, recognized with a two-day celebration the 20,000 men and women who worked on the Gemini space program. NASA had

---

[83] NASA, "Revolutionary New Launch Vehicle Renamed for Space Pioneer," news release 96-114, 7 June 1996; Associated Press, "Reusable Rocket Ends Flight," 7 June 1996.
[84] John Noble Wilford, "Whiff of Organic Stuff Found Between Stars," *New York Times*, 11 June 1996; Associated Press, "Vinegar Found in Far Away Space," 11 June 1996.
[85] NASA, "NASA Begins Study on Reliability of Space Life Support System," news release 96-120, 12 June 1996.
[86] Rowan Scarborough, "Engen To Head Air and Space," *Washington Times*, 13 June 1996.

launched 10 Gemini missions between 1965 and 1966, each spacecraft carrying two astronauts. Many staff members and astronauts involved in the Gemini program went on to work on the Apollo Moon program. Gemini astronauts included Neil A. Armstrong, Jim Lovell, and Edwin E. "Buzz" Aldrin Jr. The event organizers celebrated the achievements of the Gemini missions, in addition to raising funds to establish a permanent monument to the Gemini program.[87]

*20 June*
Shuttle *Columbia* lifted off from KSC in Cape Canaveral, Florida, embarking upon Mission STS-78. The Shuttle carried U.S. astronauts Terrence T. Henricks, Kevin R. Kregel, Richard M. Linnehan, Susan J. Helms, and Charles E. Brady Jr.; French astronaut Jean-Jacques Favier; and Canadian astronaut Robert B. Thrisk. NASA planned to use an orbiting research laboratory to conduct a variety of experiments during the mission. In addition to its human cargo, the Shuttle transported the payload for the Office of Life and Microgravity Sciences, used to conduct research on musculoskeletal physiology and on the responses of living organisms to a low-gravity environment.[88]

*21 June*
Russian officials announced that the necessity of cutting costs at the Russian Space Agency had caused a 40-day delay in returning cosmonauts Yuri I. Onufrienko and Yury V. Usachev from the *Mir* space station to Earth. Russian budget cuts had slowed the production of the Soyuz booster rockets needed to fuel a return trip for the cosmonauts. Because *Mir* was deteriorating, Russia could not leave it without a crew. This was the second time that financial and technical difficulties had led Russia to delay the return of the two men. The episode signaled a deepening crisis for the Russian space program, the United States' key partner in building the International Space Station (ISS).[89]

*24 June*
Astronauts aboard Shuttle *Columbia* conducted an overnight experiment on the effects of space travel on the body's biological clock. Four astronauts slept with special caps equipped with electrodes to monitor their brain and muscle activity. Astronauts in space have no exposure to the once-daily rising and setting Sun, which helps trigger a regular sleep pattern on Earth. In space the Sun rises and sets more than a dozen times each "day." NASA administrators planned the experiment because they recognized the value of restful sleep for astronauts working in space and were concerned that sleep loss would become a greater disadvantage the longer an astronaut stayed in space. With the ISS on the horizon, NASA researchers considered the possible effects on astronauts of record-setting stays in space a significant issue. "Sleep disruption is something that has to be taken seriously," said Timothy H. Monk, the lead researcher on the project.[90]

*26 June*

---

[87] Marilyn Meyer, "History-Making Gemini Team To Reunite," *Florida Today* (Brevard, FL), 8 June 1996.
[88] NASA, "Life and Microgravity Science Research Highlight Shuttle Mission STS-78," news release 96-116, June 1996; NASA, *Aeronautics and Space Report of the President, Fiscal Year 1996 Activities* (Washington, DC, 1997), p. 103.
[89] *USA Today*, "Russian Cosmonauts," 24 June 1996; Associated Press, "Moscow," 21 June 1996.
[90] Reuters, "Shuttle Mission Probes Body Clock," 24 June 1996.

Astronomers using the Hubble Space Telescope (HST) identified what they believed to be the most distant objects observed to date. Researchers based at State University of New York at Stony Brook found several dozen galaxies so far away that they might have existed when the universe was less than 5 percent of its present age. The team of astronomers made its calculations on the distances of the galaxies from the Earth by relying on the relationship between speed and distance in the expanding universe. According to the scientific understanding of the speed-distance relationship, "the expansion of the universe causes the light from distant galaxies to be 'redshifted'," meaning that, because of the expansion of space, light leaving a distant galaxy as blue arrives at the HST as red light. The redshifting phenomenon allows researchers to tentatively measure the distance the light has traveled. Noting that other factors, such as dust, also might make a galaxy's light red, Mark E. Dickinson of the Space Telescope Science Institute categorized the study as "enticing," but also clarified, "it's not a proof."[91]

*27 June*
NASA's Galileo spacecraft completed a flyby of Jupiter's largest moon, Ganymede, passing only 519 miles (835 kilometers) above the moon's surface. NASA had equipped Galileo with 10 scientific instruments to gather data from the expedition. The instruments began making measurements on 23 June, as the probe approached Ganymede. Images revealed an icy surface, with bright, clean ice covering part of the moon and darker, dirty ice covering the rest. This was the first data gathered during a collection mission planned for 1996 and 1997, in which Galileo would pass Ganymede before completing 11 orbits of Jupiter and surveying the planet's other moons. NASA monitored Galileo's progress through its Deep Space Network, using control stations in California, Spain, and Australia. Scientists and engineers at NASA's Jet Propulsion Laboratory had built the Galileo spacecraft.[92]

## JULY 1996

*2 July*
NASA successfully launched its Total Ozone Mapping Spectrometer Earth Probe (TOMS-EP) aboard the Japanese Advanced Earth Observing Satellite, via an Orbital Sciences Pegasus XL launch vehicle from Vandenberg Air Force Base. A jet flying roughly 7 miles above the California coast released the rocket that was carrying the satellite, placing the TOMS-EP in a successful orbit. The McMurdo Sound Tracking Station in Antarctica acquired the satellite almost immediately and determined that the spacecraft was functioning properly. After it reached a circular orbit, the TOMS-EP was ready to conduct its scientific mission.[93]

Paul F. Holloway, Director of Langley Research Center (LARC) in Hampton, Virginia, announced his retirement. Holloway had served at LARC for his entire aerospace career, beginning in 1960. He became the Center's Director in 1991, overseeing 4,500 employees and a budget of more than US$600 million. During his tenure, Holloway published 42 technical articles in the fields of hypersonic aerodynamics; boundary layer transition and flow separation;

---

[91] Associated Press, "Hubble Scope Peers into Universe's Past with Aid of Starlight," 27 June 1996; NASA, "Findings from Hubble Deep Field Home in on Distant Galaxies," news release 96-123, 26 June 1996.
[92] NASA, "Galileo Readies for Close Flyby of Jupiter's Biggest Moon," news release 96-122, June 1996.
[93] NASA, "Pegasus/TOMS Launch Successful from Vandenberg AFB," news release, 2 July 1996; Associated Press, "BRF—Satellite Launch," 2 July 1996.

analysis of entry flight mechanics; and Earth orbital and planetary space missions. NASA awarded Holloway its Outstanding Leadership Medal, Exceptional Service Medal, Distinguished Service Medal, and Equal Employment Opportunity Medal. Holloway also received the Presidential Rank of Meritorious Service and two Senior Executive Service Distinguished Presidential Rank Awards. NASA Administrator Daniel S. Goldin summarized Holloway's contributions to NASA, saying that his "career spans almost the entire history of the Space Age, and his many achievements at Langley are a tribute to his talent, his professionalism, and his dedication to the importance of NASA aeronautical programs."[94]

Lockheed Martin won the contract to design and build the X-33 test vehicle, a scaled-down model of the reusable launch vehicle (RLV) that NASA planned as an eventual replacement for the Space Shuttle. NASA set a budget of US$941 million for the project. In announcing the new contract, Administrator Daniel S. Goldin expressed NASA's high hopes for the project: "We want to develop technologies that will allow industry to build a vehicle that takes days, not months, to turnaround; dozens, not thousands of people to operate; reliability ten times better than anything flying today; and launch costs that are a tenth of what they are now. Our goal is a reusable launch vehicle that will cut the cost of sending a pound of payload to orbit from [US]$10,000 to [US]$1,000." Lockheed planned to invest US$220 million in the RLV project. The X-33 plan built upon the successful engineering of the Clipper Graham (DC-XA), launched from White Sands Missile Range in New Mexico. The agreement between NASA and Lockheed Martin split the risk between government and industry. Although NASA passed over Rockwell International for the contract, Rockwell designed the vehicle's engine.[95]

*4 July*
Reuters News Service published an internal NASA memorandum written by an independent safety commission, the Aerospace Safety Advisory Panel, cautioning NASA against cutting the budget for the Shuttle program too quickly. According to the panel's analysis, NASA's attempts to reduce costs by conducting fewer Shuttle safety inspections and by transferring responsibility for day-to-day operations of the Shuttle program to United Space Alliance had demoralized NASA's workforce at Kennedy Space Center (KSC). The memo made clear the opinion of the safety inspectors: "Overall, it was the clear consensus of the team that a cooling off period is absolutely necessary if KSC operations are to continue safely . . . without this hiatus the safety risk is likely unacceptable."[96]

*5 July*
In a study released in the journal *Science*, Charles P. Sonett detailed the results of his research measuring changes in the motion of Earth. Sonett reported that tidal forces from the gravitational force of the Moon had caused the Earth's rotation to slow gradually over a period of millions of years. About 900 million years ago, according to Sonett, the length of the Earth's day was only

---

[94] NASA, "NASA Langley Director Paul F. Holloway To Step Down," news release 96-127, 2 July 1996.
[95] NASA, "Lockheed Martin Selected To Build X-33," news release 96-128, 2 July 1996; Jeff Cole, "Lockheed Wins $1 Billion Pact for the Shuttle," *Wall Street Journal*, 3 July 1996; Curt Suplee and Steven Pearlstein, "Lockheed Martin Gets Spacecraft Project," *Washington Post*, 3 July 1996.
[96] Reuters, "Warning Light for NASA's Shuttle Program—Memo," 4 July 1996.

about 18 hours. The gravitational pull of the Moon, Sonett explained, "acts like a brake on the Earth."[97]

*7 July*
Shuttle *Columbia* landed at KSC in Cape Canaveral, Florida, ending the longest Space Shuttle mission ever. The mission, lasting nearly 17 days, set a new space endurance record for the crew. The astronauts had bested their previous record by nearly 7 hours. *Columbia* had carried into space a torch for the 1996 Olympic relay, and upon landing, two members of the crew joined the Olympic torch relay team in a special ceremony. The astronauts had overcome several small malfunctions to conduct the successful mission and land the Shuttle on time. Less than 2 hours before the Shuttle's scheduled arrival, the crew had discovered a clogged cooling line, but the astronauts were able to flush ice from the system to clear the line. In addition, the mission brought to light a potential ongoing problem for *Columbia*. In studying the launch, NASA officials had observed gas and flames issuing from the rocket boosters, probably through the O-ring joints used to connect the booster's fuel segments. Although *Columbia*'s crew was not in danger, NASA officials decided that the Shuttle would not launch again until scientists had determined how well the new water-based adhesive would function under extreme heat. NASA had discontinued the adhesive used previously on the rocket boosters, because the adhesive was an ozone-depleting agent.[98]

*9 July*
NASA announced the recipients of the 1996 Software of the Year Award, honoring those individuals who had created software enhancing NASA's mission and the U.S. aerospace industry. Linked Windows Interactive Data System, developed by NASA's Jet Propulsion Laboratory, and the Tetrahedral Unstructured Software System, developed by LARC, won the 1996 awards. The former system allowed scientists to examine geophysical and climatological data gathered from satellites, and the latter aided researchers in studying problems with "spacecraft, rotorcraft, automotive, turbomachinery, and medical analysis and design."[99]

*14 July*
NASA announced the six-week delay of astronaut Shannon W. Lucid's return from her outpost aboard the *Mir* space station. NASA made the schedule change because of concerns about the reliability of the joint sealant used on the Shuttle's reusable rocket boosters during the June launch of *Columbia*. NASA planned to replace the booster before launching *Atlantis*, scheduled to retrieve Lucid. The reusable rocket boosters, packed with fuel, launch the Shuttle. After pushing the Shuttle into space, the boosters fall off and descend, attached to parachutes, into the ocean. Upon retrieving the rocket boosters, engineers disassemble them into smaller pieces, repack them with fuel, and rebuild them for a new Shuttle launch, sealing each of the rocket's three field joints with an O-ring. NASA sought to fortify these joints further, and the adhesive used to secure them, before launching another Shuttle mission. Space Shuttle Program Manager

---

[97] C. P. Sonett et al., "Late Proterozoic and Paleozoic Tides, Retreat of the Moon, and Rotation of the Earth," *Science* 273, no. 5271 (5 July 1996): 100–104; Associated Press, "Astrodynamics: A Brake on the Earth's Spin," 8 July 1996.
[98] William Harwood, "Record Shuttle Flight Ends; Problem May Delay Next Launch," *Washington Post*, 8 July 1996; Associated Press, "Columbia Ends 17-Day Mission," 8 July 1996.
[99] NASA, "NASA Announces 1996 Software of the Year Award Recipients," news release, 9 July 1996.

Thomas W. Holloway stated that NASA remained confident in the overall design of the rocket booster and in the joint, but wanted to take no avoidable risks.[100]

*15 July*
Astronaut Shannon W. Lucid established a new American record for the longest space mission. Lucid conducted a news conference from the *Mir* space station marking her record of 115 days and 44 minutes, which eclipsed the record of her predecessor aboard *Mir*, Norman E. Thagard. With the changed date for the rendezvous between Shuttle *Atlantis* and *Mir*, NASA predicted that Lucid would also break the record number of 170 days spent by a woman in space.[101]

NASA named the crew members for Mission STS-84, the sixth scheduled docking mission between the Space Shuttle and Russia's *Mir* space station. NASA selected Charles J. Precourt as Mission Commander and Eileen M. Collins as Shuttle Pilot, with Mission Specialists Edward T. Lu and Carlos I. Noriega and European Space Agency astronaut Jean-François Clervoy rounding out the crew. The scheduled *Atlantis* mission also included Mission Specialist C. Michael Foale, who planned to stay on *Mir* for four months.[102]

*18 July*
NASA released new images of Io, one of Jupiter's moons, obtained by the Galileo spacecraft. Michael J. S. Belton of the National Optical Astronomical Observatories pointed out that radical changes had occurred on the moon during the 17 years between Galileo and the first observation of Io, via the Voyager spacecraft in 1979. According to Belton, active volcanoes had altered the color of the moon's surface, forming deposits of sulfur and sulfur dioxide that painted the surface white.[103]

*20 July*
NASA commemorated the 20th anniversary of Viking 1's landing on Mars, the first time that a spacecraft had successfully descended to the surface of the planet. NASA marked the occasion with an address from Administrator Daniel S. Goldin and the celebration of Mars Day at the Smithsonian Institution's National Air and Space Museum. Viking 1 had searched for life on Mars, but at the end of its quest, scientists had found no evidence of life on Mars. As it celebrated its accomplishments, NASA continued its efforts to learn more about Mars throughout its 20th anniversary year, with the development of Mars Global Surveyor and Mars Pathfinder.[104]

*22 July*
While constructing a piping trench, two surveyors at NASA's LARC located the fossil remains of a 3.5 million-year-old, 30-foot (9-meter) baleen whale. The workers found the whale's skull, tympanic bulla, vertebrae, and rib fragments. Langley loaned the remains to the nearby College of William and Mary. Gerald H. Johnson, a geologist at the College of William and Mary, suggested that the discovery provided a clue to the history of the Langley site, indicating that the

---

[100] William J. Broad, "Astronaut's Ride Home Is Delayed for 6 Weeks," *New York Times*, 13 July 1996.
[101] NASA, "Shannon Lucid Breaks a Record; The 20th Anniversary of the Viking Mission to Mars," video advisory V96-84, 15 July 1996; Associated Press, "With One Record Down, She's Still Up," 16 July 1996.
[102] NASA, "Crew Named to Sixth Shuttle–*Mir* Docking Mission," news release 96-139, 15 July 1996.
[103] NASA, "Galileo Finds Big Changes on Jupiter's Volcanic Moon Io," news release 96-143, 18 July 1996.
[104] NASA, "Activities Planned To Commemorate 20th Anniversary of First Mars Landing," news release 96-141, 15 July 1996.

Atlantic shoreline probably used to be located 30–40 miles (48–64 kilometers) west of its present location.[105]

*24 July*
As Congress prepared to vote on NASA's budget, Thomas J. Shulz of the U.S. General Accounting Office (GAO) testified before a U.S. Senate committee that NASA would likely exceed its projected budget for constructing the International Space Station (ISS) and making it operational. NASA refuted the claim. GAO's report suggested that NASA had made unrealistic cost estimates and that its system to monitor costs and schedule was inadequate. The report also outlined GAO's concerns about Russia's continued commitment to the ISS endeavor. Shulz testified that "in a worst-case situation, these issues could threaten the future of the program, especially if they result in significantly higher cost estimates and substantial schedule delays." NASA's spokesperson on the issue Wilbur C. Trafton defended NASA's projected budget and its relationship with Russia. Trafton pointed out that Vice President Albert A. Gore Jr. and NASA Administrator Daniel S. Goldin had procured a new agreement during a July meeting in Russia, indicating the solidity of the partnership between the United States and Russia. GAO's critique occurred against the backdrop of the continuing quest during the mid-1990s to make NASA more cost effective.[106]

*31 July*
NASA Administrator Daniel S. Goldin announced the appointment of Jeremiah F. Creedon as Director of LARC. Creedon succeeded Paul F. Holloway, who had served for 36 years at LARC. With this appointment, Creedon moved up from his position as LARC's Director of the Airframe Systems Program Office to return to the research center where he had begun his engineering career 33 years earlier. During his career at NASA, Creedon had received the Presidential Rank of Meritorious Executive in the Senior Executive Service, NASA's Outstanding Leadership Medal, and the Presidential Rank of Distinguished Executive in Senior Executive Service.[107]

## AUGUST 1996

*1 August*
The Boeing Company announced that it had purchased most of the aerospace and defense holdings of its competitor Rockwell International Corporation. The US$3 billion deal restructured the two companies in drastically different ways, permitting Boeing to increase significantly its potential to win government aerospace contracts, while Rockwell International turned its attention to the lucrative commercial market in automation. Whereas Boeing solidified its established dominance in the commercial jetliner construction business, Rockwell left behind its signature industry—the company had been nearly synonymous with NASA's piloted space program. However, in the face of dwindling government contracts, Rockwell had been concentrating for years on developing its commercial product lines. Although the merger was not a surprise, the sale of the company responsible for engineering the Apollo spacecraft and the B-1

---

[105] NASA, "Ancient Whale Surfaces at NASA Langley," news release 96-145, 22 July 1996.
[106] William E. Clayton Jr., "GAO Criticizes NASA's Estimate on Cost of Building Space Station," *Houston Chronicle*, 25 July 1996; Larry Wheeler, "Report: NASA Budget Much Too Rosy," *Florida Today* (Brevard, FL), 25 July 1996.
[107] NASA, "Dr. Jeremiah Creedon Named Director, Langley Research Center," news release 96-152, 31 July 1996.

bombers marked the end of an era in the aerospace industry. The *Wall Street Journal* estimated that Boeing's share of aerospace and defense revenues would increase to US$8–9 billion annually.[108]

NASA scientist Jay R. Hermon published an article in *Geophysical Research Letters*, claiming that he had used NASA's Total Ozone Mapping Spectrometer spacecraft to demonstrate the dramatic increase, between 1979 and 1992, of solar ultraviolet (UV) radiation reaching the Earth's surface. The article, "UV-B Increases (1979–1992) from Decreases in Total Ozone," explained that the ozone necessary to protect the Earth from the Sun's harmful UV rays was declining. Herman found that annual UV-B exposure had increased 6.8 percent per decade at 55° north latitude and 9.9 percent per decade at 55° south latitude. Both latitudes had significant population centers that emitted harmful emissions destructive to the ozone layer.[109]

*2 August*
NASA and the U.S. Air Force unveiled to the public an "intelligent" aircraft, using a computerized system to "learn as it flies." Researchers had engineered the hypersonic aircraft to take inventory of flight maneuvers performed while operating under remote control, eventually using its acquired knowledge to develop the capability to fly itself. According to the flight researchers, the aircraft's control system would create a continually altering set of "control laws," to optimize flight performance. After the aircraft had perfected its system, it could assist pilots in performing such tasks as landing partially damaged planes. The intelligent aircraft could also enable flight at speeds and altitudes that human pilots could hardly ever achieve, while still maintaining adequate control of an aircraft.[110]

*5 August*
NASA convened a five-member investigation board to determine the cause of the postlanding explosion of the Clipper Graham (DC-XA) rocket on 31 July 1996. The Clipper Graham had successfully flown a 2-minute, 20-second flight profile, but had tipped over and caught fire after landing when one of its four landing gears failed to deploy. The DC-XA had tested successfully three times before the explosion destroyed it. Former astronaut Vance Brand chaired the panel, bringing together representatives from NASA's Dryden Flight Research Center, Marshall Space Flight Center, Langley Research Center, and Kennedy Space Center. Researchers hoped the investigation would illuminate any flaws in the Clipper Graham's design, allowing engineers to make necessary adjustments in future prototypes.[111]

*7 August*

---

[108] The Boeing Company, "Boeing To Acquire Rockwell Aerospace and Defense Units," news release, 1 August 1996; NASA, "NASA Comment on Boeing Corporation Acquisition of Rockwell Aerospace and Defense Businesses," news release N96-51, 1 August 1996; Associated Press, "Boeing To Buy Rockwell International's Aerospace Holdings," 1 August 1996; James F. Peltz, "Rockwell To Sell Space and Defense Divisions to Boeing," *Los Angeles Times*, 2 August 1996; Jeff Cole and Steven Lipin, "Boeing Deal Will Strengthen Company," *Wall Street Journal*, 2 August 1996.

[109] J. R. Herman et al., "UV-B Increases (1979–1992) from Decreases in Total Ozone," *Geophysical Research Letters* 23, no. 16 (1996): 2117–2120; NASA, "Surface Ultraviolet Radiation Levels Have Increased from 1979 to 1992," news release 96-153, 1 August 1996.

[110] NASA, "Intelligent Test Aircraft Unveiled in Wisconsin Today," news release 96-154, 2 August 1996.

[111] NASA, "Clipper Graham Incident Investigation Board Convenes," news release 96-158, 5 August 1996.

NASA reported that a team of scientists from Johnson Space Center and Stanford University had found evidence that "strongly suggested" that primitive life existed on the planet Mars. During a two-year study, the researchers discovered what they believed to be organic molecules of Martian origin on a potato-sized meteorite, which members of the National Science Foundation's Antarctica Meteorite Program had found in 1984. Scientists estimated that the meteorite was about 4.5 billion years old. According to the team's report, published in *Science* magazine, the remnants on the meteorite shared several mineral characteristics indicating biological activity, possibly even containing microscopic fossils of primitive, bacteria-like organisms. The minute scale of the evidence required the careful research of the team of specialists in the fields of microbiology, mineralogy, and chemistry. The largest of the possible fossils measured about $1/100^{th}$ the diameter of a human hair. Only with recently developed technological instruments, including a dual-laser mass spectrometer, were the scientists able to make any substantial observations.[112] The announcement generated immediate excitement in the scientific community and in the mass media, although some scientists responded with a multitude of critical questions. The researchers carefully qualified their report, stating, "It is very difficult to prove life existed 3.6 billion years ago on Earth, let alone on Mars."[113] NASA Administrator Daniel S. Goldin pledged to make samples of the meteorite available to scientists throughout the world, to encourage a full scientific investigation.[114]

*13 August*
NASA released images from its Galileo spacecraft, indicating that water might have existed on Jupiter's moon Europa in the past, or might still exist there. According to scientists studying the images, Europa may have "warm ice" on its surface. Finding such ice would help scientists determine whether Europa's climate could support life. "What we are really looking for is niches that could support life," Ronald Greeley, a Galileo team scientist clarified, at a press briefing releasing the images. At first glance, the pictures, shot from a distance of about 95,700 miles (154,000 kilometers), seemed to depict a series of white stripes stretched across Europa's landscape. However, the stripes were actually ice glaciers, long of interest to scientists. Researchers had long suspected that Europa might be one of the places in the universe possessing water and, therefore, an environment that could support life. NASA Administrator Daniel S. Goldin expressed "skeptical optimism" at the Europa findings and urged the scientific community to examine the pictures further, in a tempered reaction similar to his stance regarding the discovery, one week earlier, of microbiological evidence of life on Mars.[115]

*17 August*

---

[112] David S. McKay et al., "Search for Past Life on Mars: Possible Relic Biogenic Activity in Martian Meteorite ALH84001," *Science* 273, no. 5277 (16 August 1996): 924–930; NASA, "Meteorite Yields Evidence of Primitive Life on Early Mars," news release 96-160, 7 August 1996; Kathy Sawyer, "NASA Releases Images of Mars Life Evidence," *Washington Post*, 8 August 1996. See also Kathy Sawyer, *The Rock from Mars: A Detective Story on Two Planets* (New York: Random House, 2006).
[113] Malcolm W. Browne, "Planetary Experts Say Mars Life Is Still Speculative," *New York Times*, 8 August 1996; Joyce Price, "Clinton Trumpets NASA 'Triumph'; Scientists Skeptical," *Washington Times*, 8 August 1996.
[114] McKay et al., "Search for Past Life on Mars," 924–930; NASA, "Meteorite Yields Evidence"; Sawyer, "NASA Releases Images."
[115] NASA, "Statement by Administrator Daniel S. Goldin on the Release of New Galileo Spacecraft Images of Europa," news release 96-166, 13 August 1996; Reuters, "Galileo Hints at Water on Europa," 14 August 1996; NASA, "Jupiter's Europa Harbors Possible 'Warm Ice' or Liquid Water," news release 96-164, 13 August 1996; Kathy Sawyer, "Jupiter Moon May Contain Ice Floes, Sea," *Washington Post*, 14 August 1996.

The Russian spacecraft *Soyuz TM-24* launched from Baikonur Cosmodrome in Kazakhstan carrying cosmonauts Valery G. Korzun and Alexander Y. Kaleri and the first female French astronaut, Claudie André-Deshays. The crew of *Soyuz* planned to resupply *Mir*'s long-term crew.[116]

*19 August*
The Russian *Soyuz TM-24*, which launched on 17 August, arrived at *Mir*, carrying France's first female astronaut, Claudie André-Deshays. André-Deshays, a rheumatologist, joined Russian cosmonauts Valery G. Korzun and Alexander Y. Kaleri on the flight from Baikonur Cosmodrome in Kazakhstan. She planned to study the effects of weightlessness on the human body during her stay aboard *Mir*. The crew of *Soyuz* joined Shannon W. Lucid, the American astronaut who had been aboard the Russian space station since March. Reporting the safe arrival of André-Deshays at *Mir*, the head of France's space program also announced that France would likely end its short-term space missions conducted aboard Russian vehicles. France intended to focus on longer, less frequent missions, facilitating more detailed research programs. The change in policy went into effect immediately, with France canceling two French-Russian missions planned for 1998 and 1999.[117]

*21 August*
NASA launched its Fast Auroral Snapshot (FAST) Explorer from a test range near Vandenberg Air Force Base in California. Engineers had designed the FAST vehicle to travel about 1,500 miles (2,400 kilometers) above Earth, measuring energetic particles, and magnetic and electric fields, at high altitudes where auroras form, to explore the "physical processes that produce aurora borealis and aurora australis." NASA's Small Explorer Project, tasked with providing frequent and low-cost research missions for astrophysics investigations, had carried out the FAST mission as the second in a series of five missions.[118]

*23 August*
NASA's analysts at Marshall Space Flight Center (MSFC), who had investigated the causes of the *Challenger* explosion, joined the probe into the crash of TWA flight 800, which exploded near Long Island, New York, on 17 July 1996. NASA offered the services of its MSFC investigative team to analyze why the center fuel tank of the Boeing 747 had exploded. Officials from the National Transportation Safety Board led the investigation, with support from a variety of other agencies, including NASA, the U.S. Navy, and the FBI.[119]

*27 August*

---

[116] *Spacewarn Bulletin*, no. 514, 25 August 1996, *http://nssdc.gsfc.nasa.gov/spacewarn/spx514.html* (accessed 29 July 2008); *Milwaukee Journal Sentinel* (WI), "Russian Spaceship Heads for Space Station," 18 August 1996.
[117] Craig Covault and Pierre Sparaco, "French Astronaut Joins Russian/U.S. Mir Crew," *Aviation Week and Space Technology* 145, no. 9 (26 August 1996): 69; Associated Press, "Russian Spaceship Docks with Orbiting Station," 19 August 1996; Associated Press, "Fifth Joint French-Russian Mission May Be Last, Official Says," 19 August 1996.
[118] NASA Goddard Space Flight Center, "FAST Status Report #1: Spacecraft Successfully Launched," news release, 21 August 1996; Bruce Smith, "Fast Launched Successfully," *Aviation Week and Space Technology* 145, no. 9 (26 August 1996): 71.
[119] Robert Davis, "Challenger Experts Aid in TWA Probe," *USA Today*, 23 August 1996; Eric Malnic, "Experts at NASA Join Probe of TWA Crash," *Los Angeles Times*, 23 August 1996.

The Galileo spacecraft lost some of its collected data when a problem with one of its computer processors caused a shutdown of nonessential systems. Project Manager William J. O'Neil said that much of the lost data was of little consequence to researchers, but that data from Galileo's continuous study of Jupiter's magnetic field had suffered a significant loss, creating an information gap of about one week. The episode, which occurred in the midst of NASA's string of impressive research achievements, highlighted the difficulty of directing a data-gathering probe thousands of miles from Earth's surface.[120]

*28 August*
NASA and Orbital Sciences Corporation signed a contract, which was not a cooperative agreement, whereby Orbital Sciences would build a new X-34 reusable launch vehicle, smaller than that originally planned. NASA had first made public the original X-34 program in 1994. NASA had intended the program, also known as the Reusable Small Booster Program, to stimulate the development of a reusable commercial launch vehicle to place smaller payloads in low orbit. In addition, NASA planned for the X-34 program to demonstrate that an industry-led partnership between NASA and the aerospace industry could accomplish this objective within 30 to 36 months and within a fixed government budget. On 12 January 1995, NASA released a final cooperative agreement notice for the X-34 program, and in March 1995, NASA selected Orbital Sciences Corporation as its industry partner. Over the next eight months, as design of the X-34 progressed, problems arose over the selection of an appropriate rocket engine, resulting in NASA's temporarily shutting down the program on 2 November 1995. In January 1996, Orbital Sciences Corporation suggested that NASA explore the potential of a smaller vehicle. Orbital Sciences then issued stop-work orders to all its X-34 subcontractors, effectively ending the program. Because NASA wanted to use remaining federal dollars to develop a smaller vehicle, capable of demonstrating various reusable launch technologies, it rebid the contract in March 1996. On 10 June 1996, NASA selected Orbital Sciences to build the new X-34.[121]

## SEPTEMBER 1996

*1 September*
More than 200 experts met in Sioux Falls, South Dakota, to plan the design and implementation of a new fleet of radar satellites. The group emphasized the recent geological discoveries made with data from radar satellites, especially from those satellites tracking the movements and shifts in Earth's plates that have been impossible to track until recently. The U.S. Geological Survey and NASA's Jet Propulsion Laboratory sponsored the event. Geophysicist Howard A. Zebker described the satellites as a "great leap forward. It's like suddenly having [x]-rays to see inside a body."[122]

*4 September*

---

[120] Kathy Sawyer, "Spacecraft in Orbit Around Jupiter Loses Data," *Washington Post*, 27 August 1996; Associated Press, "Computer Error on Spacecraft Exploring Jupiter," 28 August 1996. See also Michael Meltzer, *Mission to Jupiter: A History of the Galileo Project* (Washington, DC: NASA, 2007).
[121] NASA History Division, "X-33 History Project, Fact Sheet #7: The Policy Origins of the X-33, Part VII: The X-34," March 25, 2000, *http://www.hq.nasa.gov/pao/History/x-33/facts_7.htm*.
[122] William J. Broad, "Satellite Radar Unveils Subtle, Slow Wrinkling of the Planet's Surface," *New York Times*, 3 September 1996.

Hurricane Fran delayed the launch of Shuttle *Atlantis*. NASA managers decided to move the Shuttle off the launchpad when it became clear that the hurricane's path lay through the southeast region of the United States. Because of the slow rollback of the vehicle from the launchpad to the assembly building, managers determined that they would not be able to meet the scheduled 14 September departure date.[123]

*5 September*
Astronomers using the Hubble Space Telescope identified 18 star clusters that they believed to be embryonic galaxies. According to the *New York Times*, because the clusters resided 11 billion light-years from Earth, essentially, the scientists had glimpsed back in time. The observations supported a theory that galaxies begin with stars grouping together. The astronomers from Arizona State University and the University of Alabama called the discovery "the first page in an otherwise blank book."[124]

NASA named astronaut Michael Lopez-Alegria to replace astronaut Wendy B. Lawrence as NASA's Director of Operations at Gagarin Cosmonaut Training Center in Star City, Russia. The job of the director was to coordinate the training of U.S. astronauts headed for the *Mir* space station and to act as liaison between NASA and the Russian Space Agency.[125]

*6 September*
NASA awarded grants to fund 19 studies that researchers had proposed in response to a NASA research announcement. NASA awarded a total of US$2.9 million for the studies, which focused on innovative mission concepts in space physics. NASA expected most of the studies to last one or two years and to cost about US$100,000 each year. NASA planned to use the studies in forming its space science strategic plan and in considering future missions.[126]

*7 September*
Because of delays in launching Shuttle *Atlantis*, astronaut Shannon W. Lucid broke Elena V. Kondakova's record for longest single flight in space by a woman. NASA engineers had delayed the launch because they suspected that the O-rings on the solid rocket boosters were faulty and wanted to replace them. During Lucid's stay in space aboard *Mir*, she and the two Russian cosmonauts had conducted a variety of scientific experiments.[127]

*11 September*
NASA announced that engineers from NASA and Lockheed Martin had completed successful tests on a prototype, lightweight, external fuel tank for the Shuttle. The tests measured the tank's strength and reliability under conditions exceeding recommended flight certification. NASA's External Tank Project Manager Parker V. Counts reported that the tests had demonstrated that the unique composition of the very lightweight tanks did not reduce their strength. NASA had

---

[123] NASA, "Atlantis Moved to VAB; STS-79 Delayed," launch advisory 96-181, 4 September 1996.
[124] Lee Bowman, "Possible View of Galaxies Being Born," *Washington Times*, 5 September 1996; John Noble Wilford, "Looking Back 11 Billion Light-Years, a Glimpse at Galaxy Birth," *New York Times*, 5 September 1996.
[125] NASA, "Lopez-Alegria To Replace Lawrence as NASA Manager to Russia," news release 96-182, 5 September 1996.
[126] NASA, "NASA Selects 19 Innovative Space Physics Mission Concepts," news release 96-184, 6 September 1996.
[127] NASA, "Lucid To Break Record; Latest Images of Hurricane Fran," video advisory V96-111, 6 September 1996.

awarded Lockheed Martin the contract to produce the lighter tanks, to enable the Shuttle to carry cargo to the high inclination orbit where the International Space Station (ISS) would reside. To build the tanks, NASA had used aluminum lithium, a material both lighter and stronger than the metal alloy used in previous tank construction. NASA had designed the tanks to weigh 7,500 pounds (3,400 kilograms) less than the models used on earlier Shuttles.[128]

*12 September*
Scientists at NASA's Goddard Space Flight Center (GSFC) received the first images from the United States' Total Ozone Mapping Spectrometer (TOMS), as the instrument orbited aboard the Japanese Advanced Earth Observing Satellite (ADEOS). GSFC scientists noted the superior quality of the initial images. ADEOS continued a long line of satellites launched to aid scientists in observing total levels of ozone and volcanic sulfur dioxide. Because of data obtained from the TOMS instruments deployed by NASA, millions of Americans became aware of the *ozone hole* and began to use *ozone* as a household word. The latest TOMS instrument joined a series of instruments aboard ADEOS, including the Improved Limb Atmospheric Sounder, the Interferometric Monitor for Greenhouse Gases, and the Retroreflector in Space.[129]

*13 September*
In Marshall Space Flight Center's (MSFC's) underwater weightless simulator, NASA began testing a remotely controlled robot created by a team of researchers from the University of Maryland. Team leaders nicknamed the robot Ranger, predicting that it would one day assist astronauts aboard the Shuttle or the ISS. Project Manager Joseph Parrish pointed out that astronauts often use valuable spacewalk time completing mundane tasks, such as collecting tools, establishing footholds, and cleaning up. Scientists hoped Ranger could perform such tasks under the command of an operator located in space or even that of an operator at NASA's command center. The 8-foot-long (2.4-meter-long), 1,700-pound (770-kilogram) robot used gas thrusters, robotic arms, lights, and cameras. The academic project filled a void in experimental robot design, a program affected by NASA's budget cuts. Allowing university researchers and professors to design and test the robot cost NASA about US$8 million, far less than NASA would pay for similar projects contracted through an aerospace firm. During the tests, the University of Maryland's team operated Ranger remotely from its campus hundreds of miles away from MSFC's facility in Huntsville, Alabama.[130]

*16 September*
Shuttle *Atlantis* launched from Kennedy Space Center (KSC) in Cape Canaveral, Florida. Hurricane Fran had delayed the launch and the subsequent Shuttle–*Mir* rendezvous. The mission was the fourth *Mir* docking mission for a U.S. Space Shuttle. William F. Readdy commanded the six-astronaut crew aboard *Atlantis,* including Pilot Terrence W. Wilcutt and Mission Specialists Jay Apt, Thomas D. Akers, Carl E. Walz, and John E. Blaha. During Mission STS-79, the Shuttle crew planned to retrieve astronaut Shannon W. Lucid from *Mir*, leaving astronaut John Blaha as her replacement. The Shuttle carried 4,600 pounds (2,090 kilograms) of supplies to *Mir*.

---

[128] NASA, "Shuttle Super Lightweight Fuel Tank Completes Tests," news release 96-186, 11 September 1996.
[129] NASA, "First Global Image of Total Atmospheric Ozone Obtained from NASA Instrument Aboard Japanese Satellite," news release 96-188, 16 September 1996.
[130] Martin Burkey, "Ranger Gets a Workout," *Huntsville Times* (AL), 13 September 1996.

Mission STS-79, the 17th flight of Shuttle *Atlantis*, entailed the largest transfer of supplies ever made to *Mir*.[131]

*19 September*
The White House released a new national space policy, the first such directive since the Cold War. White House policy advisors constructed the plan over an 18-month period, amidst government-wide calls for tightening NASA's budget. The policy statement called for controlling costs by using private-sector industry to build, develop, and operate selected space technologies. The directive also advocated drawing together civilian and military space activities, and forging closer ties with foreign countries conducting space research. The Clinton administration reversed several decisions of President George H. W. Bush, most noticeably the former President's call to land an astronaut on Mars. Space experts called the initiative too costly and dangerous. Instead, the Clinton plan relied on robots to explore Mars and targeted the ISS as the hub of human-based space exploration, as well as enshrining the "faster, better, cheaper" mantra of NASA Administrator Daniel S. Goldin. Congress planned to hold a funding summit in the months following the release of the policy, to bring together key political, military, and science officials.[132]

*26 September*
Astronaut Shannon W. Lucid returned to Earth after her record-setting space sojourn on Russia's *Mir* space station. Space Shuttle *Atlantis* touched down at KSC in Cape Canaveral, Florida, delivering Lucid home to a celebration honoring her 188 days spent in space, most of it aboard *Mir*. During her mission, Lucid had traveled 75 million miles (121 million kilometers), the equivalent of 157 trips to the Moon and back. Doctors stood by to greet the Shuttle, expecting Lucid to experience some muscle weakness as her body re-accustomed itself to the effects of gravity. President William J. Clinton called to congratulate Lucid, saying, "We are all so proud of you."[133]

*28 September*
During a top-level meeting between officials of the United States and Russia, NASA made public a shift in its policy, announcing that U.S. astronauts would not always have command aboard the ISS. In a compromise meant to encourage Russia to build its portion of the station, NASA conceded that astronauts from other countries would have the opportunity to command the ISS for significant periods.[134]

*30 September*

---

[131] NASA, "Space Shuttle Mission STS-79—Press Kit," news release 96-179, September 1996; William Harwood, "Shuttle Arrives for Astronaut's Long-Awaited Pickup," *Washington Post*, 19 September 1996.
[132] Kathy Sawyer, "White House Releasing New National Space Policy," *Washington Post*, 19 September 1996; Associated Press, "Space Policy," 19 September 1996; The White House National Science and Technology Council, "Fact Sheet: National Space Policy," 19 September 1996, *http://history.nasa.gov/appf2.pdf*.
[133] Associated Press, "Space Shuttle," 26 September 1996; Associated Press, "Remarks by President Clinton in Conversation with Astronaut Lucid," 26 September 1996; United Press International, "Lucid's Six-Month Mission Ends," 26 September 1996.
[134] Knight-Ridder News Service, "Command of Space Station Will Be Shared by Other Nations," 28 September 1996.

Dryden Flight Research Center (DFRC) at Edwards Air Force Base in California commemorated its 50th anniversary with a celebration called Discovery Through Flight Research. The Center got its start when five aeronautical engineers from Langley Research Center arrived to explore flight beyond the speed of sound, in preparation for the X-1 tests, featuring pilot Charles E. "Chuck" Yeager at the controls. Researchers at DFRC contributed to a wide variety of aeronautical achievements, including breaking the sound barrier and helping launch the Space Shuttle program. At the age of 50, DFRC employed 900 government and civilian contractor employees.[135]

With the push of a button at GSFC, NASA discontinued use of the International Ultraviolet Explorer satellite, which had been in orbit since 1978. When NASA sent a signal to the satellite to empty its fuel, the satellite effectively went to sleep, spinning off into space. Although NASA had originally designed the satellite to spend only three years in space, scientists continued to use it to gather data far beyond the spacecraft's anticipated lifespan. NASA ended control of the satellite to save between US$1 and US$2 million annually in operating costs. GSFC had operated the satellite as part of a three-way agreement with the European Space Agency and the Particle Physics and Astronomy Research Council of the United Kingdom. More than 2,000 researchers from around the world had made observations from the satellite, research that had directly led to the publication of more than 3,200 papers in peer-reviewed scientific journals.[136]

## OCTOBER 1996

*1 October*
United Space Alliance (USA), a joint venture of Lockheed Martin Corporation and Rockwell International, took over daily operations of NASA's Space Shuttle fleet. NASA had previously used private contractors. NASA Administrator Daniel S. Goldin marked the occasion as "the first day of a new space program in America." Financial experts forecast that the government-private industry pact might be worth as much as US$12 billion by the end of its 10-year venture. According to the agreement, USA shouldered the responsibility for the 12 Shuttle-related contracts carried out at Kennedy Space Center (KSC) in Cape Canaveral, Florida, and Johnson Space Center (JSC) in Houston, Texas. NASA stated that USA might take over additional tasks in the future. NASA had decided to consolidate the Shuttle operations with USA to improve cost efficiency, a vital step in light of congressional budget cuts. Kent Black, the CEO of USA, predicted that the move would save NASA as much as 20 percent of its operating budget for the Shuttle in the first year of the contract alone. Even with the turnover, NASA maintained complete control over planning Shuttle missions, selecting astronauts, and approving launches of the Shuttle. NASA planned to take on an oversight role similar to that of the Federal Aviation Administration, rewarding USA with bonuses for savings and imposing monetary penalties if the company failed to achieve "very good" safety ratings.[137]

---

[135] *Antelope Valley Press* (Palmdale, CA), "Dryden Marks Half-Century of Achievements," 25 September 1996.
[136] Associated Press, "Ancient Spacecraft," 1 October 1996.
[137] Associated Press, "NASA Turns Shuttle Operations Over to Private Industry," 30 Sept 1996; Reuters, "NASA Signs $7 Billion Shuttle Pact," 30 September 1996; United Press International, "NASA signs $12 Billion Shuttle Pact," 1 October 1996.

NASA introduced a new technology with the potential to make much smaller and more efficient electronic devices: the Thin-Layer Composite-Unimorph Piezoelectric Driver and Sensor (THUNDER). Researchers at Langley Research Center (LARC) recognized the potential of piezoelectric material, because of its well-known capability to generate movement when subjected to an electric current. NASA hoped that the THUNDER technology would improve devices in "electronics, optics, jitter (irregular motion) suppression, noise cancellation, pumps, valves and a variety of other fields." LARC's interdisciplinary team had improved upon already available commercial-grade piezoelectric material, producing a more durable, cheaper, and more effective product. *R&D Magazine* announced that it would honor THUNDER as one of its top 100 most technologically significant new products of the year. Six companies signed agreements with NASA to develop THUNDER, and a dozen other companies expressed interest in negotiating similar pacts.[138]

*3 October*
China announced it would participate in an international collaborative space research project for the first time in its history. Chinese scientists planned to assist their colleagues from Russia and the United States in the future *Discovery* mission. Although China had no plans to send an astronaut on the Shuttle, Chinese aerospace engineers volunteered to construct the sophisticated magnets needed to build a magnetic spectrometer. The project was the first time the usually secretive Chinese space science program had opened itself to outside scrutiny.[139]

NASA discovered the second of two misplaced hand tools left inside the propulsion system of the recently landed Shuttle *Atlantis*. Although the Shuttle had experienced no problems during its 10-day flight, NASA formed an investigative panel to determine who had left the tools inside the Shuttle's engine compartment after completing work. Inspectors had discovered the first missing tool inside the electronic control compartment of the Shuttle's reusable rocket booster when they retrieved the booster from the Atlantic Ocean shortly after the launch. Workers discovered the second tool during a routine postflight inspection of *Atlantis*'s rocket engine compartment. Although the tools likely posed no threat to the safety of the Shuttle crew, a NASA representative said that, for safety reasons, NASA intended to find out how workers had made the error and, if possible, to identify the Shuttle contractor responsible.[140]

*8 October*
President William J. Clinton signed legislation ending funding for the federal helium reserve. The United States, through the Bureau of Mines, had bought and stockpiled helium since the 1960s—expending more than US$250 million. Experts estimated that the U.S. government's helium reserve could supply the world for more than 10 years and the federal government for 80. The legislation to end the reserve, spearheaded by U.S. Representative Christopher Cox (R-CA), closed down the helium facility outside Amarillo, Texas. NASA planned to sell the helium slowly, over a period of 18 years, so that the sales would not destabilize the worldwide helium market. The closure had the potential to affect NASA, the federal government's largest user of

---

[138] NASA, "NASA Rolls Out Award-Winning 'Thunder'," news release 96-197, 1 October 1996.
[139] United Press International, "China To Assist U.S. Space Program," 3 October 1996.
[140] Mark Carreau, "Another Wrenching Experience for Shuttle," *Houston Chronicle*, 3 October 1996.

helium. At the time of the reserve's closure, NASA was using about 70 million cubic feet (2 million cubic meters) of helium annually to pressurize Shuttle fuel tanks, among other uses.[141]

*10 October*
NASA researchers at Dryden Flight Research Center (DFRC) completed the testing of the F-16XL at Edwards Air Force Base. The tests had taken more than one year to complete. Scientists were studying how the modified F-16 flew with US$14 million laminar-flow gloves on its wings. The data they collected demonstrated that the glove had reduced the turbulent layer of air that passes over an airplane's wing during flight. Project Manager Carol A. Reukauf called the tests successful. NASA hoped the results would provide manufacturers with a new technology that would improve aeronautical design, decreasing air drag on airplanes and, consequently, reducing the fuel required to fly.[142]

*11 October*
NASA marked the 50-year anniversary of the Bell X-1 aircraft's first flight, which took place at the site of Edwards Air Force Base, home of NASA's DFRC. According to Jay Miller, the author of *The X-Planes: X-1 to X-33*, the X-1 had given flight researchers and engineers their first full-scale tool to study transonic aerodynamics. The Bell Aircraft Corporation, the U.S. Army Air Forces (predecessor of the U.S. Air Force), and NASA's predecessor, the National Advisory Committee for Aeronautics, had collaborated to get the first test plane off the ground. The X-1 test not only provided aeronautical data, but also pioneered the test methods later adopted by DFRC flight researchers.[143]

*15 October*
Russia postponed the scheduled launch of Progress M-33, a cargo vehicle loaded with supplies for the two Russian cosmonauts and the one American astronaut aboard the *Mir* space station. The delay highlighted once again the severe budget crisis of the Russian Space Agency. A lack of funding held up the production of the rocket booster needed to send the craft into space. The crew on *Mir* had sufficient supplies aboard the space station, but missed the visit from Earth. Besides supplying food, regular planned visits provided needed contact for the space dwellers. The postponement was the fourth time in 1996 that Russia had delayed the scheduled launch to *Mir* because of problems paying for the production of *Soyuz*-U boosters. Reuters announced that because of the delay, a group of laboratory monkeys, already wired to monitor the effects of space travel, would die without ever leaving Earth.[144]

*18 October*
NASA announced the completion of a major benchmark in the development of its Advanced X-ray Astrophysics Facility (AXAF), as engineers completed the delicate task of assembling the high-resolution, cylindrical mirrors needed for the facility's telescope. Engineers aligned and cemented the mirrors into place at the facility in Rochester, New York. Unlike optical telescopes,

---

[141] Associated Press, "Clinton Signs Law Killing Federal Helium Reserves," 9 October 1996.
[142] Jay Levine, "Laminar Flow Tests Successful," *Antelope Valley Press* (Palmdale, CA), 10 October 1996.
[143] NASA, "The X-Planes: 50 Years of High Desert 'Right Stuff'," news release 96-204, 10 October 1996.
[144] Reuters, "Russia Postpones Mir Supply Rocket Mission," 16 October 1996.

the AXAF used a series of shallow mirrors shaped like cylindrical cones to produce its image. AXAF's mirrors were the largest set of such mirrors ever constructed.[145]

*19 October*
For the first time in its 30-year history, KSC in Cape Canaveral, Florida, welcomed the public for an open house. Nearly 40,000 visitors came to KSC for the occasion. NASA allowed visitors to drive directly down the road used by the astronauts and up to Shuttle *Columbia*, sitting on the launchpad. Many visitors commented on how worn the Shuttle looked, after its many launches and reentries. The details of the Shuttle's appearance were not visible to viewers watching the Shuttle on television. NASA officials also opened the Vehicle Assembly Building, one of the largest structures in the world. Astronauts were available to greet the public.[146]

*22 October*
NASA announced the retirement of Jay F. Honeycutt, Director of KSC. A search for Honeycutt's successor began immediately following the announcement. Honeycutt had joined NASA in 1966 after working as an engineer for Redstone Arsenal. He had begun as an engineer in flight operations and had worked at JSC, later training Apollo astronauts for the lunar landings, and had moved to NASA Headquarters in 1981. Honeycutt had transferred to KSC in 1989 as Director of Shuttle Management and Operations and had become Director of KSC in 1995. Honeycutt had won two Exceptional Service Medals, the Special Achievement Award, NASA's Outstanding Leadership Medal, NASA's Equal Employment Opportunity Award, and the Meritorious Executive Presidential Rank Award.[147]

*24 October*
DFRC test pilot Edward T. Schneider received the American Institute of Aeronautics and Astronautics Chanute Flight Test Award for his work testing the F-18 High Alpha Research Vehicle at extremely high angles of attack. The Institute honored Schneider at the World Aviation Congress and Exposition, held in Los Angeles, California. Schneider had participated in NASA's Attack Research Program, meant to improve the control and maneuverability of high-performance aircraft. He tested the thrust-vectoring control system installed in the F-18, which allowed pilots to make high-angle motions and still maintain control of their planes. Schneider spent nine years testing the F-18 High Alpha Research Vehicle during his distinguished 18-year career as a test pilot. Schneider had received NASA's Exceptional Service Medal in 1996.[148]

*30 October*
A NASA research team published in the *Journal of Geophysical Research* its findings on urban-like pollution over the tropical South Atlantic Ocean. The team used data gathered from an experiment called the Transport and Chemistry near the Equator of the Atlantic (TRACE-A), conducted with the Brazilian Space Agency and scientists from South Africa. The scientists suspected that the high concentrations of ozone, like those found in densely populated urban

---

[145] NASA, "Completion of Mirror Assembly Marks Milestone for NASA's Advanced X-ray Astrophysics Facility," news release 96-212, 18 October 1996.
[146] Reuters, "Kennedy Space Center Opens Doors to Fans," 21 October 1996.
[147] NASA, "KSC Director Jay Honeycutt To Retire," news release 96-215, 22 October 1996.
[148] *Antelope Valley Press* (Palmdale, CA), "Dryden Pilot Receives Flying Award," 25 October 1996, http://www.nasa.gov/lb/centers/dryden/news/Biographies/Pilots/bd-dfrc-p014.html (accessed 5 August 2007).

areas, resulted from seasonal burnings in South America and Africa. According to the NASA researchers, the presence of pollution over the tropical ocean confirmed that humans had extended their negative impact on "traditionally pristine air" far beyond reaches directly above land. According to one researcher, the results of TRACE-A greatly improved the scientific community's understanding of the atmospheric chemistry of the Southern Hemisphere.[149]

## NOVEMBER 1996

*4 November*
NASA released images taken with the Hubble Space Telescope, revealing a dust storm on Mars nearly the size of the state of Texas. The storm appeared in the pictures of the planet as a 600-mile-long (965-kilometer-long), salmon-colored patch. Scientists believed this was the first time that one of the huge storms had neared the planet's northern polar ice cap.[150]

*6 November*
A team of scientists from the United States, the United Kingdom, and Australia discovered new evidence that pushed back considerably the scientifically estimated date of the origin of life on Earth. The researchers projected that the Earth's age might be 350 million years older than previously thought. The research team, led by Gustaf Arrhenius, published its findings in the 7 November issue of *Nature*. The scientists argued that the presence of apatite (basic calcium phosphate) in Earth's oldest known sediment sequences suggested that life had emerged on Earth 3.85 billion years ago. The scientific community met these findings with both excitement and skepticism.[151]

*7 November*
NASA spacecraft managers reported the failure of two scientific satellites. NASA had designed the Scientific Applications Satellite-B (SAC-B), launched three days earlier from NASA's Wallops Flight Facility, to survey solar flares and gamma-ray bursts. The project was a cooperative effort between NASA and Argentina. NASA scientists had intended the second satellite, HETE (High Energy Transient Experiment), to remain dormant until it detected sunlight, at which time it would deploy, transmitting signals. Scientists doubted this would ever occur, because the satellite had fallen into a tumbling pattern. Officials hypothesized that the satellites had not deployed properly and, therefore, had orbited uselessly, lacking electrical power. NASA valued both satellites at more than US$20 million. The launch failure was the third miscue in two years for the Pegasus rocket program. The Orbital Sciences Corporation of Dulles, Virginia, developed and ran the program.[152]

---

[149] NASA, "NASA Team Finds Urban-Like Pollution in Tropical South Atlantic," news release 96-220, 30 October 1996.
[150] Reuters, "Hubble Shows Texas-Size Dust Storm on Mars," 4 November 1996.
[151] S. J. Mojzsis et al., "Evidence for Life on Earth Before 3,800 Million Years Ago," *Nature* 384, no. 6604 (7 November 1996): 55; Malcolm W. Browne, "Evidence Puts Date for Life's Origin Back Millions of Years," *New York Times*, 7 November 1996.
[152] John Mintz, "Orbital Sciences Fails in Another Launch," *Washington Post*, 5 November 1996; Associated Press, "NASA Reports Failure of Scientific Satellites," 7 November 1996; NASA, "SAC-B/HETE Spacecraft No Longer Operational," news release 96-231, 7 November 1996.

NASA launched into space its unpiloted Mars Global Surveyor, the first in a series of missions to explore the Red Planet further. The Mars Global Surveyor Mission and vehicle embodied a shift in the United States' approach to space exploration. With the expressed intent to make exploration as affordable and reliable as possible, NASA had developed the smaller Mars Global Surveyor, measuring 5 by 5 by 10 feet (1.5 by 1.5 by 3 meters) and designed to carry less fuel than Mars Observer. Mars Global Surveyor cost NASA US$150 million to develop, whereas Observer had cost nearly US$1 billion and had failed to reach Mars after its 1992 launch. Surveyor's ability to function with less fuel than previous models heightened the chances that it would succeed in its mission. NASA predicted that after ascending to the desired 235-mile-high (378-kilometer-high) orbit, the robotic Global Surveyor would have to travel nearly halfway around the Sun before catching up with Mars in September 1997. The spacecraft would then begin mapping the Martian atmosphere and surface, a survey planned to last 687 days. Mission objectives included searching for evidence of life on Mars, gathering data about the Martian climate, learning more about the planet's geology, and determining what resources were necessary to support future human missions to Mars. According to NASA, the search for water was the unifying theme of all missions to Mars. The Mars Global Surveyor Mission continued the exploration of Mars that NASA had begun with the Mariner and Viking missions of the 1960s and 1970s.[153]

*12 November*

In its annual contest, *Popular Science* magazine honored NASA's experimental flight-control system as one of the best technological developments of 1996. NASA had installed the system, developed for the use of NASA and the U.S. Air Force, in a remotely piloted aircraft called LoFLYTE (Low Observable Flight Test Experiment). On board the aircraft, the system used techniques involving neural networks to "learn" to fly, actually acquiring flight skills by mimicking the actions of the pilot controlling the plane from the ground. Neural networks are systems that learn skills by doing them. Scientists hoped the technology might eventually have an impact on commercial and military aviation.[154]

A malfunction in an unpiloted test plane forced NASA officials to detonate the plane high above Edwards Air Force Base in California. The lightweight plane, dubbed Theseus, veered out of control during a routine test flight. Aurora Flight Systems had developed the US$5 million plane for long-term flights to observe climatic conditions from altitudes as high as 60,000 feet (18,300 kilometers). A spokesperson for Aurora indicated that the company had insufficient data to determine the cause of the failure. NASA had contracted with Aurora Flight Systems to develop the plane as part of its Mission to Planet Earth initiative. At the time of the forced detonation, NASA had not designated funds to develop any subsequent Theseus models.[155]

*15 November*

---

[153] Associated Press, "Mars Probe Heads for Mapping," 8 November 1996; Paul Hoversten, "Surveyor Sets Off for Mars," *USA Today*, 8 November 1996; NASA, "1996 Mars Mission—Press Kit," news release 96-207, November 1996.
[154] NASA, "Aircraft Flight-Control System Wins 'Best of What's New'," news release 96-233, 12 November 1996.
[155] David Colker, "Pilotless Test Plane Blown Up After It Veers Off Course," *Los Angeles Times*, 13 November 1996; Jim Skeen, "Pilotless Aircraft Crashes at Edwards," *Antelope Valley News* (Palmdale, CA), 13 November 1996.

NASA appointed Samuel L. Venneri as Chief Technologist at NASA Headquarters, reporting directly to Administrator Daniel S. Goldin. The appointment placed Venneri in charge of all of NASA's technology policy initiatives and programs. Venneri had served at NASA since 1981, after working as an aerospace consultant for Swales and Associates and as an engineer for Fairchild Space Electronics.[156]

*17 November*
A Russian nuclear-powered Mars probe crashed into the South Pacific Ocean, about 500 miles (800 kilometers) southeast of New Zealand. The U.S. Space Command, located in Colorado Springs, Colorado, monitored the descent of the disabled craft, but was unsure exactly when and where the probe had hit the ocean. Russian space scientists assured NASA that no danger of nuclear contamination existed. A handful of experts, however, cautioned that in an extremely unlikely, worst-case scenario, the impact could result in a small but lethal plutonium cloud. President Clinton, vacationing in Australia at the time, offered the services of U.S. teams trained to locate and recover stray nuclear materials. Russian officials believed that the probe, named Mars-96, had failed to reach orbit because of a malfunction during the fourth stage of the Proton-K rocket's ascent. Mars-96 carried plutonium pellets the size of eraser heads, designed to withstand the pressures of entering Mars's atmosphere. The probe's failure was a setback for the international effort to explore Mars, as well as for the Russian space program. Donna L. Shirley of NASA's Jet Propulsion Laboratory, the chief of the U.S. Mars effort, called the failure a "terrible, terrible tragedy" for researchers. Some people speculated that the crash would prompt a rebirth of protests against using nuclear materials in spacecraft.[157]

NASA officials confirmed that tense negotiations with Russia over the composition of the first crew to travel to the International Space Station (ISS) had resulted in Russia's withdrawing one of its most senior and well-trained cosmonauts from consideration. NASA had selected U.S. astronaut William M. Shepherd to lead the 1998 mission and veteran space traveler Anatoly Y. Solovyev to accompany Shepherd. Each country strongly insisted that its own astronaut lead the international venture. In a letter to Russian Prime Minister Viktor S. Chernomyrdin, Russian space officials stated, "Despite the Russian efforts to make experienced cosmonaut Solovyev the crew commander, the American side is [pre]emptorily insisting on the candidacy of American astronaut Shepherd." Because the United States had insisted that Shepherd command the mission, Russian officials threatened that Russian Space Agency cosmonauts would not cooperate, claiming, "experienced Russian crew commanders doubt the expediency of their participation." The standoff highlighted the diplomatic difficulty of holding together the international coalition necessary to construct the ISS.[158]

*19 November*
Space Shuttle *Columbia* launched from Kennedy Space Center in Cape Canaveral, Florida, at 2:55 p.m. (EST). In addition to a five-day delay because of the weather, NASA had delayed the mission for a week so that its engineers could conduct a careful check of possible problems with the rocket nozzles. News agencies reported that the fifteen-year-old *Columbia*, the nation's

---

[156] NASA, "Venneri Named NASA Chief Technology Officer," news release 96-240, 15 November 1996.
[157] Warren P. Strobel, "Russian's Mars Probe Crashes," *Washington Times*, 18 November 1996; Richard Boudreaux and K. C. Cole, "Failure of Mars Mission Called Scientific Tragedy," *Los Angeles Times*, 19 November 1996.
[158] Reuters, "Tug of War Knocks Cosmonaut off Crew," 18 November 1996.

oldest Space Shuttle, was carrying into space sixty-one-year-old F. Story Musgrave, the oldest astronaut. With his sixth spaceflight, Musgrave became the first astronaut to fly on all five of NASA's Shuttles. In Mission STS-80, NASA's final Shuttle flight of 1996, the crew planned to deploy and retrieve two free-flying spacecraft, conduct spacewalks, and carry out a number of microgravity research experiments. Additionally, NASA and the National Institutes of Health planned to collaborate on tests further exploring how the human body reacts to the space environment. Only hours after *Columbia*'s launch, Shuttle crew released the U.S.-German Orbiting Retrievable Far and Extreme Ultraviolet Spectrometer, designed to make as many as 300 observations of stars and interstellar gas during the Shuttle's mission.[159]

*21 November*
The European Telecommunications Satellite Organization (Eutelsat) launched from Cape Canaveral, Florida, aboard a Lockheed Martin Atlas 2A rocket, a satellite aimed at serving the burgeoning European television market. The satellite *Hot Bird 2* cost US$100 million to develop and was powerful enough to beam television images throughout Europe and the Middle East. Eutelsat led a consortium of co-owner companies from the 45 different countries that co-own the satellite. Eutelsat assumed responsibility for operating *Hot Bird 2*, calling it "the most powerful commercial satellite launched to date."[160]

*25 November*
NASA returned the record-setting, solar-powered, research aircraft Pathfinder to flight-testing at Dryden Flight Research Center. Pathfinder had set an altitude record of more than 50,000 feet (15,000 meters) during a flight on 11 September 1995. Wear and tear during previous operations, as well as a ground accident inside a hangar, had damaged the craft. NASA had hired AeroVironment to repair the remotely piloted plane. AeroVironment made several improvements and rebuilt the damaged portions of the aircraft. The company installed stronger rib structure and higher-efficiency solar cells. NASA had adopted Pathfinder in the 1980s after the aircraft's release from a classified military program.[161]

*28 November*
NASA canceled a spacewalk for the astronauts aboard Shuttle *Columbia* when the astronauts could not open the outer air lock. After communicating for more than 2 hours with Mission Control in Houston, with officials instructing them to "put as much force as you feel comfortable applying to the handle," *Columbia*'s crew ended preparations for the spacewalk. During the spacewalk, the astronauts had intended to test a large crane and various power tools planned for the ISS.[162]

*29 November*

---

[159] Reuters, "Shuttle Crew with Oldest Astronaut Is Set To Fly," 19 November 1996; John Noble Wilford, "Shuttle Blasts Off on Mission with 2 Scientific Satellites," *New York Times*, 20 November 1996; NASA, "Space Shuttle Mission STS-80—Press Kit," news release 96-206, November 1996; Associated Press, "Astronauts Release Ultraviolet Telescope," 20 November 1996.
[160] Todd Halverson, "Atlas Sends $100 Million Satellite into Orbit," *Florida Today* (Brevard, FL), 22 November 1996.
[161] NASA, "Solar-Powered Pathfinder Returns to Skies," news release, 25 November 1996.
[162] William Harwood, "Spacewalk Called Off as Air Lock Hatch Won't Open," *Washington Post*, 28 November 1996.

The Russian-built Tu-144LL supersonic passenger jet, which U.S. and Russian aerospace industries and NASA had modified for commercial purposes, performed well in its first test flight after refurbishment. The aircraft took off from the Zhukovsky Airfield south of Moscow. During the Tu-144LL flight, researchers studied the impact of excessive speeds on the aircraft's internal and exterior surfaces, engine temperature, and handling. The aircraft's body, constructed primarily from light aluminum alloy, held up well during the initial test. The Tu-144LL project supported NASA's High-Speed Research Program, which had begun in 1990, to encourage the development of technologies supporting supersonic commercial travel. The Boeing Company had led the U.S. industry involvement in the project, with support from McDonnell Douglas Aircraft, Rockwell International Corporation, Pratt & Whitney, and General Electric Company. The project's aim was to prepare flight-research facilities to adapt to market demands of the future. Aviation experts predicted that a commercial market for advanced supersonic transport would evolve early in the $21^{st}$ century.[163]

## DECEMBER 1996

*2 December*
President William J. Clinton awarded the Congressional Space Medal of Honor to astronaut Shannon W. Lucid, the first woman to receive the award. Praising Lucid as a "determined visionary," President Clinton presided over the Oval Office ceremony, as the fifty-three-year-old astronaut recounted her 188 days aboard the *Mir* space station. Lucid recalled fondly her time spent with the two Russian cosmonauts, a period characterized by "working together, laughing together, and having a good time together." Lucid's husband Michael attended the ceremony, along with U.S. Senators John H. Glenn Jr. (D-OH) and Conrad Burns (R-MT), NASA Administrator Daniel S. Goldin, and Russian Ambassador Yuli Vorontsov. The U.S. Congress had created the Space Medal of Honor in 1969, awarding it to eight astronauts before Lucid. The award is distinct from the Medal of Honor, awarded for the highest acts of military service and extraordinary heroism on the field of combat.[164]

Scientists announced the discovery of frozen water on the Moon, calling into question the long-held belief that the Moon lacked any form of hydration. The Clementine spacecraft had found the ice while using radar signals to examine the depths of the Moon's craters. The discovery was the by-product of a US$75 million mission, co-sponsored by NASA and the U.S. Department of Defense, to test the Ballistic Missile Defense Organization's "Star Wars" sensors developed to detect and track missiles. The possible ice deposit, composed of the suspected ice crystals mixed with dirt and spread over a vast landscape, was termed a *dirty lake*. The scientists discovered the ice in a massive, extremely deep crater. One scientist described the crater as twice the size of Puerto Rico and deeper than the height of Mount Everest. Scientists discussing the significance of the discovery postulated that the presence of water on the Moon might be a boon to future space exploration, perhaps enabling the construction of a "filling station" on the Moon. Other

---

[163] NASA, "First Flight of U.S.-Russian Supersonic Testbed Scheduled for Nov. 29," news release N96-78, 26 November 1996; "Russian Tu-144LL SST Joint NASA Flying Laboratory—Flight November 29, 1996," *http://nix.larc.nasa.gov/info;jsessionid=1f4kro1asktqm?id=EC96-43859-2&orgid=7* (accessed 1 August 2007).
[164] Associated Press, "President Clinton's Remarks in Awarding Medal to Shannon Lucid," 2 December 1996; Associated Press, "Clinton Gives Medal to Record-Setting Astronaut," 2 December 1996. For more information, see NASA History Division, "Congressional Space Medal of Honor," 28 April 2006, *http://history.nasa.gov/spacemedal.htm*.

scientists offered a more sober opinion of the findings, referring to reports that the presence of water on the Moon might someday allow for colonization as "hyperbole."[165]

*4 December*
NASA launched its second unpiloted craft, Mars Pathfinder, less than a month after the first, continuing its concentrated program of Mars exploration. The US$196 million Pathfinder lifted off from Cape Canaveral, Florida, aboard a Delta II rocket, on a mission planned to cover more than 300 million miles (480 million kilometers). NASA researchers expected the six-wheeled craft to land on Mars, on 4 July 1997, and to journey across the planet's surface, gathering images of rocks and soil. Depending on how useful the images were, scientists planned for Pathfinder to roam Mars for at least one week and perhaps for months, straying no farther than approximately 60 feet (18 meters) away from its landing site.[166]

*5 December*
NASA officials announced that a funding crisis in Russia would delay the human occupation of the International Space Station (ISS) up to eight months. Before the announcement, NASA had learned that the financial problems of the Russian Space Agency had virtually halted Russia's work on the orbiting laboratory's nerve center. Without Russia's contribution to the station, the crew of the ISS would have nowhere to stay during the construction of the station in space. NASA officials remained hopeful that Russia would release funding for the project and keep its commitment to the effort. Nevertheless, NASA officials began revisiting contingency plans to address reductions in funding. Russian Space Agency officials planned to meet with their NASA counterparts to discuss the situation.[167]

*6 December*
Scientists at the University of Texas at Dallas revealed their discovery of an ancient river channel buried beneath the Sahara Desert in Africa, publishing their findings in *Science* magazine. The discovery offered an answer to a question that had long perplexed the scientific community, why does the Nile River, as it flows through the world's largest desert (in Sudan) make a huge, looping bend in its course? The Nile generally flows toward the north, except at the Great Bend's turning, where the river flows toward the southwest for more than 200 miles (320 kilometers). After this detour, the Nile resumes its course toward the Mediterranean Sea. The researchers had used the images taken by the Spaceborne Imaging Radar C/X-Band Synthetic Aperture Radar (SIR-C/X-SAR), which had flown with Shuttle *Endeavour*'s Mission STS-59 in 1994, to study the ancient structures in the Precambrian rock, guiding the river's course. The SIR-C/X-SAR radar waves had penetrated the sand to reveal unseen structures, not visible without the satellite images. Diane L. Evans, SIR-C Project Manager at NASA's Jet Propulsion Laboratory (JPL), called the discovery "one of the most exciting discoveries from the SIR-C/X-

---

[165] John Noble Wilford, "The Moon May Have Water, and Many New Possibilities," *New York Times*, 4 December 1996; Associated Press, "Deep Within Lunar Crater, Small Frozen Lake Is Found," 3 December 1996; Joyce Price, "Prospect of Ice on Moon Fires Hopes of Scientists," *Washington Times*, 4 December 1996; Curt Suplee, "The Moon's 'Dirty Lake' Makes Waves," *Washington Post*, 4 December 1996.

[166] John Noble Wilford, "U.S. Launches Second Craft on a Mission To Study Mars," *New York Times*, 5 December 1996.

[167] William Harwood, "Funding Crisis To Delay Occupation of Space Station," *Washington Post*, 6 December 1996.

SAR mission to date." JPL was managing the program as a part of NASA's Mission to Planet Earth initiative.[168]

*7 December*
Shuttle *Columbia* landed at Kennedy Space Center in Cape Canaveral, Florida. Forced to remain in space an additional day because of bad weather, *Columbia* and its crew had set an endurance record of nearly 18 days in space, only one day short of the previous record of 17 days. Heavy fog in Florida was the cause of the landing delay.[169]

*9 December*
NASA announced the completion, on time and on budget, of the first major component of the ISS. Russia's Khrunichev Industries and the Boeing Company had combined efforts to construct Functional Cargo Block (known by its Russian abbreviation FGB), a 20-ton (18,000-kilogram or 18-tonne) pressurized module. Members of the international space team planned to launch the spacecraft from Russia in November 1997. FGB formed the initial building block of the ISS, featuring computer equipment, thermal controls, fire detection tools, and navigational apparatus. Khrunichev State Research and Production Space Center and Boeing Defense and Space Group received US$190 million for developing and building the craft. According to the plan, the ISS team would add many other components to FGB.[170]

*11 December*
In a postflight inspection, NASA engineers discovered that one loose screw had caused the hatch on Shuttle *Columbia* to jam, preventing the astronauts from taking their scheduled spacewalks. The screw had probably dislodged during takeoff, falling into the gears of the hatch. NASA had decided to cancel the spacewalks, rather than force the hatch open and risk damaging it permanently. With the Shuttle scheduled to launch again in one month, NASA faced the question of whether to remove and replace the gearboxes on all the hatches inside the Shuttle as a precautionary measure.[171]

*12 December*
Researchers at NASA's Marshall Space Flight Center announced that tests aimed at improving the efficiency of Space Shuttle engines would likely translate into substantial savings for commercial airlines as well. Researchers had learned that locating an engine's turbine airfoils in the "optimum position" could cut down the flow of fluttering wakes of gases. When wakes occur, an engine requires additional energy to run and generates higher temperatures. NASA researchers predicted that the findings would have special application for the Boeing 777. Improving efficiency by a half percent would save hundreds of gallons of fuel per flight. The researchers had based the new positioning of the turbines upon the simple principle evident when a bike racer drafts behind a competitor—because the first rider takes the full brunt of the wind,

---

[168] Robert J. Stern and Mohamed Gamal, "The Origin of the Great Bend of the Nile from SIR-C/X-SAR Imagery," *Science* 274, no. 5293 (6 December 1996): 1696–1698; NASA, "Space Radar Unearths Secrets of the Ancient Nile," news release 96-251, 6 December 1996.
[169] Associated Press, "Space Shuttle Sets a Record for a Mission," 7 December 1996.
[170] NASA, "Station's First Module Assembled; Ready for Testing," news release 96-253, 9 December 1996.
[171] Associated Press, "Loose Screw Caused Stuck Hatch on Shuttle," 12 December 1996; Associated Press, "NASA Hatching Up Plan in Case Emergency Spacewalk Is Needed," 2 December 1996.

the second can travel equally as fast with less exertion. The engineers found that by aligning turbines in the most efficient positions, the engine derived a similar drafting.[172]

New audio recordings, gathered from the excursion of NASA's Galileo spacecrafts to Jupiter's large moon Ganymede, contained a "soaring whistle and hissing static," suggesting that the moon possesses a planet-like magnetosphere. According to Donald A. Gurnett, who had conducted the experiments using Galileo's plasma wave instrument, the data "is kind of like looking at a musical score." The sound patterns mimic those found on Earth, Saturn, and Jupiter. Gurnett suggested that the new findings would have broad significance for the scientific community, possibly allowing scientists to draw further conclusions about the interplay between magnetic forces and matter throughout the universe.[173]

NASA announced the successful harvest of the first crop of healthy plants grown completely in space. Grown aboard the Russian space station *Mir*, the super-dwarf wheat grew robustly in the microgravity of space. Researchers at NASA's Ames Research Center termed the successful harvest a step toward a greater human presence in space. "The development of plant-based, regenerative life support systems is critical to sustaining a crew during long-duration missions such as the Mars exploration," NASA's David Bubenheim explained. The space station crew had raised the plants aboard *Mir* in a greenhouse designed by a team of Russian and Bulgarian engineers. Florescent lights had provided sufficient light for the plants, and the astronauts had injected water directly into the soil.[174]

*13 December*
The Aerospace Safety Advisory Panel reported its findings on the safety of the Space Shuttle program, concluding that aggressive cost-cutting measures had not increased risks. The review responded to critics who had suggested that measures, such as outsourcing flight operations, had been too drastic, causing a precipitous decline in NASA workers' morale. Administrator Daniel S. Goldin, responsible for making NASA a more fiscally lean organization, praised NASA's "clean bill of health." However, the Aerospace Safety Advisory Panel warned NASA about challenges that would arise during the assembly of the ISS, cautioning those in favor of cutting the budget against trimming NASA's funds too drastically.[175]

*15 December*
The Boeing Company and the McDonnell Douglas Corporation announced that McDonnell Douglas would merge with Boeing in a stock-for-stock transaction. The merger, the biggest in aerospace history, positioned Boeing to compete with Europe's Airbus Industrie for nearly all commercial airline contracts. The merger also strengthened Boeing's position as one of the most influential companies, both domestically and internationally, in the United States. The US$13.3 billion deal would place under a single umbrella nearly 200,000 employees, producing annual

---

[172] NASA, "Applying NASA Shuttle Engine Test Findings May Save Airlines Millions in Fuel Costs," news release 96-254, 12 December 1996.
[173] D. A. Gurnett et al., "Evidence for Magnetosphere at Ganymede from Plasmawave Observations by the Galileo Spacecraft," *Nature* 384, no. 6609 (12 December 1996): 535; NASA, "Big Icy Moon of Jupiter Found To Have a 'Voice' After All; Europa Flyby Next for Galileo," news release 96-255, 12 December 1996.
[174] NASA, "NASA Harvest of Mir Space Wheat Marks U.S.-Russian First," news release 96-256, 12 December 1996.
[175] NASA, "Panel Review Finds Space Shuttle Safe," news release 96-257, 13 December 1996.

revenues in excess of US$48 billion. The Boeing Company also emerged from the deal with improved chances of winning major military contracts, whereas, previously, the company had few ventures involving the military aerospace industry. However, McDonnell Douglas had been the main supplier to the military of F/A-18 and F-15 Eagle-model fighter jets to the military. Thus, the merger brought together leaders in the commercial and military sectors of the aerospace industry, creating what one industry analyst called an "800-pound gorilla." McDonnell Douglas stock rose 20 percent on the first day of trading following the announcement.[176]

*16 December*
NASA Administrator Daniel S. Goldin hosted members of Congress and other government and aviation industry leaders at a ceremony marking the start of NASA's General Aviation Propulsion program. NASA launched the endeavor to encourage the development of technologies and manufacturing practices that would create more cost-effective and environmentally friendly aviation-propulsion systems. NASA selected as its partners William International and Teledyne Continental Motors, along with the Federal Aviation Administration and the Industry Advanced General Aviation Transport Experiments Consortium. Aviation leaders praised the collaboration as the beginning of a movement to revolutionize aircraft engines and, consequently, the aviation industry.[177]

*20 December*
Astronomer and Pulitzer Prize winner Carl Sagan died after a long fight against a bone marrow disease; he was 62 years old. Since 1971, Sagan had directed Cornell University's Laboratory for Planetary Atmospheres and Surfaces, where he forged his reputation as an expert in exobiology. Sagan helped plan the Mariner, Viking, and Voyager missions, receiving NASA's Medal for Exceptional Scientific Achievement in 1972. Sagan also helped popularize modern science with his widely viewed television series *Cosmos*. Sagan described the television series as an endeavor "to show that science is a delight and to end people's artificial alienation from it." Sagan's *The Dragons of Eden* (1977) won the Pulitzer Prize. Throughout his work, Sagan had also argued for the possibility of extraterrestrial life.[178]

*24 December*
Russia launched a satellite carrying two monkeys and a collection of insects, snails, and plants. The satellite *Bion-11* launched from the Plesetsk Cosmodrome, 600 kilometers (373 miles) north of Moscow. The Russian Space Agency planned to study the effects of weightlessness on the animal "crew."[179]

**JANUARY 1997**

*7 January*

---

[176] The Boeing Company, "McDonnell Douglas To Merge with Boeing, Combination To Be the World's Largest Aerospace Company," news release, 15 December 1996; Peter Kaplan, "Mission Inevitable," *Washington Times*, 17 December 1996; David E. Sanger, "A Giant in Jets and Foreign Policy," *New York Times*, 17 December 1996.
[177] NASA, "Signing Ceremony To Initiate Development of Revolutionary Aircraft Engines," news release N96-80, 10 December 1996.
[178] Cornell University, "Carl Sagan, Cornell Astronomer, Dies Today (Dec. 20) in Seattle," news release, 20 December 1996; United Press International, "Astronomer Carl Sagan Has Died," 20 December 1996.
[179] Reuters, "Russia Launches Satellite with Monkeys, Insects," 24 December 1996.

The Clipper Graham (DC-XA) Incident Investigation Board released its final report concerning the explosion of the vehicle at the White Sands Missile Range in New Mexico. The Board, chaired by former astronaut Vance Brand, determined that a disconnected brake line on the secondary landing gear had prevented the gear from fully extending. The DC-XA, a prototype developed by McDonnell Douglas for NASA, had flown successfully in its fourth test flight on 31 July 1996, before failing on landing. NASA's director of space transportation noted that the failure, which occurred as part of ongoing research into reusable launch vehicles, had strengthened NASA's resolve to reduce costs and to achieve efficient reusability, safety, and reliability, ideally using a combination of automation and human control.[180]

*9 January*
NASA reported that one of the two monkeys sent into space aboard the Russian *Bion-11* flight in December 1996 had died upon completing postmission tests in Moscow. Both NASA and the Russian Space Agency announced that they would investigate the death. According to observers, both monkeys aboard the mission returned to Earth alert and active. The purpose of the flight was to investigate further the effects of spaceflight on the musculoskeletal system.[181]

*12 January*
In the first spaceflight of 1997, Shuttle *Atlantis* launched for a linkup mission with the *Mir* space station. Mission STS-81 was the fifth Shuttle voyage to bring supplies and new crew members to the Russian space station. *Atlantis* carried the SPACEHAB module in the payload bay of its orbiter, with experiments for *Atlantis*'s crew, as well as equipment for *Mir*. Michael A. Baker, making his fourth Shuttle flight, commanded the mission, with Brent W. Jett Jr. serving as Pilot. Astronaut Jerry M. Linenger was also aboard, set to replace astronaut John E. Blaha as the representative of the United States aboard *Mir*. Blaha had spent nearly four months aboard the orbiting station, roughly the same tenure planned for Linenger. So that *Atlantis* could rendezvous with *Mir*, the Shuttle had launched at precisely 4:27 a.m. (EST). *Atlantis* would spend two days orbiting, periodically firing its engines to draw closer to *Mir*, approximately 240 miles (386 kilometers) above Earth. The United States and Russia planned for the Shuttle–*Mir* missions to contribute to the international team's development of procedures and techniques for use on the International Space Station (ISS). NASA and the Russian Space Agency referred to the missions as Phase I of the ISS effort.[182]

*13 January*
An international team of astronomers announced the discovery of three black holes in three nondescript galaxies, suggesting that nearly all galaxies may in fact have massive black holes. Using NASA's Hubble Space Telescope (HST), the researchers took a census of 27 galaxies to find a series of black holes massive enough to consume millions of Sun-like stars. The findings, which the scientists presented in full at the 189th meeting of the American Astronomical Society, provided insight into the origin and evolution of galaxies, as well as clarifying the role of quasars

---

[180] NASA, "Clipper Graham Incident Report Release," news release 97-3, 7 January 1997.
[181] Associated Press, "Russian Space Monkey Multik Dies," 12 January 1997; NASA, "Monkey Dies After Completing 14-Day Bion Mission," news release 97-8, 9 January 1997.
[182] NASA, "Fifth Shuttle–*Mir* Docking Flight Highlights STS-81 Mission," news release 97-2, January 1996; William Harwood, "Atlantis Roars into Orbit, Begins Chasing *Mir* for Tuesday Linkup," *Washington Post*, 13 January 1997; Associated Press, "Atlantis en Route to Russian Station After 'Gorgeous Launch'," 12 January 1997.

in the galaxy. Douglas O. Richstone of the University of Michigan, who led the research team, stated that two of the black holes had "weighed in" at 50 million and 100 million solar masses, respectively. Scientists remained divided on why black holes exist so abundantly in space. One theory suggested was that, at the center of most or all galaxies, supermassive black holes exist, where gases ignite to the hottest temperatures known, sending nearby stars spiraling in new directions. Ralph Narayan of the Harvard-Smithsonian Center for Astrophysics called the black holes "the ultimate victory of gravity."[183]

*14 January*
Henry C. Ferguson, an astronomer at the Space Telescope Science Institute in Baltimore, Maryland, announced the discovery of stars residing outside of any defined galaxy. Scientists had long suspected that such stars existed, but Ferguson and his team were the first to confirm the theories. They did so using images captures by the HST, which revealed as many as 600 stars in the seemingly blank spaces among the Virgo cluster of galaxies, approximately 60 million light-years away from Earth. Ferguson suspected that many more stars—perhaps as many as 1 trillion Sun-like stars—are adrift in the galaxies of Virgo. The astronomers theorized that galactic mergers, or the tidal forces of nearby galaxies, had displaced the stars from their home galaxies.[184]

*15 January*
Shuttle *Atlantis* safely docked with the Russian space station *Mir*, while both spacecraft orbited at speeds of more than 17,500 miles (28,000 kilometers) per hour. Russian cosmonauts greeted the six-person American crew with the traditional Russian welcoming gift of bread and salt. Communication problems delayed the opening of the connection hatch temporarily, before American astronaut John E. Blaha, who had been aboard *Mir* for four months, was able to greet his fellow compatriots and his replacement. Blaha reported that the arriving Shuttle looked like a "shiny star" as it approached *Mir*. On hearing *Atlantis*'s approach, Blaha, ready to go home, reported, "All bags are packed. Ready for transfer." NASA Administrator Daniel S. Goldin took the opportunity to congratulate Blaha on his contributions to space exploration and to improved international relations. "We spent 50 years aiming weapons at each other," Goldin stated, and he praised the United States and Russia for cooperating in sustaining *Mir* and planning for the ISS.[185]

*16 January*
Johnson Space Center (JSC) awarded BRSP Inc. a US$128 million, five-year contract to provide base-operation support services, including plant maintenance and operations, logistics support, and security services for JSC's Houston, Texas, facility. BRSP, a joint venture of Brown and Root Services and Pioneer Contract Services, had formed to bid for the JSC contract.[186]

---

[183] Paul Recer, "Astronomers Find Stronger Evidence of Massive Black Holes," *Huntsville Times* (AL), 14 January 1997; NASA, "Massive Black Holes Dwell in Most Galaxies, According to Hubble Census," news release 97-9, 12 January 1997.
[184] John Noble Wilford, "Hubble Detects Stars that Belong to No Galaxy," *New York Times*, 15 January 1997.
[185] Associated Press, "Atlantis and *Mir* Make Docking Astronauts Meet for Crew Swap," 15 January 1997; Associated Press, "Atlantis Closes in for Late-Night Docking with Russian Station," 14 January 1997.
[186] NASA, "BRSP Chosen for JSC Base Operation Contract," news release C97-b, 16 January 1997.

Fred C. Adams and Gregory P. Laughlin, astrophysicists at the University of Michigan, presented their theory of a "Dying Universe," at the American Astronomic Society conference. Adams and Laughlin had projected the future of the universe based upon quantitative theory, concluding that the Earth would die out with a whimper. According to their theory, the Sun will eventually die out, and the light of all stars will vanish. The projection was so far off in the distance (10,000 trillion trillion trillion trillion trillion trillion trillion trillion years from now) that it drew a mixture of criticism, amusement, and intrigue from the scientific community. Adams and Laughlin insisted that recently acquired insights into cosmic evolution had allowed them to draw conclusions about the end of the universe, albeit in broad terms. Previously, scientists had lacked even the basic data and theoretical framework necessary to undertake the task. The two researchers emphasized that they made no claim to foresee when the universe would end, but rather offered the first long-term, science-based deduction on how the universe would evolve through four periods of expansion—the star-filled era (the present), the degenerate era, the black-hole era, and the dark era. The ultimate result of such a progression was too difficult to estimate, according to the scholars.[187]

NASA announced that planned major modifications of Space Shuttle *Atlantis*, as well as routine inspections, would take place at the Boeing Company's facility in Palmdale, California, rather than at Kennedy Space Center (KSC) in Cape Canaveral, Florida. United Space Alliance, the company contracted for the maintenance and operation of the Shuttle program, had recommended the changes. According to estimates of the U.S. General Accounting Office, completing the renovations in California, rather than in Florida, would cost NASA an additional US$20 million because of higher labor costs. NASA explained that if the engineers had done the work in Florida, the launch schedule at KSC, along with the need for *Atlantis* to make a prolonged stay for extensive modifications, would have threatened future Shuttle flights. The changes planned, including rerouting the Shuttle's fuel lines, involved more than routine maintenance. NASA also pointed out that its most experienced engineers worked at the Palmdale facility, and that workers at the facility had already made similar modifications to Shuttles *Discovery* and *Endeavour*. Moreover, NASA maintained that permitting KSC to focus on Shuttle launches and missions would be the most efficient use of its resources. Nevertheless, immediately after the announcement, some observers criticized NASA's continued use of Boeing's private facility in preference to that of the government-owned KSC. NASA's previous intimation that the Palmdale facility would be "mothballed" for Shuttle repair and maintenance intensified critics' reaction to the announcement.[188]

*17 January*
NASA released images taken from Galileo as it soared just 430 miles (690 kilometers) above the surface of Jupiter's moon Europa. The images revealed the apparent traces of massive ice volcanoes. According to early analysis of the pictures, ice volcanoes and the movement of tectonic plates had reshaped the surface of Europa, one of the bodies in the universe that

---

[187] John Noble Wilford, "At Other End of 'Big Bang', Finale May Be Big Whimper," *New York Times*, 16 January 1997; Tim Friend, "We've Got 100 Trillion Years Until 'Lights Out' in Universe," *USA Today*, 16 January 1997; Associated Press, "End of It All," 16 January 1996.

[188] Seth Borenstein, "Despite Boost in Cost, Atlantis To Get Tuneup in California," *Orlando Sentinel* (FL), 17 January 1997; NASA, "Space Shuttle Atlantis Modification Work To Be Performed at Palmdale Facility," news release 97-11, 16 January 1997; Todd Halvorson, "Atlantis Overhaul To Be Done Out West," *Florida Today* (Brevard, FL), 17 January 1997.

scientists believed might once have hosted life. The presence of water, organic compounds, and adequate heat, all of which would be present if ice volcanoes existed on the moon, had led scientists to focus on Europa as a location for possible development of life. According to Galileo team member Robert J. Sullivan, the apparent traces of ice volcanoes supported the even more significant possibility that an ocean might exist below the surface of Europa. Because NASA's Jet Propulsion Laboratory (JPL) managed the Galileo Mission, the scientific community credited JPL with providing the vital new images of the surfaces of Europa and Jupiter that had led to further discussion about the possibility of life beyond Earth.[189]

*21 January*
Astronaut John E. Blaha, just returning from a four-month stay aboard the *Mir* space station, reported that he had experienced feelings of depression and anxiety during his long deployment in space. Blaha indicated that the conditions aboard *Mir*, rather than any elements of the space environment, had caused his difficulty. According to Blaha, the lack of private quarters for U.S. astronauts, who unlike the Russian cosmonauts did not have even small personal quarters, was one of the most difficult aspects of his stay aboard *Mir*. He also reported that, although relations with his Russian counterparts always had remained respectful, the tension of getting to know men who spoke a different language in such tight quarters created "another element of psychological pressure and stress," aside from the inevitable feelings of isolation. Blaha reported that he had experienced psychological depression, especially during his first month aboard the Russian space station. He made these remarks aboard Shuttle *Atlantis* as it returned him to Earth, in the course of radio communications with four people living inside an airtight chamber at JSC in Houston, who were taking part in a 60-day experiment for future ISS deployments.[190]

*22 January*
John E. Blaha became the first astronaut to leave the U.S. Space Shuttle carried by medical personnel. Blaha was experiencing weakness upon his return to Earth's gravitational conditions, and NASA's doctors wanted to gauge immediately the effects of long-term weightlessness on his body. Blaha remarked upon landing that he could hardly move, saying that he was "absolutely stunned" at the difficulty of returning to a gravity-controlled environment. NASA's doctors focused specifically on Blaha's dizziness and on his weakened bones and muscles, hoping to learn how to improve health plans for astronauts living aboard the ISS. Blaha had maintained an arduous schedule of exercise aboard *Mir*, to mitigate the weakening effects of the weightless environment. In addition, he had made the return flight aboard *Atlantis* lying down in a Shuttle seat, to minimize the crush of gravity.[191]

*23 January*
NASA announced the promotion of Gretchen W. McClain to the position of Acting Director of Space Station Requirements. As Director, McClain took over responsibility for establishing

---

[189] NASA, "Ice Volcanoes Reshape Europa's Chaotic Surface," news release 97-12, 17 January 1997.
[190] Associated Press, "Astronaut Tells of Down Side to Space Life," *New York Times*, 22 January 1997.
[191] William Harwood, "Atlantis Astronaut Wobbly on His Return to Gravity," *Washington Post*, 23 January 1997; Associated Press, "Astronaut Carried off Shuttle, a First for NASA," 23 January 1997; Associated Press, "Returning Astronaut Sheds Right Stuff Image, Is Carried off Shuttle," 22 January 1997.

policy and standards for the ISS, as NASA continued to plan for the space station, with ongoing support from the Russian Space Agency, the European Space Agency, Japan, and Canada.[192]

*24 January*
NASA Administrator Daniel S. Goldin named Roy D. Bridges Jr., retired U.S. Air Force Major General and Shuttle astronaut, as Director of KSC. Goldin called Bridges "the right person to take KSC into the next century." As Director of KSC, Bridges assumed responsibility for the only site for launches of the Shuttle. At the time of Bridges's appointment, about 2,000 employees worked at KSC, along with 14,000 contractors. Bridges's appointment followed years of highly decorated service in the U.S. Air Force and NASA. Graduating with distinction from the U.S. Air Force Academy, he had served in several leadership positions in the Air Force and had received recognition as a distinguished graduate of the U.S. Air Force Pilot Training and Test Pilot Schools. Among other awards, he had received the Distinguished Service Medal, the Defense Service Medal with oak leaf cluster, the Legion of Merit with oak leaf cluster, the Distinguished Flying Cross with two oak leaf clusters, the Meritorious Service Medal, the U.S. Air Force Commendation Medal, and NASA's Flight Space Medal.[193]

## FEBRUARY 1997

*3 February*
*Space News* reported that the Italian Space Agency had decided not to participate in construction of the International Space Station (ISS) because of NASA's insistence that Italy meet previously established deadlines. The Italian firm Alenia Aerospazio had expressed interest in building, on behalf of Italy, two nodes, the pieces of hardware used to connect different sections of the orbiting laboratory. The Italian Space Agency had requested more time to investigate the feasibility of participating, after it determined that funding the $150 million project, solely in return for additional future use of the ISS, might strain its budget for other scientific projects. Gretchen W. McClain, Acting Director of Space Station Requirements, said that, although NASA regretted the loss of Italy's participation in the ISS, placing the project on hold while the Italian Space Agency made its decision would result in significant schedule shortfalls. The development highlighted the difficulty of fostering international participation and, at the same time, maintaining the target delivery date of the ISS.[194]

*5 February*
*Flight International* published a story claiming that, more than 10 years after the explosion of *Challenger*, aerospace engineer Ali Abu Taha had discovered new evidence in Time-Life photographs demonstrating that a breach and a fire in *Challenger*'s right solid rocket booster (SRB) had caused the explosion. One photograph, taken about 20 seconds after liftoff, revealed a 3-meter-long (10-foot-long) flame issuing from an SRB joint, and another clearly showed a white object separating from the same SRB seconds later. The discovery seemed to support Taha's controversial theory that the booster had caught fire at liftoff and burned continuously until the explosion occurred. In contrast to Taha's findings, which focused on dynamic liftoff

---

[192] NASA, "NASA Names Acting Space Station Director," news release, 97-15, 23 January 1997.
[193] NASA, "Roy Bridges Selected as KSC's New Center Director," news release 97-17, 24 January 1997.
[194] Peter B. de Selding and William Harwood, "Italian Nodes for Station Unlikely," *Space News*, 3–9 February 1997.

loads, the congressionally mandated Presidential Commission on the Space Shuttle Challenger Accident (Rogers Commission) had reported that faulty O-rings caused the tragedy. Automatic cameras set up around the Cape Canaveral, Florida, launch site had taken the photographs that Taha used as evidence for his theory, images never released to the public. Taha's photographic analysis also revealed that the explosion had propelled *Challenger*'s crew compartment thousands of meters away from the explosion, partially explaining why investigators had taken 40 days to locate the shell. Taha suggested that a shock wave had killed the crew instantly, although NASA had never found evidence of a shock wave. News of the photographs and of Taha's research kept alive the debate surrounding the decade-old accident.[195]

*6 February*
President William J. Clinton released his budget proposal, which included a request for US$75 million to develop the Space Infrared Telescope Facility. NASA had argued that the infrared telescope would aid scientists in answering some of the universe's most fundamental questions including, what is the energy source for the universe's brightest stars? and, where is the universe's missing dark matter? Presidential support for the initiative led NASA to hope that it might finally receive the funding necessary to begin building the last of four space-based telescopes proposed as a part of the Great Observatories program. In addition to the highly publicized Hubble Space Telescope (HST), NASA had successfully placed in orbit the Compton Gamma Ray Observatory and had immediate plans to launch the nearly completed X-ray Astrophysics Facility. NASA expected the four telescopes, using various kinds of cosmic radiation to survey the atmosphere, would provide an unprecedented view of the universe. The four telescopes operating together would allow us "to look at the universe at all ages at once," explained Michael Werner, a scientist at NASA's Jet Propulsion Laboratory (JPL). Upon the program's conception in 1991, the National Academy of Sciences had called the Great Observatories the "highest priority for any major new program in space-based astronomy," but a steady stream of budget cuts had made it increasingly difficult to realize the project.[196]

NASA Administer Daniel S. Goldin praised the Clinton administration's budget proposal for 1998 as a sign that NASA's deep budget cuts were leveling off. NASA had braced itself for a budget as low as US$13.1 billion, and, therefore, found President William J. Clinton's US$13.5 billion proposal a pleasant surprise. The request still entailed a US$200 million budget cut, and Congress might reduce the budget further. Nevertheless, Goldin responded to the proposed budget exuberantly, exclaiming, "Holy mackerel—this is a great program," and remarking that for the first time during his five-year tenure, NASA could hope for budgetary stability. The proposal also projected that NASA's budget in 2000 would be approximately US$13.2 billion, rather than the US$11.6 billion forecast in previous Clinton administration proposals. NASA planned to absorb the small reduction in funding by increasing efficiency, especially in the Shuttle program, having recently transferred management of the program to the privately held United Space Alliance. NASA officials believed that NASA could continue sending up seven or eight Shuttle missions annually, if budget trends continued to stabilize. On the day of the budget announcement, however, a NASA advisory panel warned once again that a drastic decrease in

---

[195] Tim Furness, "New Evidence Reveals Fire on Doomed *Challenger*'s Booster," *Flight International*, 5–11 February 1997.
[196] Paul Hoversten, "NASA Seeks $75 Million for Another Eye in the Sky," *USA Today*, 6 February 1997.

the Shuttle's funding would compromise its safety. Nevertheless, Goldin maintained that the proposal would allow NASA to continue fostering "faster, better, cheaper" science.[197]

Vice President Albert A. Gore Jr. met with Russia's Prime Minister Viktor S. Chernomyrdin to discuss, among other topics, the United States' concern that the Russian Space Agency's lack of funding might derail plans for the ISS. U.S. officials declared that funding was the primary issue in the construction of the ISS. "We have no doubts about their science and engineering proficiency. Where we have concerns relates to adequate funding on the Russian side," said one American space leader. Although the ISS had received contributions from the United States, Russia, Japan, Europe, and Canada, Russian construction stoppages on a crucial ISS component already had delayed the project. Before the Gore-Chernomyrdin meeting, NASA received assurances from the Russian Space Agency that, despite financial difficulties, Russia would maintain its commitment to the project. During the meeting, the leaders agreed upon a joint statement committing the Russians to a new deadline. Gore stated that he believed the summit's reaffirmation of previous commitments would satisfy members of Congress who were critical of Russia's trustworthiness.[198]

*7 February*
NASA announced that it would join an international consortium of space agencies to support the launch of a Japanese satellite designed to create the largest astronomical instrument ever. The launch of the Very Long Baseline Interferometry (VLBI) Space Observatory would create a radio telescope more than twice the diameter of the Earth, giving astronomers their sharpest view yet of the universe. Astronomers praised the Japanese-led launch as a significant advance for space research. VLBI satellites would allow radio astronomers to link together widely separated radio telescopes, so that they would function as a single instrument with extraordinarily sharp resolving power. The farther apart the telescopes, the greater would be the image's resolution. NASA equated the power of the new tool with the ability to see a grain of rice in Tokyo from Los Angeles. Scientists hoped that the new tool would allow for further exploration of active galaxies with massive black holes. Researchers also hoped that the instrument would foster understanding of the mysterious quasars, which pour out tremendous amounts of energy and host or create black holes. Creating the enormous VLBI network involved 40 radio telescopes from 15 nations. Japan's Institute of Space and Astronautical Science led the international consortium of science agencies, including NASA's JPL, the U.S. National Science Foundation's National Radio Astronomy Observatory, the Canadian Space Agency, the Australia Telescope National Facility, the European VLBI Network, and Europe's Joint Institute for VLBI.[199]

U.S. astronaut Jerry M. Linenger, NASA's temporary resident aboard Russia's *Mir* space station, became the second American to ride aboard the three-person *Soyuz* capsule that had delivered the cosmonauts to *Mir*. Cosmonauts Valery G. Korzun and Alexander Y. Kaleri, Linenger's Russian

---

[197] Anne Eisele, "Budget Plan Brings Stability by 2000," *Space News*, 10–16 February 1997; Larry Wheeler, "NASA Makes Do on Slimmer Budget, *Florida Today* (Brevard, FL), 7 February 1997; Sean Holton, "NASA Cheers Budget that Cuts Agency Only a Bit," *Orlando Sentinel* (FL), 7 February 1997.
[198] Associated Press, "US-Russia," 6 February 1997; Reuters, "Gore-Chernomyrdin Talks on NATO, Space, Summit," 5 February 1997; Cragg Hines, "Russia Pledges To Pay Its Share of Space Station," *Houston Chronicle*, 9 February 1997.
[199] NASA, "Launch Will Create a Radio Telescope Larger than Earth," news release 97-24, 7 February 1997; NASA, *Aeronautics and Space Report of the President: Fiscal Year 1996 Activities* (Washington, DC, 1997), p. 25.

partners aboard *Mir*, took Linenger for a short ride around *Mir*, to free a docking station for the new *Soyuz* vehicle that Russia would soon launch. In allowing an American astronaut aboard the Russian craft, Russia took another step with the United States to foster a cooperative postcommunist relationship between the two countries.[200]

*10 February*
The Russian Space Agency launched *Soyuz TM-25*, carrying a crew of two Russian Space Agency cosmonauts, Vasili V. Tsibliyev and Alexander I. Lazutkin, and German astronaut Reinhold Ewald to the *Mir* space station, from Baikonur Cosmodrome in Kazakhstan. The cosmonauts would become *Mir*'s new, long-term crew, and the German astronaut would return with the retiring crew in two weeks. Germany had paid the Russian Space Agency millions of dollars to carry Reinhold Ewald aboard the Shuttle to *Mir*.[201]

NASA researchers, astronauts, and scholars gathered in Washington, DC, for a conference of the National Academy of Sciences, disclosing and discussing details of experiments conducted aboard the Shuttle during two 1996 missions. The event commemorated the one-year anniversaries of the second U.S. Microgravity Laboratory and the third U.S. Microgravity Payload. Scientists predicted that discoveries made aboard the Shuttle flights in 1996 would eventually lead to technological advances. Research on numerous topics, from the climate of the universe to human biology, could lead to production of cheaper metals and alloys or new synthetic drugs. Highlights from the research presentations included the announcement of the discovery that space-grown crystals are of much higher quality than those grown on Earth; the description of experiments intended to uncover the effect of space on the production of metal and alloys; and the presentation of evidence that the microgravity environment of space had enabled scientists to make more precise measurements of the physical properties of elemental gas.[202]

*11 February*
Shuttle *Discovery*, carrying a crew of seven astronauts, launched from Kennedy Space Center (KSC) in Cape Canaveral, Florida, on a mission to service the HST. NASA planned the 10-day Shuttle Mission STS-82 to upgrade significantly the scientific capabilities of HST, as well as to perform standard maintenance on the US$2 billion instrument. The spaceflight was the second servicing mission for HST since its deployment in April 1990 aboard Shuttle *Discovery*'s Mission STS-31. The astronauts planned to conduct at least four spacewalks to install new components, such as the Space Telescope Imaging Spectrograph and the Near Infrared Camera and Multi-Object Spectrometer. A NASA scientist described the improvements as "replacing 1970's technology with 1990's technology." In addition to the work conducted on the HST, *Discovery*'s crew planned to move the telescope a few miles farther into the atmosphere, improving its chances for a longer operating life. Astronomers hypothesized that the potential gains from refurbishing the HST outweighed the risks of tampering with its already spectacular abilities, justifying the US$800 million cost of the mission. The head of the Harvard-

---

[200] Association Press, "Astronaut Rides Soyuz," 8 February 1997.
[201] Reuters, "Russians Launch 3 into Space," 11 February 1997.
[202] NASA, "Significant Discoveries from Space Shuttle Experiments To Be Presented at Conference," news release 97-23, 5 February 1997.

Smithsonian Center for Astrophysics speculated that the refurbished HST could lead to "a whole different kind of science."[203]

*12 February*
The White House Commission on Aviation Safety and Security, chaired by Vice President Albert A. Gore Jr., released its recommendations for reforming the United States' aviation industry. The Commission challenged NASA, the Federal Aviation Administration (FAA), and the U.S. Department of Defense to combine their efforts, with the goal of reducing aircraft accident rates by 80 percent over the next five years. NASA Administrator Daniel S. Goldin voiced immediate support for the challenge, explaining, "We're looking for solutions that will save lives." He pledged NASA's commitment to interagency cooperation and committed US$500 million in NASA funds, to achieve a major reduction in aircraft-related fatalities. Both NASA and the FAA cited previous successful collaboration efforts between their agencies. The Gore Commission's most controversial recommendation was the use of computerized databases to identify potential terrorists. Critics quickly called the plan an exercise in profiling, which would foster biased treatment, especially toward Arab Americans and passengers with Arab-sounding names. Other safety proposals included requiring children under the age of two to sit in their own seats, rather than on the lap of an adult, as the airlines had long permitted. The Commission further suggested that user fees, rather than ticket taxes, fund air-traffic control; that airports deploy a new satellite-based, air-traffic-control system, as soon as possible; and that airports make all airmail packages weighing more than 1 pound (0.5 kilograms) subject to inspection. In total, the Commission recommended 53 changes to the current system.[204]

*19 February*
*Discovery*'s crew released the HST from the Shuttle's cargo bay after nearly a week of maintenance and refurbishing work. The astronauts had conducted a fifth spacewalk, one more than NASA had originally planned, to place a makeshift patch on the HST's peeling insulation. "It's not the Hubble Space Telescope. It's really Hubble Space Telescope II," remarked NASA's chief Hubble scientist upon releasing the telescope. NASA deemed the mission a complete success.[205]

*20 February*
Astronaut and lawmaker U.S. Senator John H. Glenn Jr. (D-OH), on the 35th anniversary of his journey as the first American to orbit Earth, announced that he would retire from the U.S. Senate at the conclusion of his fourth term. On 20 February 1962, Glenn had orbited Earth three times, symbolizing a hard-fought achievement in the United States' effort to catch up with the Soviet Union in the space race. Glenn, first elected to the Senate from Ohio in 1974, made the

---

[203] NASA, "Space Shuttle Mission STS-82—Press Kit," news release 97-18, February 1997; John Noble Wilford, "Shuttle Starts Space Telescope Mission," *New York Times*, 12 February 1997; Associated Press, "Improved Hubble Telescope To Give Astronomers New View of the Universe," 8 February 1997.

[204] NASA, "NASA, FAA, and DOD Challenged To Achieve White House Commission Goal To Reduce Aircraft Accidents," news release 97-26, 12 February 1997; John Mintz, "Gore Panel Proposes Aviation Safety Steps," *Washington Post*, 13 February 1997; Robert Davis, "FAA Ready To Depart on 'Clear Path of Action'," *USA Today*, 13 February 1997; Robert Davis, "53 Air Safety Changes Urged," *USA Today*, 13 February 1997.

[205] Associated Press, "Hubble Released To Search Universe," 20 February 1997; Associated Press, "Looking for Answers," 19 February 1997.

announcement at Muskingum College in New Concord, Ohio, joking that since "there is still no cure for the common birthday," he intended to retire at age 75.[206]

*21 February*
NASA announced that a plan for collaborative research and development between NASA and five other government agencies might result in the creation of an Internet connection a million times faster than home computer modems. The Next Generation Internet (NGI) initiative dedicated US$300 million and three years to developing the NGI, to fulfill the promise President William J. Clinton had made in his State of the Union address: "We must build the second generation of the Internet so that our leading universities and national laboratories can communicate at speeds one thousand times faster than today; to develop new medical treatments, new sources of energy, new ways of working together." NASA sites designated for early trials of NGI connections included Ames Research Center, Goddard Space Flight Center, Langley Research Center, Lewis Research Center, and Jet Propulsion Laboratory.[207]

Shuttle *Discovery* ended Mission STS-82 with a safe landing at KSC in Cape Canaveral, Florida. The early morning landing occurred later than NASA had planned because of cloudy weather in Florida. NASA praised *Discovery*'s crew for its work refurbishing the HST.[208]

*23 February*
A small fire broke out aboard the *Mir* space station, raising new questions about the fitness of the aging station for occupation and research. The six crew members on board quickly extinguished the fire, the result of a faulty air purification unit. The astronauts and cosmonauts estimated that the fire had burned for about 90 seconds, forcing the crew to don respirator masks. *Mir* continued to function normally after the fire, although the heat had melted several cables.[209]

## MARCH 1997

*2 March*
Two Russian cosmonauts and a German astronaut returned safely to Russia after tenures aboard the Russian space station *Mir*. The cosmonauts had spent more than six months in space, and Reinhold Ewald of Germany had spent two weeks aboard *Mir* as a paying guest of the Russian Space Agency. Ewald conducted geophysical, astrophysical, and medical experiments during his stay. The German Space Agency paid about US$60 million for Ewald's trip, providing a much-needed infusion of capital for Russia's space program.[210]

*3 March*

---

[206] Paul Souhrada, "Glenn, 75, Will Retire, Citing 'No Cure for Common Birthday'," *Washington Times*, 21 February 1997; NASA, "Glenn Orbits the Earth 35 Years Ago Today," video advisory V97-12, 20 February 1997.
[207] NASA, "Next Generation Internet a Million Times Faster than Home Computer Modems," news release 97-29, 21 February 1997.
[208] William Harwood, "Discovery Lands: Shuttle Back from Hubble Repair Mission," *Chicago Sun-Times*, 22 February 1997.
[209] NASA, "Small Fire Extinguished on *Mir*," news release 97-30, 24 February 1997; Associated Press, "Fire on Russian Space Station Doused, but Raises Concern," 25 February 1997.
[210] Associated Press, "Russia-Space," 2 March 1997.

NASA commemorated the 25th anniversary of the still-orbiting Pioneer 10 space probe. Launched 2 March 1972, Pioneer 10 was the functioning probe farthest from Earth, at approximately 6 billion miles (9.6 billion kilometers) away. NASA originally had intended the probe to orbit for 21 months, but the spacecraft had continued to perform remarkably well and efficiently long after its designated mission. The probe completed its planned objective of surveying Jupiter in 1972; after that, NASA launched Pioneer 10 into space at 86,000 miles (138,000 kilometers) per hour. Since then, scientists had tracked the probe, learning more about magnetic fields, solar wind, cosmic particles, and ultraviolet glow. In June 1983, Pioneer 10 became the first human-made object to leave the solar system. On the day marking Pioneer 10's 25th year in space, the probe still broadcast "crystal-clear data" to the NASA control room on Earth. NASA did not plan a 26th anniversary for the probe, explaining that Pioneer 10's pulse was finally growing so weak that it could no longer facilitate significant scientific experiments.[211]

*Design News* honored as its Engineer of the Year, Bernard P. Dagarin, the key design engineer behind the Galileo space probe, which had successfully navigated Jupiter's atmosphere in 1996. Dagarin had begun working on prototypes for the Jupiter probe in 1978, when NASA awarded Hughes Space and Communications Company the initial Galileo contract. According to *Design News*, "Never before had a spacecraft been asked to survive such a long and difficult mission." Galileo had achieved the near impossible. Not only was Galileo's Jupiter mission a complete success, achieving the desired data, but the mission had also become the capstone of Dagarin's decorated career.[212]

*4 March*
The unpiloted cargo spacecraft Progress M-33 failed in its planned docking with the *Mir* space station, forcing Russian space officials to decide whether to allow the cargo craft to burn up in the atmosphere or to try again to connect with *Mir*. The docking of Progress M-33, which only had some surplus fuel aboard, was not of critical importance to *Mir*, but Russian Space Agency scientists had planned to use the craft's docked mass to adjust the trajectory of *Mir*. News of the failure of the Progress's docking tempered Russia's celebration of its first satellite launch from a new cosmodrome in Svobody, occurring on the same day. On 6 March, the saga ended, when Russian officials decided to let the Progress M-33 drift off to burn up in the atmosphere.[213]

*6 March*
NASA researchers at Marshall Space Flight Center (MSFC) in Huntsville, Alabama, announced a plan to collaborate with Lawrence Berkeley National Laboratory in Berkeley, California, to explore the potential of the substance Aerogel, which has tremendous insulating capabilities. Scientists said that they were optimistic that they could manipulate the material, rendering it transparent and, therefore, especially useful in constructing energy-efficient windows for homes and automobiles. During the 1930s, a researcher at Stanford University had discovered Aerogel,

---

[211] Associated Press, "Pioneer 10," 3 March 1997; USA Today, "NASA To Retire Oldest Interplanetary Explorer," 3 March 1997.
[212] Lawrence D. Maloney, "Galileo Probe's Guardian Angel," *Design News* 52, no. 5 (3 March 1997): 74–83.
[213] Reuters, "Russia May Try New Space Docking or Let Craft Burn," 5 March 1997; Reuters, "Russian Space Launch Marred by Failed Docking," 4 March 1996; Reuters, "Russia To Scrap Spacecraft after Docking Snag" 6 March 1997.

often called frozen smoke because of its hazy appearance. NASA engineers had used Aerogel as an insulating agent aboard Mars Pathfinder in 1996. Although the substance is the lightest solid known—only three times the density of air, a block of Aerogel weighing less than 1 pound (0.5 kilograms) can support 0.5 tons (450 kilograms or 0.45 tonnes). Aerogel's large internal surface area, which disperses heat throughout its structure, causes the material's extreme insulating capability—a 1-inch-thick (2.5-centimeter-thick) Aerogel window would offer the same insulation value as 15 panes of glass. John M. Horack, Assistant Laboratory Director for Space and Science Communications at MSFC, called the development of Aerogel "a great example of how NASA space research generates scientific knowledge that can be used to improve the quality of life on Earth."[214]

A panel of scientists at the National Research Council issued a report warning NASA of the slight possibility that a mission to Mars might actually bring microbes back to the United States. Although their report clearly stated that the so-called Microbe Peril was unlikely, the panel argued, nevertheless, that a probe returning from an exploratory mission to Mars could possibly serve as a conduit to bring Martian germs back to Earth. The panel issued its warning because of NASA's plan to send a mission to Mars to obtain samples of substances on the planet, and because NASA scientists had recently identified signs of possible microscopic life on a meteorite. The panel's report stated, "If life forms ever existed on Mars, either by having been formed in an independent origin or having been transferred there from Earth, it is possible that they could have continued to exist up to the present time."[215]

NASA announced the selection of four industry partners to study and develop a new Earth-imaging radar system. The proposed LightSAR system would use advanced technologies to reduce the cost and increase the efficiency of radar-based satellites, which gather scientific data and provide commercial remote sensing. Project Manager Steven Bard of NASA's Jet Propulsion Laboratory (JPL) pointed out the unique team approach planned for the project. NASA intended to work with industry partners from the start, to assess the potential commercial market for LightSAR products and services and to share the cost of developing the expensive new technology with private companies. The four primary contractors selected were DBA Systems Inc., Lockheed Martin Astronautics, Research and Development Laboratories, and Vexcel Corporation.[216]

*7 March*
Edward M. Purcell, 1952 Nobel Prize winner in physics and long-time researcher at Harvard University, died. Purcell discovered a means to detect the extremely weak magnetism of the atomic nucleus, making it possible to "listen to the whisperings of hydrogen through the universe." Purcell had served as President of the American Physical Society. In 1967 Purcell had won the Oersted Medal of the American Association of Physics Teachers, and in 1979 he had won the National Medal of Science. At Harvard, Purcell held the position of Gerhard Gade

---

[214] NASA, "NASA Research in Space May Redesign Household Windows," news release 97-34, 6 March 1997.
[215] Warren E. Leary, "Microbe Peril from Mars Is Possibility, Panel Warns," *New York Times*, 7 March 1997; Associated Press, "Scientists Warn on Mars Bugs," 7 March 1997; Mark Carreau, "Fear of Martians," *Houston Chronicle*, 7 March 1997.
[216] NASA, "Teams Selected for Studies of Potential Partnership with NASA To Develop New Earth Imaging Radar System," news release 97-35, 6 March 1997.

University Professor, retiring from the university in 1977. Of his Nobel Prize–winning research Purcell commented, "We are dealing not merely with a new tool but with a new subject, a subject I have called simply nuclear magnetism."[217]

*9 March*
Professional and amateur astronomers in northern China and in the eastern reaches of Russia witnessed the near-simultaneous occurrence of a total solar eclipse and an unimpeded view of Comet Hale-Bopp. According to news reports, tens of thousands of people crowded city streets and braved freezing temperatures to observe the spectacle, many using pieces of smoked glass to protect their eyes. Mohe County in China's northern tip had banned the use of smoke-producing stoves and other heating devices, to keep the air as clear as possible for the view. Chinese television stations broadcast the event, only the third time in recorded history when a full eclipse and the passing of a nearby comet had occurred simultaneously.[218]

*10 March*
NASA announced an agreement with the European Space Agency (ESA) for the European construction of two docking nodes for the International Space Station (ISS). In exchange, NASA agreed to launch the ESA's *Columbus* laboratory module. The "launch-offset barter agreement" was typical of the arrangements in progress to bring the ISS closer to fruition. The ESA announced immediate plans for the Italian Space Agency to carry out the actual construction of the two nodes, slated for mid-2000 launches. In addition to the construction arrangement, the ESA agreed to provide a variety of minor hardware, including refrigeration units, for the astronauts' living space and for research areas of the ISS.[219]

*12 March*
NASA announced the completed construction of the Lunar Prospector spacecraft, scheduled for launch in September 1997. Engineers had designed the craft to obtain the first compositional and gravity maps of the Moon. G. Scott Hubbard, NASA's Lunar Prospector Mission Manager at Ames Research Center, praised contractor Lockheed Martin for its efficient construction and for keeping project costs at forecasted levels. NASA projected that the total cost to build Lunar Prospector and send it to the Moon would be about US$63 million. Scientists explained the need to return to the Moon, even though it was "conquered" decades ago, reminding the public that many important questions about the Moon's history and its fundamental composition remained unexplored. Researchers hoped that Lunar Prospector's planned one-year mission orbiting the Moon would create a detailed map of the Moon's surface composition and gravitational and magnetic fields. Lunar Prospector itself was compact—only 4.5 feet (1.37 meters) high, 4 feet (1.22 meters) in diameter, and weighing 660 pounds (300 kilograms). Even before its launch, scientists hailed Lunar Prospector as a new type of pathfinder, which had "made history in terms of management style, technical approach, cost management, and focused science." Prospector used a gamma-ray spectrometer to provide maps of the composition of the Moon, thereby allowing scientists to understand better the elements present in the surface layer of the Moon.[220]

---

[217] "Edward Purcell, 84, Physicist Who Shared a Nobel Prize," *New York Times*, 10 March 1997.
[218] Associated Press, "Comet, Eclipse Dazzle Chinese, Russians," 10 March 1997.
[219] NASA, "NASA-ESA Agreement Enhances Station with Additional Node," news release 97-36, 10 March 1997.
[220] NASA, "Lunar Prospector Spacecraft Construction Complete," news release 97-38, 12 March 1997.

*13 March*
As the public anticipation of the appearance of Comet Hale-Bopp heightened, NASA announced that NASA-supported researchers would use its vast resources—including spacecraft and the Hubble Space Telescope (HST)—to study the comet. Scientists hoped that studying Hale-Bopp would lead to a greater general understanding of comets. Because comets are composed of loosely packed dirt and ice, experts consider them the best-preserved remnants of the early solar system. NASA disclosed that the Wallops Flight Facility would launch four rockets using ultraviolet wavelengths, to study Comet Hale-Bopp. In addition, the Ulysses spacecraft, the joint project of NASA and the ESA, which was already in orbit at the time, would chart the effects of solar-wind conditions on comets. However, scientists also disclosed that the close proximity of Hale-Bopp to the Sun posed a danger to HST's sensitive detectors; therefore, they had determined to wait until a few months after the comet's approach to Earth before tracking it with the telescope. JPL planned to host a public event, called Comet Chasers: On the Trail of a Comet, to bring scholars and the public together to discuss the significance of Comet Hale-Bopp.[221]

New support emerged for NASA scientists' claim that a meteorite discovered in August 1996 held fossil evidence of primitive Martian life. In an article in *Science,* a group of researchers from California Institute of Technology and McGill University used magnetic studies to demonstrate that the findings on the meteorite were not, as some critics had contended, the result of a high-temperature environment that would have made life impossible. "What we are able to show from the magnetic studies is that these things [the fossil remnant that had excited the interest of Johnson Space Center geologists] couldn't have been heated even to the boiling point of water." Although the scientists did not explicitly support NASA's earlier claim that the remnant provided evidence for bacterial life, the team's discovery opened the intriguing possibility that remnants of primitive life could migrate from one planet to another.[222]

NASA announced that, through its Small Business Innovative Research program, researchers had successfully used recycled plastic milk bottles to create a more effective, lightweight insulation for clothing and blankets. The material had the same honeycomb structure as that of the metal heat barriers used in spacecraft. According to the principal investigator, Steven D. Miller of S. D. Miller and Associates, "The blankets are better than wool or fleece because they are non-allergenic, and they dry five times faster. The new material is also four times warmer than wool in cold and damp conditions." The research team hypothesized that agencies such as the Red Cross, or other emergency medical personnel, eventually would be able to use the new material to warm patients more quickly. After developing the product for space and commercial uses, NASA planned to allow Miller's company to pursue further commercialization opportunities without NASA funding.[223]

---

[221] NASA, "NASA Plans Comet Hale-Bopp Observing Campaign, Activities," news release N97-17, 13 March 1996.
[222] Joseph L. Kirschvink et al., "Paleomagnetic Evidence of a Low-Temperature Origin of Carbonate in the Martian Meteorite ALH84001," *Science* 275, no. 5306 (14 March 1997): 1629–1634; Mark Carreau, "California Scientists Back NASA's Theory of Life on Mars," *Houston Chronicle*, 14 March 1997; John Noble Wilford, "Study Backs Idea that Meteorite Hints of Life on Mars," *New York Times*, 14 March 1997.
[223] NASA, "NASA Technology Creates Market for Recycled Milk Bottles," news release 97-39, 13 March 1997.

NASA announced the establishment of a National Microgravity Center, formed in conjunction with Case Western Reserve University and the Universities Space Research Association. NASA selected Lewis Research Center (LERC) in Cleveland, Ohio, to lead its research efforts in the project. Plans for the nation's first center dedicated exclusively to microgravity research focused on five areas: 1) research and technology development; 2) science program outreach and development; 3) scientific support; 4) technology transfer to industry; and 5) public education initiatives. NASA Administrator Daniel S. Goldin announced the endeavor, stating, "The National Center for Microgravity Research on Fluids and Combustion represents a commitment to our goal to strengthen the partnership between NASA and our nation's research community in universities and industry so that together we can increase the scientific and economic payoffs from NASA's Microgravity Science Program." NASA selected Simon Ostrach, a distinguished professor of engineering at Case Western University, as Director of the National Microgravity Center. As Director, Ostrach would be responsible for managing the Center's staff of more than 30 people and for shaping an agenda of research that would take advantage of opportunities aboard the ISS.[224]

*20 March*
Japan's National Space Development Agency (NASDA) unveiled its key contribution to the ISS, two satellites named after stars—*Orihime* and *Hikoboshi*, which Japan planned to launch aboard an H-2 rocket. The satellites would make it easier to perform unpiloted docking experiments on the ISS. Japan's interest in robotic space research fueled the proposal, which complemented NASDA's plan to build an unpiloted space shuttle to ferry Japan's experiments to the ISS.[225]

NASA awarded its 1996 Government Invention of the Year to a patented, high-temperature seal developed for the National Aerospace Plane project. Bruce M. Steinetz and Paul J. Sirocky of NASA's LERC directed the research. The seal used ceramic and superalloy fibers braided into a malleable structure to seal high-temperature joints on airplanes and Shuttles. The seal maintained its form in temperatures up to 2,000°F (1,090°C).[226]

*22 March*
NASA marked the one-year anniversary of the United States' continuous presence in space, a tenure that began with the mission of astronaut Shannon W. Lucid aboard the *Mir* space station. Astronauts John E. Blaha and Jerry M. Linenger had followed Lucid's stay. NASA planned to maintain a continuous presence aboard *Mir* until 1998.[227]

**APRIL 1997**

*1 April*
The famed Comet Hale-Bopp made its closest approach to the Sun, accelerating the comet's process of shedding the gargantuan ice shards that, scientists hypothesized, might contain the basic ingredients necessary for life to begin. Based on data gathered by researchers during the 20 months between the discovery of the comet and its approach to the Sun, scientists theorized that

---

[224] NASA, "NASA Establishes New National Microgravity Center," news release 97-40, 13 March 1997.
[225] Reuters, "Japan Unveils Experimental Docking Satellites," 20 March 1997.
[226] NASA, "NASA Selects 1996 Government Invention of the Year," news release 97-47, 20 March 1997.
[227] NASA, "Anniversary Marks Milestone of U.S. Presence in Space," news release 97-45, 17 March 1997.

the outer layers of Comet Hale-Bopp probably contained methanol, formaldehyde, carbon monoxide, hydrogen cyanide, hydrogen sulfide, and other carbon compounds. Because of its proximity to the Sun, many telescopes, most significantly the Hubble Space Telescope (HST), had to stop tracking the 25-mile-wide (40-kilometer-wide) comet to avoid damage to their sensitive instruments. The approach of the comet was a scientific opportunity occurring once "every 200 years or so," according to Brian G. Marsden of the Harvard-Smithsonian Center for Astrophysics. Comet sightings thrilled amateur and professional astronomers alike, as well as causing some paranoia. In one tragically irrational response to its approach, 39 members of the Heaven's Gate religious community committed suicide, under the delusion that they were about to catch a ride on a spaceship hidden behind the comet's tail.[228]

NASA announced that the newly refurbished and improved HST had allowed astronomers to observe the fading, visible-light counterpart of a gamma-ray burst (GRB), a process that researchers called "one of the most energetic and mysterious" in the universe. The scientists speculated that the observable counterpart signified the existence of an unobservable, catastrophic burst of gamma rays, unleashing as much energy in a few seconds as the Sun does in 10 billion years. The New Technology Telescope and the W. M. Keck Telescope, added to the HST in 1997, had enabled scientists to make the observation. According to Gerald J. Fishman of Marshall Space Flight Center, the lead investigator of NASA's Compton Gamma Ray Observatory, astronomers could not overestimate the significance of the find, even if the visible-light phenomenon was only an indicator of another unobservable phenomenon. "This [discovery] opens up a whole new era in GRB research," Fishman explained. "We now know that it is possible to see the fading optical emission by rapid follow-up observations with powerful telescopes. With several more of these, we should be able to narrow the models of what could be causing these gigantic outbursts."[229]

*4 April*
Shuttle *Columbia* launched from Kennedy Space Center (KSC) in Cape Canaveral, Florida, embarking on a mission primarily to conduct research on microgravity. To understand the effect of the space environment on fire, and the potential danger of a fire to the astronauts, the mission crew planned to light more than 200 small, controlled fires during the mission. James D. Halsell Jr. commanded the Mission STS-83 crew, and Susan L. Still piloted Shuttle *Columbia*. Still was the second American woman to pilot a Space Shuttle. Mission Specialists Michael L. Gernhardt and Donald A. Thomas, Payload Commander Janice E. Voss, and Payload Specialists Roger K. Crouch and Gregory T. Linteris completed the crew. The Shuttle carried the Microgravity Science Laboratory in its payload, and the crew planned dozens of experiments, to "serve as a bridge to America's future in space." In the course of conducting these experiments, focused on processes necessary to carry out the long-duration research planned for the International Space Station (ISS), the astronauts hoped to create new research procedures and protocols. The STS-83 marked the 22$^{nd}$ mission of *Columbia*. The Shuttle's launch took place after a one-day delay, so that engineers could install thermal insulation on two exposed Shuttle pipes.[230]

---

[228] William J. Broad, "The Comet's Gift: Hints of How Earth Came to Life," *New York Times*, 1 April 1997.
[229] NASA, "Hubble Tracks the Fading Optical Counterpart of a Gamma-Ray Burst," news release 97-63, 1 April 1997.
[230] NASA, "Microgravity Research Highlights Mission STS-83," news release J97-8, April 1997; Associated Press, "Shuttle Leaves on a Mission To Study Fire," 5 April 1997.

*6 April*
Astronauts aboard *Columbia*, consulting with NASA ground support, decided to cut short their planned 16-day mission because of the potential for an explosion in the Shuttle's electronic generator. NASA decided to abort the mission after the crew had attempted for two days to fix the balky generator. Despite the fact that the Shuttle's two other generators could provide enough power for the Shuttle, NASA's procedure required the Shuttle's return, to prevent subsequent problems. Space Shuttle Program Manager Thomas W. Holloway clarified that the astronauts faced no imminent danger and that terminating the mission was a precaution rather than an emergency. "The conservative thing to do is land on Tuesday," Holloway said at a news conference. *Columbia*'s early return to Earth in Mission STS-83 was only the third time in the history of the Shuttle program that NASA had ended a mission early. The astronauts had time to complete only a few of the planned 33 experiments on the mission, which cost nearly US$500 million.[231]

*8 April*
Satellite-based monitoring instruments, owned by NASA and used by the National Oceanic and Atmospheric Administration (NOAA), documented startlingly low levels of ozone over the Arctic North Pole. The measurements indicated ozone levels of nearly 40 percent below the levels found in 1979 and 1982. Scientists attributed the low ozone levels to a series of unusual meteorological conditions, although they did not yet know the principal causes of the change.[232]

Shuttle *Columbia* landed safely at KSC in Cape Canaveral, Florida, less than three days after its liftoff. The mission, planned for 16 days, ended early, and the Shuttle completed the return flight without incident. *Columbia* weighed more than 236,000 pounds (107,000 kilograms) at landing, the heaviest Shuttle ever to land, because of its load of unused supplies for experiments. Almost immediately after the safe landing, NASA officials began to consider repeating the flight as soon as possible, to allow the Shuttle crew to accomplish its original objectives and to avoid the substantial cost of launching an entirely new mission.[233]

*9 April*
Michael H. Carr of the U.S. Geological Survey announced that the Galileo spacecraft had captured images revealing iceberg-like structures, suggesting the presence of frozen water on Jupiter's moon Europa. The discovery provided strong support for the theory that an ocean once existed on the moon's surface. Astronomers at NASA's Jet Propulsion Laboratory called the discovery "mind-blowing" and "the clearest evidence to date that there is liquid water and melting close to the surface of Europa." Ronald Greeley, the Arizona State University geologist who managed the Galileo imaging team, described the structures in the Europa images as "blocks of ice, similar to those seen on the Earth's polar seas during springtime thaws." Greeley concluded that a "thin icy layer covering water or slushy ice" might possibly exist on Europa's surface. The images revealed one possible ice formation that spread across Europa's surface in

---

[231] William Harwood, "Shuttle Flight Cut Short as Risk Persists," *Washington Post*, 7 April 1997; Warren Leary, "Power Problem on Shuttle Forces a Tuesday Landing," *New York Times*, 7 April 1997; Associated Press, "Deteriorating Generator Cuts Short Mission," 6 April 1997.
[232] NASA, "Low Ozone Measured over North Pole," news release 97-64, 8 April 1997.
[233] William Harwood, "Columbia Lands After Aborted Mission," *Washington Post*, 9 April 1997.

patches as wide as 4 miles (6.4 kilometers). Scientists disagreed on exactly how to interpret the images and on whether future exploration would be able to prove the presence of an ocean. Nevertheless, excitement over the findings spread throughout the research community. In its use of the Galileo probe to explore Europa, NASA continued its mission to search the universe for other environments suitable for life. The Galileo probe almost lost the images when an antenna on the craft jammed, slowing data transmission to Earth. Galileo came within 363 miles (584 kilometers) of Europa to capture the highly celebrated pictures.[234]

At a tense hearing of the U.S. House Committee on Science and Technology, NASA announced the necessity of a significant delay in the on-orbit assembly of the ISS. NASA Administrator Daniel S. Goldin explained the delay, "We knew from the outset that building the International Space Station was going to be tremendously challenging. Space exploration is not easy or predictable. We will work through this schedule issue." NASA had scheduled the on-orbit construction to begin in November 1997, after Russia had launched Functional Cargo Block, known by its Russian abbreviation FGB. Inadequate funding for the Russian Space Agency, however, had delayed the construction of this ISS building block, pushing back the overall schedule. Russia had also delayed building the service module, another key component of the ISS.[235] NASA's announcement invoked further criticism from members of Congress opposed to the ISS initiative. Representative F. James Sensenbrenner Jr. (R-WI), long a vocal opponent of the ISS, said simply, "I told you so." Before the official announcement of the delay, Sensenbrenner had argued that NASA needed to consider alternative plans, in case the Russian Space Agency proved unable to meet its obligations. NASA Associate Administrator for Space Flight Wilbur C. Trafton countered criticism, noting that the international contribution to the effort had surpassed US$6 billion and that NASA and its partners still expected to complete the overall project on time in 2002. Because the enormous undertaking involved "15 countries, dozens of companies, and thousands of workers," keeping production to preestablished deadlines had proven difficult.[236]

*12 April*
The Russian Space Agency celebrated the 36th anniversary of sending the first Russian, Yuri Gagarin, into space. At the same time, two cosmonauts and U.S. astronaut Jerry M. Linenger, who were on board *Mir*, reported that they had almost had to abandon the Russian space station twice over the past few months. Russia's successful history of space exploration contrasted with the current struggles of the Russian Space Agency to obtain adequate funding for *Mir* and other space initiatives. When Russia launched its first piloted mission in 1961, the Cold War was driving an intense space competition between the Soviet Union and United States. In the days leading up to the anniversary, Russian President Boris N. Yeltsin expressed his support for increased funding to the Russian Space Agency in the future.[237]

*16 April*

---

[234] Kathy Sawyer, "Signs of Ocean Found on a Jupiter Moon," *Washington Post*, 10 April 1997; Associated Press, "Scientists Spot Water on Moon of Jupiter," 10 April 1997; NASA, "New Images Hint at Wet and Wild History for Europa," news release 97-66, 9 April 1997.
[235] NASA, "NASA Revises International Space Station Schedule," news release 97-65, 9 April 1997; Kathy Sawyer, "NASA Delays Start of Space Station," *Washington Post*, 10 April 1997.
[236] William J. Broad, "Panel Upset by New Delay in Russian Space Module," *New York Times*, 10 April 1997.
[237] Reuters, "Mir Crew Has Little To Celebrate on Space Day," 11 April 1997.

NASA and the U.S. Air Force Space Command announced an agreement to collaborate on several projects of mutual interest. NASA Administrator Daniel S. Goldin explained that the agreement was part of NASA's ongoing mission to become a more efficient and cost-effective agency, and that sharing information would lead to "greater efficiencies in our respective missions." Goldin and U.S. Air Force Space Commander General Howell M. Estes III signed the pact, establishing seven teams to explore areas of potential cooperation. The areas of research included the feasibility of launching defense satellites from the Shuttle; the use of the Shuttle for U.S. Air Force technology payloads; and development of a plan to meet the dual space needs of NASA and the U.S. Air Force. The two agencies also planned to examine ways they might share their common infrastructure and facilities.[238]

In the first study to observe directly a change in the growth cycles of a large swath of the Northern Hemisphere, scientists reported that spring was arriving a full week earlier in the Earth's coldest regions than only a decade earlier. The report contributed to growing concern over the prospect of global warming. Using images obtained from NOAA satellites, a team of five scientists from Boston University demonstrated that plants budding in early spring had used 10 percent more carbon than previously, indicating the spring's earlier onset. The scientific community reacted to the report with interest, although most scientists cautioned that understanding climate change would be a long and difficult process.[239]

*23 April*
NASA scientist Addison Bain released a study concluding that the cause of the highly publicized explosion of the airship Hindenburg in 1937 was a special paint used to protect the aircraft from sunlight—not the ignition of inflammable hydrogen as historians had previously thought. Bain and a team of researchers used reels of archived film, models, and computer simulations to reach the conclusion that the outside of the Hindenburg had caught fire first. Because the Nazi government viewed the Hindenburg accident as an embarrassment, Germany had restricted the investigation into the explosion, which caused the deaths of the 35 passengers. The cover-up had led researchers to propose the scientifically unconvincing theory of a hydrogen fire.[240]

*25 April*
NASA announced that it had rescheduled for July 1997 Space Shuttle *Columbia*'s Microgravity Science Laboratory (MSL) mission, cut short in early April because of a fuel cell problem. The quick repair and relaunch plan used the same crew and set the same objectives as the initial Mission STS-83. Space Shuttle Program Manager Thomas W. Holloway stated, "we are now in the position to do everything possible to complete the MSL mission with minimal impact to downstream flights." He also pointed out that the attempt to complete the MSL mission provided NASA with a "unique opportunity to demonstrate our ability to respond to challenges." In its decision to insert *Columbia*'s Mission STS-83 into an already arranged Shuttle docket, NASA

---

[238] NASA, "NASA and Air Force Space Command Announce Cooperative Efforts," news release 97-68, 16 April 1997.
[239] Joby Warrick, "Spring Sprouting Earlier than a Decade Ago," *Washington Post*, 17 April 1997.
[240] Reuters, "Paint Led to Hindenburg AirshipTragedy," 23 April 1997.

acknowledged that it would have to push back slightly other flights, already staffed and planned.[241]

*28 April*
At the fourth Compton Symposium on Gamma Ray Astronomy and Astrophysics, scientists from Northwestern University and the U.S. Naval Research Laboratory announced the unexpected discovery of two clouds of antimatter in the Milky Way Galaxy. The researchers termed the material "antimatter annihilation radiation." The team of researchers used NASA's Compton Gamma Ray Observatory to discover the clouds, which scientists could not fully explain. "The origin of this new and unexpected source of antimatter is a mystery," William R. Purcell of Northwestern University explained. The scientists postulated that the clouds might have resulted from starbursts, jets of material from a nearby black hole, or from the merger of two neutron stars. Attempting to explain the startling nature of the discovery, Charles D. Dermer of the U.S. Naval Research Laboratory said, "It is like finding a new room in the house we have lived in since childhood. And the room is not empty—it has some engine or boiler making hot gas filled with annihilating antimatter."[242]

*29 April*
NASA astronaut, American Jerry M. Linenger, and Russian Space Agency cosmonaut, Ukrainian-born Vasili V. Tsibliyev, made the first joint U.S.-Russian spacewalk in the history of space exploration. Linenger, more than three months into his four-month stay aboard the *Mir* space station, joined Tsibliyev for nearly 5 hours outside the station. The two men gathered cosmic dust samples and installed a radiation meter. It was Linenger's first spacewalk and Tsibliyev's third. The men both wore Russian spacesuits and spoke in Russian as they cooperated to complete the exercise. Scientists planned to use the samples gathered from the spacewalk to continue tests to find the materials best suited for long-term space occupation, a question of preeminent importance with the ISS nearing fruition. Because of the orbiting position of the space station, Russian Mission Control was unable to communicate with Linenger and Tsibliyev while they were outside *Mir*, somewhat heightening tensions during the spacewalk.[243]

## MAY 1997

*2 May*
NASA announced that John C. Mather, a senior astrophysicist at Goddard Space Flight Center, had been elected to the National Academy of Sciences, often considered the highest recognition possible for scientists and engineers. Mather's work at NASA had begun in 1974, specifically, on the Cosmic Background Explorer (COBE) spacecraft. He had served as a project scientist when NASA launched the spacecraft in 1989, and in 1992 he had participated in the COBE

---

[241] NASA, "Microgravity Science Laboratory Mission Set for July; Remaining 1997 Shuttle Manifest Adjusted Slightly," news release 97-81, 25 April 1997; Seth Borenstein, "Space Shuttle Columbia Gets Chance To Finish What It Started," *Orlando Sentinel* (FL), 18 April 1997.
[242] NASA, "Antimatter Clouds and Fountain Discovered in the Milky Way," news release 97-83, 28 April 1997; Malcolm W. Browne, "Enormous Plume of Antimatter Alters View of the Milky Way," *New York Times*, 29 April 1997.
[243] NASA, "U.S. Astronaut Ready for Milestone Spacewalk," news release 97-80, 25 April 1997; Associated Press, "American, Russian Take a Spacewalk," 30 April 1997; Associated Press, "Astronauts Make First Joint U.S.-Russian Spacewalk," 29 April 1997.

project, which effectively mapped primordial hot and cold spots in the cosmic microwave background radiation. The map demonstrated that radiation from the Big Bang conformed to theoretical predictions, providing new evidence for the long-standing Big Bang theory. Mather received his PhD in physics from the University of California at Berkeley in 1974.[244]

*6 May*
NASA announced that it had granted Hitco Technologies exclusive rights to use a NASA-developed, heat-resistant material to produce high-performance pistons. NASA had developed the carbon-carbon composite in the 1960s, for use as a heat shield for missile applications. Researchers speculated that pistons composed of carbon-carbon, a material with virtually no thermal expansion, would allow engines to perform more efficiently, holding their shape and rigidity in temperatures well above 2,500°F (1,370°C). According to G. Burton Northam of Langley Research Center, the carbon-carbon composite was "the material of choice for the most demanding applications." Allowing industry to use one of NASA's patented technologies continued the trend of government-industry collaboration that Administrator Daniel S. Goldin had fostered.[245]

*8 May*
At Baltimore's Space Telescope Science Institute, NASA scientists described the long-duration-exposure pictures taken by the Hubble Space Telescope (HST) as the "best picture we have yet of the early universe," announcing that, approximately 12 billion years ago, the universe had been completely barren of galaxies. Richard S. Ellis of England's Institute of Astronomy emphasized the significance of science's new ability to use the HST to view "a time before galaxy formation." The HST had opened its lens for many hours of exposure, focusing on a tiny wedge of the sky to capture its most probing picture ever, known as the Hubble Deep Field Survey. Focusing only on a small speck of the sky had enabled astronomers to look deeper into the universe than ever before. Mario Livio of the Space Telescope Science Institute estimated that the HST had captured a view of the universe at only about 10 percent of its present age.[246]

*12 May*
Hubble scientists released their first reviews of the newly upgraded HST, concluding enthusiastically that the refurbished space telescope offered exciting new possibilities. The upgrades to the HST significantly improved the range and precision of the instrument. Scientists reported peering approximately 50 million light-years away and zeroing in on previously undetectable black holes. According to Edward J. Weiler, the head of NASA's Hubble team, the added Space Telescope Imaging Spectrograph, installed in February 1997, reduced the length of time it took to confirm the presence of black holes in the universe. Weiler called the HST a "census bureau" for black-hole hunting, allowing researchers to survey these objects as a population rather than as individual phenomena. Another new instrument installed in February, the Near Infrared Camera and Multi-Object Spectrometer (NICMOS), enabled the HST to capture near-infrared wavelengths, beginning to penetrate the "dusty veil" that had prevented astronomers from studying the birth and death of stars. NASA engineers also disclosed that an early focusing problem with one of the three cameras used by the NICMOS system seemed to be

---

[244] NASA, "Goddard Scientist Selected for National Academy of Sciences," news release 97-88, 2 May 1997.
[245] NASA, "Heat-Resistant Material Licensed for High-Performance Pistons," news release 97-91, 6 May 1997.
[246] Associated Press, "Hubble-Look Back," 9 May 1997.

correcting itself. Although the flaw could limit the life of the system, scientists were confident that NICMOS would still be able to gather and preserve all scientific data. During early testing of the NICMOS system, scientists had uncovered a region of the universe in which young stars eject material into a molecular cloud, a process that researchers had long hypothesized but had been unable to observe.[247]

*13 May*
The *New York Times* reported a soon-to-be released study by *Research Policy*, providing strong evidence that publicly funded scientific research often fueled industrial advancements in the United States. According to the study, prepared for the National Science Foundation, 73 percent of all American industrial patents filed during the two-year study cited scientific advancements made possible by government funding. The report had particular relevance because the Clinton administration and its congressional allies, both Democratic and Republican, had considered paring down the nation's science outlays to balance the federal budget. The Council of Scientific Society Presidents called the report "a wake-up call for [f]ederal investment policies." The study joined an ongoing debate over exactly what contribution general scientific research made to the national economy. The National Science Foundation and NASA were two of the agencies most cited in industrial patent requests.[248]

For the first time, NASA deployed its ER-2 aircraft, a civilian version of the U-2 reconnaissance plane, over the North Pole. The vehicle's flight supported the Photochemistry of Ozone Loss in the Arctic Region in Summer (POLARIS) project. The ER-2 flew at 70,000 feet (21,000 meters), an ideal height for atmospheric research. Michael J. Kurylo, Manager of NASA's Upper Atmosphere Research Program, emphasized the importance of aircraft such as the ER-2 to fill this specific research niche: "It is really critical that we have access to consistent measurements at this key altitude, which is an intermediate region between aerosol particle-driven processes measured by standard aircraft-based sensors and gas-phase processes monitored by orbiting satellites." Scientists said they were optimistic that the ER-2 would aid significantly the POLARIS endeavor to understand why the ozone layer over the North Pole had experienced reductions during each Arctic summer.[249]

*14 May*
NASA and Japan's Institute of Space and Astronautical Science announced an agreement to combine their efforts to collect and study samples from the surface of the asteroid Nereus. According to the plan, a Japanese launch vehicle would carry a NASA-engineered rover onto the surface of the asteroid. The planned rover, with a visible-imaging camera and a near-infrared point spectrometer, would weigh less than 3 pounds (1.3 kilograms), the smallest spacecraft ever sent into space. A recovery capsule would return to Earth the samples the rover gathered from Nereus. The project would have a lengthy timeframe, with the return of the samples to Earth scheduled for 2006. NASA and the Institute of Space and Astronautical Science were hopeful

---

[247] NASA, "Hubble's Upgrades Show Birth and Death of Stars; Discover Massive Black Hole," news release 97-93, 12 May 1997; Kathy Sawyer, "Improved Hubble Telescope Gets Rave Reviews at NASA," *Washington Post*, 13 May 1997; Reuters, "Hubble Captures Star Birth," 13 May 1997; Associated Press, "Upgrade Hubble Telescope Dazzles Astronomers," 12 May 1997.
[248] William J. Broad, "Study Finds Public Science Is Pillar of Industry," *New York Times*, 13 May 1997.
[249] NASA, "NASA Earth Science Research Aircraft Soars to New Heights," 13 May 1997.

and excited at the prospect of joining forces. "This ambitious mission is an opportunity for two spacefaring nations to combine their expertise and achieve something truly fantastic," said Jurgen H. Rahe, NASA's Director of Solar System Exploration.[250]

*15 May*

Space Shuttle *Atlantis* took off at dawn from Kennedy Space Center (KSC) in Cape Canaveral, Florida, carrying a crew of seven. Mission STS-84 was the fourth mission of 1997 and the sixth of nine planned missions to the *Mir* space station. *Atlantis* carried astronaut C. Michael Foale, who was to replace Jerry M. Linenger as the United States' long-term resident aboard *Mir*. Charles J. Precourt served as Commander of Mission STS-84 and Eileen M. Collins as Pilot for the Shuttle. In addition, the crew included U.S. astronauts Carlos I. Noriega and Edward T. Lu and Mission Specialists Elena V. Kondakova of the Russian Space Agency and Jean-François Clervoy of the European Space Agency (ESA). Kondakova was the first Russian woman to fly aboard a NASA Space Shuttle. With the recent problems aboard *Mir*, many viewed the mission with special urgency. Concerned about the numerous breakdowns and the fire aboard the Russian space station, NASA even considered refusing to allow Foale to take his place aboard *Mir*. The Shuttle carried to *Mir* a new oxygen generator, which astronauts planned to unload almost immediately upon docking. *Atlantis* took into space more than 7,000 pounds (3,200 kilograms) of cargo for *Mir* and its crew.[251]

European astronomers unveiled a new, more accurate map of the stars. Scientists immediately praised the 17-volume celestial guide as a milestone in astronomy. The ESA's Hipparcos project had produced the map over 17 years, at a cost approaching US$1 billion. Astronomers called the release of the Hipparcos data the beginning of a new and fruitful debate. Hipparcos Project Scientist Michael Perryman said, "It's a massive leap forward in our understanding. But the science of Hipparcos doesn't stop here. For the scientific community it is just beginning." In total, the map indicated the positions and motions of more than 118,000 stars.[252]

*19 May*

NASA unveiled a new facility, called the Chemical Crib, at Dryden Flight Research Center (DFRC). Officials hoped the new facility would reduce chemical waste by as much as 50 percent in three years. Waste disposal had become an increasingly expensive problem for NASA. "For every dollar spent to buy chemicals," NASA's Hazardous Materials Officer reported, "we spend three dollars to dispose of them." The Chemical Crib would use a variety of techniques and technologies to compress waste, such as breaking down photographic-waste chemicals into a sludge, possessing as little as 5 percent of the mass of the original waste. NASA planned to construct similar facilities at its other sites. Staff of DFRC's Safety, Health, and Environmental Office believed that the new facility would not only benefit the environment, but also the Center's budget.[253]

---

[250] NASA, "NASA and Japan To Cooperate on Asteroid Sample Return Mission," news release 97-95, 14 May 1997.
[251] NASA, "Atlantis Ready To Fly Sixth Shuttle–*Mir* Mission," news release J97-15, May 1997; Associated Press, "Space Shuttle," 15 May 1997.
[252] Association Press, "Mapping the Heavens," 15 May 1997.
[253] NASA, "New NASA Facility To Reduce Chemical Wastes; May Provide Better Storage for Hazardous Materials," news release 97-102, 19 May 1997.

Astronaut Shannon W. Lucid, who had spent more than six months aboard *Mir* in 1996, received the Order of Friendship Medal from Russian President Boris N. Yeltsin in a ceremony at the Kremlin. The medal was the highest honor available to a non-Russian citizen.[254]

*20 May*

NASA researchers announced the discovery of wildly oscillating weather patterns on Mars. In yet another use for the HST images, scientists tracked much colder, cloudier, harsher weather conditions than they had expected. R. Todd Clancy of the Space Science Institute attributed Mars's chaotic weather to a combination of factors, including the planet's thin atmosphere and elliptical orbit and the ice and dust clouds surrounding it. When the planet's orbit places it closest to the Sun, large windstorms push dust into the atmosphere; the dust absorbs sunlight, causing the air to heat, and Mars's temperature to rise as high as 30°F (-1°C). HST images seemed to demonstrate further that, when the planet is farthest from the Sun, the dust storms remain at low altitudes, and ice clouds surround the planet, causing temperatures to plunge precipitously. The findings were of particular importance to NASA's ongoing Mars Pathfinder Project, with Pathfinder's entrance into the Martian atmosphere planned for July 1997.[255]

A Russian Zenit-2 rocket carrying a military satellite exploded moments after taking off from Baikonur Cosmodrome in Kazakhstan. According to Russian space officials, the failure occurred 38 seconds after liftoff, when an engine inexplicably shut off. The Russian Space Agency suspended all further launches until an investigation could determine what had caused the explosion. NASA believed that the destroyed satellite was a spy satellite intended to replace in orbit one of Russia's many outdated satellites. Russia had not insured the satellite, valued at US$17 million.[256]

*21 May*

NASA announced that an American astronaut would be the first commander of the planned International Space Station, settling a long-standing disagreement between NASA and the Russian Space Agency. NASA selected veteran U.S. astronaut William M. Shepherd to lead the three-person crew, which also included Russian cosmonauts Sergei K. Krikalev and Yuri P. Gidzenko. Gidzenko said that he had no problem participating in a U.S.-led mission: "It doesn't matter who will be commander, who will be the flight engineer or pilot. They all work together, and they will try to do their best." NASA had selected Russian cosmonaut Anatoly Y. Solovyev as a member of the crew, but Solovyev had refused to take part in an American-led expedition, opening the door to Gidzenko's participation.[257]

*23 May*

---

[254] NASA, "Astronaut Shannon Lucid Receives Russian Order of Friendship Medal," news release 97-103, 19 May 1997.
[255] Warren E. Leary, "Hubble Space Telescope Finds Big Shifts in Weather on Mars," *New York Times*, 21 May 1997; Paul Hoversten, "Kaleidoscopic Mars Shows Its True Colors," *USA Today*, 21 May 1997; Associated Press, "Hubble Finds Climate on Mars Harsher than Scientists Thought," 21 May 1997.
[256] Richard C. Paddock, "Russian Rocket Bearing Military Satellite Fails," *Los Angeles Times*, 21 May; Associated Press, "Russian Booster Fails, Explodes During Satellite Launch," 20 May 1997.
[257] Todd Halverson, "American Astronaut Will Lead First Space Station Crew in '99," *Florida Today* (Brevard, FL), 22 May 1997.

The U.S. Capitol's Statuary Hall installed a bronze statue of *Apollo 13* astronaut John L. Swigert Jr., a native of Denver, to represent the state of Colorado. Each state chooses two statues for the hall to honor its most prominent citizens. Swigert was, perhaps, most famous for uttering the words, "Houston, we've had a problem." Colorado had elected Swigert to the U.S. House of Representatives in 1982, but he had died of cancer before his swearing-in. Swigert's statue took its place alongside that of fellow Coloradoan Florence Rena Sabin, who led the crusade for women's admission to medical schools. Astronaut Thomas Kenneth Mattingly II, whom Swigert replaced only three days before the *Apollo 13* launch, attended the statue dedication, along with NASA Administrator Daniel S. Goldin and other dignitaries.[258]

*24 May*

Astronaut Jerry M. Linenger surprised onlookers at KSC when he bounded off Shuttle *Atlantis* only minutes after its landing. Physicians credited Linenger's excellent condition after spending more than four months aboard the *Mir* space station to his regimen of orbital exercise. NASA Administrator Daniel S. Goldin was on hand to greet Linenger and the other astronauts returning home on *Atlantis*. The Shuttle's safe landing returned Linenger to Earth after 132 days in space—the second longest stint by an American.[259]

*25 May*

Educators, scientists, and politicians gathered to celebrate a new nationwide observance—Space Day—meant to carry on President John F. Kennedy's call to the nation for space exploration. On 25 May 1961, Kennedy had urged the United States to "commit itself to achieving the goal before this decade is out of landing a man on the moon and returning him safely to Earth," sparking NASA's successful development of human space travel. Advocates of Space Day argued that a day of celebrating previous achievements and calling for further advances would inspire future space explorers, particularly children. More than 1,900 schools across the nation participated in Space Day activities.[260]

## JUNE 1997

*2 June*

Lockheed Martin Corporation announced a joint satellite venture with the Moscow-based Intersputnik consortium. Financial analysts estimated the pact could produce more than US$1.5 billion in annual revenue, capitalizing on the increasing demand for satellite transmission services for telecommunication signaling. The agreement, which cemented ties between the U.S. corporation and Russian space interests, combined Lockheed's capital resources with Intersputnik's rights to prime satellite positions. Lockheed assumed responsibility for constructing and launching the satellites, whereas Intersputnik took the lead in marketing and sales. The Intersputnik group represented Russia and 21 other nations. The newly formed Lockheed Martin Intersputnik Ltd. Company planned to focus on providing services to South

---

[258] Paul Leavitt, "New Statue," *USA Today*, 23 May 1997; Associated Press, "Newest Congressional Statue: Apollo 13 Astronaut John Swigert," 22 May 1997.
[259] *Los Angeles Times*, "Returning Astronaut Surprises Even Himself; Shuttle: Four Months on Mir Sap Less Strength than Traveler Jerry Linenger Expected. He Credits the Exercising He Did in Orbit," 25 May 1997.
[260] Martha Woodall for Knight-Ridder News Service, "Space Day Today To Renew Interest in Outer Space," 22 May 1997.

Asia and Eastern Europe—burgeoning markets with great potential for technological and economic growth.[261]

*3 June*
NASA announced the start of a month-long research endeavor to explore the levels of radiation present at high altitudes, using a refitted ER-2 aircraft. The project focused on altitudes between 52,000 and 70,000 feet (15,900 and 21,300 meters), where radiation typically occurs because of cosmic and solar rays. The project had both scientific and public policy applications; researchers intended to use the data gathered to assess the safety of public, supersonic travel. The campaign, funded by NASA's High-Speed Research Program, specifically investigated the plausibility of the High-Speed Civil Transport, still in the conceptual stage of planning, a commercial jetliner that would carry passengers at 2.4 times the speed of sound and at altitudes above 60,000 feet (18,300 meters). According to NASA Project Manager Donald L. Maiden, the speed of supersonic flight could mitigate any ill effects of high altitudes on passengers, because "Even though the exposure levels are higher at the higher cruise altitude, the typical flying public will actually receive less radiation exposure than on today's subsonic transports because of the higher speed of the High-Speed Civil Transport." NASA, the U.S. Department of Energy's Environmental Measurements Laboratory, the Boeing Company, and the space agencies of Canada, Germany, and the United Kingdom collaborated on the project.[262]

*4 June*
The *Los Angeles Times* reported that the students of Glassell Park Elementary School had sold an astonishing 32,000 candy bars, to pay the way for all 115 of its fifth graders to attend the U.S. Space Camp at NASA's Ames Research Center in Mountain View, California. Most of the school's 951 students participated in selling candy to send their older classmates to the educational camp. Principal Beatrice LaPisto said proudly of her students, "This is a poor neighborhood. Most children here, for example, are on the free lunch program. But these are the brightest, best, most motivated students you'll find anywhere." This year was the first in which the nonprofit U.S. Space Camp Foundation had expanded its operations to the California site.[263]

*5 June*
NASA researchers announced that data from the Galileo probe indicated that Jupiter has distinct wet and dry regions similar to the Earth. The finding refuted the previous supposition that Jupiter was mostly dry. When it descended to Jupiter in December 1995, Galileo landed in an extremely dry and hot portion of the planet—an area one member of the team called the Sahara Desert of Jupiter. Scientists assumed that the area was representative of the entire planet, until they were able to analyze more long-range images taken by the probe. The Galileo team concluded that Jupiter was remarkably similar to Earth, with wet, dry, hot, and cold regions, as well as rain, snow, and thunderstorms. Before the creation of Galileo, scientists had been unable to observe much of the planet because of surface clouds of frozen ammonia. According to Robert W. Carlson of NASA's Jet Propulsion Laboratory, NASA's most fundamental conclusion was that

---

[261] Jeff Cole, "Lockheed To Unveil Intersputnik Satellite Venture," *Wall Street Journal*, 2 June 1997; Associated Press, "Lockheed Planning Russia Venture," 2 June 1997; Greg Schneider, "Lockheed To Link Ex-Soviet Markets," *Baltimore Sun* (MD), 3 June 1997.
[262] NASA, "NASA Studies High Altitude Radiation with Upgrade ER-2," news release 97-118, 3 June 1997.
[263] Bob Pool, "Space Race," *Los Angeles Times*, 4 June 1997.

"There is weather on Jupiter." However, scientists pointed out that, in contrast to Earth, Jupiter faces weather developments that are far more extreme and long lasting. Storms on the planet can last for years or even centuries.[264]

An article in *Nature* confirmed the discovery of a small icy planet orbiting the Sun well beyond Pluto. The discovery confirmed the theory that the solar system extends far beyond previously verifiable limits. The planet, known as 1996TL66, with a surface area roughly the size of Texas, was the brightest solar-system object that astronomers had discovered beyond Neptune since 1978. Astronomers using a University of Hawaii telescope discovered the planet as its orbit passed closest to the Earth, an occurrence that happens every 800 years. In the past, some researchers had hypothesized, but had been unable to prove, that planets might exist in the area where astronomers discovered 1996TL66.[265]

*9 June*
NASA's AeroVironment Pathfinder, a remote-controlled, ultra-lightweight aircraft, set a new world record for highest altitude for a solar-powered aircraft. AeroVironment Pathfinder reached an altitude of 67,350 feet (20,500 meters), shattering the previous record of 50,500 feet (15,400 meters). The experiment took place at the U.S. Navy's Pacific Missile Range Facility in Kauai, Hawaii. Engineers had designed Pathfinder to fly at extremely high altitudes, to supply researchers with atmospheric and environmental data that other planes could not gather. In traveling to such heights, Pathfinder experienced temperatures as low as -100°F (-73°C). After climbing for about 6 hours, the aircraft ascended into record-breaking territory, flying for more than 90 minutes at an altitude above 60,000 feet (18,300 meters). A partnership formed under the Environmental Research Aircraft and Sensor Technology Alliance, involving NASA, aerospace companies, and members of the scientific community, was responsible for developing AeroVironment Pathfinder.[266]

*10 June*
Astrophysicists William P. Blair, Robert A. Fesen, and Eric M. Schlegel presented to the American Astronomical Society the first images ever recorded of two supernovas colliding. The team used the Hubble Space Telescope to capture the images of the supernovas, more than 17 million light-years from Earth. Astronomers had suspected that collisions occurred between exploding stars, but had been unable to capture an image of the process, because it was so short-lived. Although the scientific community would continue to debate the implications of the image, Blair offered this assessment of its significance: "It indicates that not only is there a lot of star formation going on, but a lot of those stars are massive. They are evolving quickly, and they are exploding as supernovas."[267]

*18 June*

---

[264] NASA, "Galileo Finds Wet Spots, Dry Spots and New View of Jupiter's Light Show," news release 97-123, 5 June 1997; K.C. Cole, "Scientists Detect Rain on Jupiter," *Los Angeles Times*, 6 June 1997; Associated Press, "Jupiter Wet and Dry," 6 June 1997.
[265] Jane Luu et al., "A New Dynamic Class of Object in the Outer Solar System," *Nature* 387, no. 6633 (5 July 1997): 573–575; Associated Press, "Astronomers Find Icy Miniplanet Beyond Pluto," 5 June 1997.
[266] Michael A. Dornheim, "Pathfinder Surpasses Propeller Altitude Record," *Aviation Week and Space Technology* 146, no. 25 (16 June 1997): 53; NASA, "Pathfinder Sets New World Record," news release 97-132, 12 June 1997.
[267] NASA, "Hubble Is First To Spot Colliding Supernovas," news release 97-129, 10 June 1997.

The U.S. General Accounting Office (GAO) released a report detailing the cost overruns for the development of the International Space Station (ISS). U.S. Senator Dale Bumpers (D-AK), a longtime opponent of the expensive ISS, requested the report. Upon seeing it, Senator Bumpers commented, "we don't need to search outer space for black holes. We have one right here on Earth. It's called the International Space Station." GAO's report found that Boeing Company, hired by NASA to build the ISS, had already incurred nearly US$300 million in cost overruns. The report also found that the cost-effectiveness of the project had deteriorated severely from the time of last accounting. In 1995 Boeing had been US$27 million under budget as opposed to the severe overages of 1997. GAO Associate Director Thomas J. Schulz recommended in the report that Congress should rethink its commitment to the program if the overruns continued to grow. The estimate of the cost, just to get the program back on schedule, was US$129 million. However, an ongoing debate on a disaster relief bill for the Midwest prevented the scheduled congressional hearing on the report.[268]

NASA announced that Lockheed Martin Missiles and Space Company had won the contract to develop and deliver solar-imaging instruments for U.S. weather satellites. The US$54 million contract provided funding for the creation of a model instrument and two flight instruments. Scientists believed that the solar x-ray images captured by the instrument would improve the ability of the National Oceanic and Atmospheric Administration and the U.S. Air Force to forecast special weather events. The proposed imager also had applications for civilian companies dependent upon predicting weather patterns.[269]

Jurgen H. Rahe, NASA's Science Program Director for Exploration of the Solar System, died when a tree collapsed on the car he was driving during a severe storm in Maryland. Rahe, 57 years old, had enjoyed a distinguished career in the field of astronomy and at NASA. At the time of his death, he was responsible for overseeing visionary NASA programs, such as the Galileo Mission to Jupiter and the much-anticipated July 1997 landing of Mars Pathfinder. Rahe had guided NASA's efforts to make more frequent and cost-effective exploratory missions of the solar system. According to one colleague, Rahe had presided over the planetary exploration program's "unparalleled period of major discoveries." Before his tenure at NASA, Rahe held tenured positions at California Institute of Technology and Germany's University of Erlangen-Nuremberg.[270]

*24 June*
As the 50th anniversary of the alleged alien sighting neared, the U. S. Air Force released a 231-page report entitled *The Roswell Report: Case Closed*. The report explained that the U.S. Air Force had recovered test dummies from the Roswell, New Mexico, crash site in 1947, rather than the bodies of aliens. The U.S. Air Force had issued a report on the Roswell matter in 1994, in which researchers argued that the presumed spacecraft that had crashed in 1947 was actually an Air Force balloon used in a top-secret program called Project Mogul. When officials discovered

---

[268] U. S. General Accounting Office, "Space Station Cost Control Problems Continue To Worsen" (report no. T-NSIAD-97-177, Washington, DC, 18 June 1997); Seth Borenstein and Tamara Lytle, "Space Lab Deep in Red, Report Says," *Orlando Sentinel* (FL), 11 June 1997; Associated Press, "Space Station," 11 June 1997.
[269] NASA, "Lockheed Martin Selected To Build Solar X-ray Imaging," news release C97-g, 18 June 1997; Grant Jerding, "NASA Brings Its X-ray Vision to the Universe," *USA Today*, 17 June 1997.
[270] Associated Press, "Obituary: Jurgen Rahe, 57, NASA Official," 20 June 1997; NASA, "NASA Mourns Dr. Jurgen H. Rahe, Solar System Exploration Science Program Director," news release 97-137, 19 June 1997.

evidence that the tests had used parachute dummies, however, the U.S. Air Force compiled an additional report with the new information. Military officials hoped that the Air Force's explanation for the supposed UFO reported in 1994, as well as for the supposed bodies of aliens discovered in the crash, would temper the controversy surrounding the issue. U.S. Air Force Colonel John Haynes, presiding over the Pentagon news conference at the release of the report, showed reporters footage of a NASA test craft that, indeed, resembled a flying saucer. Colonel Haynes explained that during testing in the 1950s, Air Force balloons had transported dummies to altitudes of 98,000 feet (29,900 meters), releasing them to fall to the ground. Since the testing was secret, the sight of falling dummies "easily could have been mistaken for something they were not."[271]

In the most serious of a string of problems for the Russian *Mir* space station, the 7-ton (6,400-kilogram or 6.4-tonne) robotic cargo spacecraft Progress accelerated out of control while docking and crashed into *Mir*, damaging the *Spektr* module, tearing a solar-power array, and crumpling an exposed radiator. When the craft crashed into *Mir*, U.S. astronaut, British-born C. Michael Foale was inside the *Spektr* module, which he had been using as his sleeping and research quarters. Foale immediately abandoned the damaged module and slammed shut a hatch, sealing off the leaking *Spektr*. *Spektr* depressurized completely, and later the entire space station suffered a slight depressurization. When the collision occurred, *Mir* cosmonaut Vasili V. Tsibliyev was controlling the Progress manually, attempting to re-dock the craft, to test the updated manual-control system. The Progress had already docked with *Mir* a few days earlier, when the crew had unloaded cargo and filled the craft's hold with trash. In addition to damaging *Mir*, the collision temporarily jarred the station off its normal flying orbit. Although the crew quickly remedied this problem, the three-man team had to work inside a darkened, partially powered station. Russian and American space officials estimated that *Mir* had lost half of its operating power from the collision. The collision heightened tensions between Russian and American space officials, as questions mounted about the safety of the crew aboard the failing space station. U.S. House Committee on Science and Technology Chairperson F. James Sensenbrenner Jr. (R-WI) met with NASA Administrator Daniel S. Goldin, urging NASA not to send American astronauts to *Mir* until the station's safety standards had improved significantly. Yuri N. Koptev, Director General of the Russian Space Agency, responded angrily to Sensenbrenner's suggestion.[272]

*26 June*
Vladimir Lobachyov, head of Russian Mission Control, explained to reporters that the collision of the robotic cargo craft Progress with the *Mir* space station had probably punctured one of the space station's solar batteries, damaging *Mir*'s power supply. The crew of the damaged *Mir*

---

[271] U. S. Air Force Headquarters, *The Roswell Report: Case Closed* (Washington, DC, 1997); Associated Press, "Air Force Says Dummies Used in Parachute Tests Were Mistaken for Aliens," 24 June 1997; Richard Leiby, "Alien Autopsy," *Washington Post*, 25 June 1997.
[272] Michael R. Gordon, "Russian Space Station Damaged in Collision with a Cargo Vessel," *New York Times*, 26 June 1997; Kathy Sawyer, "Docking Crash Cripples *Mir* Space Station," *Washington Post*, 26 June 1997; Associated Press, "Russians Slow To Announce *Mir* Crash, but More Forthcoming than in Past," 25; David Hoffman, "U.S. Questions on *Mir*'s Safety Anger Russia," *Washington Post*, 27 June 1997; Paul Hoversten, "Future of U.S.-Russian Mission Shaky," *USA Today*, 26 June 1997.

turned the space station to face directly toward the Sun, so that *Mir* would get the jolt of energy necessary to boost its solar-power supply.[273]

*27 June*
NASA's Near Earth Asteroid Rendezvous (NEAR) spacecraft engaged in a high-speed flyby of asteroid 253 Mathilde, capturing startlingly clear images of the huge space rock. Scientists had discovered Mathilde more than 100 years ago but, until the successful NEAR mission, they had been unable to take useful pictures of the 33-mile-around (53-kilometer-around) asteroid. NEAR Mission Director Robert W. Farquhar of the Applied Physics Laboratory praised the mission as "one of the most successful flybys of all time." In the color images, the asteroid resembled a massive gray potato with deep gashes scarring its surface. Some of the craters on the asteroid were large enough to "swallow the District of Columbia," according to one scientist who viewed the images as they were taken. The pictures also revealed that the rock was smaller than scientists had previously estimated and that it reflected only 3 percent of the Sun's light, making it twice as dark as a piece of charcoal. Scientists hoped that the new data might reveal why the asteroid rotates at the extremely slow rate of one rotation every 17.4 days. NEAR's multispectral imager took the pictures using very little of the spacecraft's solar-based power, a significant accomplishment, since NEAR was approximately 186 million miles (300 million kilometers) from the Sun when it captured the images.[274]

## JULY 1997

*1 July*
Space Shuttle *Columbia* lifted off from Kennedy Space Center (KSC) in Cape Canaveral, Florida. Commander James D. Halsell Jr. called Mission STS-94 a "once-in-a-career opportunity" for the seven-person crew, the same crew that had embarked upon Mission STS-83 in April 1997. NASA had cut short Mission STS-83 because of a fuel cell problem. The reflight gave NASA officials and the Shuttle crew a chance to undertake the experiments planned for the original mission, including testing hardware and procedures planned for the International Space Station. NASA was able to launch *Columbia* just 84 days after bringing the Shuttle home, at a cost of about 20 percent of the amount needed to set off a new Shuttle mission. The crew planned to carry out the original 33 experiments over the course of the 16-day science mission. Mission STS-94, NASA clarified, involved the same vehicle, crew, Microgravity Science Laboratory payload, and experiment schedule as the recalled mission. The mission was the first in the 36-year history of unpiloted spaceflight in which the same crew flew together in space more than once.[275]

*2 July*

---

[273] Reuters, "Mir Crew Turns to Sun To Help Damaged Space Station," 26 June 1997.
[274] NASA, "Asteroid Mathilde Reveals Her Dark Past," news release 97-147, 30 June 1997; Associated Press, "Spacecraft Sends Astonishing Images of Asteroid," 27 June 1997; Associated Press, "In Close Flyby Spacecraft Sends Back Photos of Battered Asteroid," 1 July 1997.
[275] NASA, "Columbia and Crew Ready for Reflight of Microgravity Sciences Laboratory-1 Payload on Mission STS-94," news release J97-21, June 1997; William Harwood, "Shuttle Begins Second Attempt at Mission," *Washington Post*, 2 July 1997; Warren Leary, "Shuttle Roars Aloft To Complete Mission Interrupted in April," *New York Times*, 2 July 1997; Associated Press, "Space Shuttle Blasts Off on Repeat Science Mission," 1 July 1997.

NASA announced its plan to compensate for the failure of Japan's Advanced Earth Observing Satellite (ADEOS) with NASA's weather instruments aboard, admitting that the loss of the spacecraft was "a real blow to NASA's science program." Japan's National Space Development Agency (NASDA) had declared the ADEOS lost on 30 June 1997. NASA scientists had been using the satellite to gather data about weather patterns and climate change. NASA resolutely pledged to continue working with NASDA despite the setback, highlighting plans to send up a second scatterometer aboard ADEOS II in 1999. Michael B. Mann, Deputy Associate Administrator of the Mission to Planet Earth Strategic Enterprise, qualified the failure saying, "The collaboration between NASDA and NASA on this mission has been outstanding . . . space operations is a risky business; those of us involved in the business strive to limit the risk but sometimes mishaps do occur."[276]

*4 July*
NASA successfully landed the Mars Pathfinder spacecraft on Mars, exactly seven months after its launch from Earth. Project managers at NASA's Jet Propulsion Laboratory (JPL) called the landing "near perfect," celebrating the beginning of the "second era in the exploration of Mars." The Mars Pathfinder Mission was the first to attempt a planetary landing on initial orbit, and Pathfinder's safe landing followed a journey in which the spacecraft traveled at speeds of up to 16,600 miles (26,800 kilometers) per hour. Although NASA made its first successful flyby of Mars in 1965, in recent years it had taken a new approach to exploring the Red Planet. NASA achieved the development and deployment of Mars Pathfinder for US$266 million, a bargain compared to the US$1 billion cost of the failed Mars Observer. Because of its tightening budget, NASA had adopted a course of planning several simple, but targeted excursions, using low-cost spacecraft, rather than mounting a few complex and expensive missions. Pathfinder landed almost exactly at the point aimed for and immediately transmitted signals back to Earth announcing its safe landing. Officials at JPL noticed a slight complication, when early images revealed that some of the airbags used to cushion the craft's landing had bunched up around the door that Sojourner—the rover set to explore the planet—would use to exit Pathfinder. NASA's engineers clarified, however, that the temporary problem would not compromise the rover's planned exploration.[277]

*5 July*
Russia launched Progress M35, a supply spacecraft headed for *Mir*, filled with repair equipment needed to fix problems caused by the crash of another Progress cargo craft and to restore full power to the space station. According to Russian space officials, repairs would entail a lengthy spacewalk. The repair tasks appeared so daunting that the Russian Space Agency engineers pushed back the date of the work, to allow the *Mir* cosmonauts more time to train. C. Michael Foale, the American astronaut aboard *Mir*, was responsible for operating the *Soyuz* spacecraft

---

[276] NASA, "NASA's Earth Science Program Adjusts to Loss of Data from Japanese ADEOS Satellite," news release 97-149, 2 July 1997.
[277] John Noble Wilford, "Landing Called Near Perfect by NASA," *New York Times*, 5 July 1997; John Noble Wilford, "Spacecraft Lands Today on the Planet of Dreams," *New York Times*, 4 July 1997; Tony Knight, "Pathfinder Zeroes in on Red Planet," *Los Angeles Daily News*, 4 July 1997; NASA, *Aeronautics and Space Report of the President: Fiscal Year 1996 Activities* (Washington, DC, 1997), p. 3.

docked at the station, so that in the event of a problem, the crew would be able to make a safe return to Earth.[278]

*6 July*
Sojourner, the roving vehicle carried to Mars aboard Pathfinder, began prospecting the surface of the Red Planet approximately two days after Pathfinder's safe landing. NASA engineers and enthralled scientists described the scene of the six-wheeled, microwave-sized Sojourner, slowly descending the ramp of Pathfinder, as similar to Neil A. Armstrong's "giant leap" in 1969. Richard A. Cook, managing the mission from JPL, set off a raucous celebration of NASA engineers when he announced, "The rover is on the surface of Mars. We've got some great images back, and all the scientists are in heaven." Weather reports gathered by the Pathfinder reported that Sojourner was working in temperatures as high as 8° below 0°F (-22°C), as well as in wind and dust storms. Sojourner, named for black abolitionist Sojourner Truth, headed for a dark rock near Pathfinder, as its first object of close study. NASA scientists, who had named the rock Barnacle Bill because of its craggy appearance, used similarly whimsical terms to keep track of the various geological objects studied by Sojourner. Sojourner's success, after some minor complication upon landing, encouraged mission scientists and drew wide acclaim from the media. Scientists began to speculate almost immediately, about whether the mission could possibly confirm or deny the existence of life on the planet.[279]

*7 July*
NASA's AeroVironment Pathfinder—distinct from Mars Pathfinder—broke the record for highest-flying, propeller-powered aircraft in a flight from the U.S. Navy's Pacific Missile Range Facility on Kauai, Hawaii. After climbing for nearly 7 hours, AeroVironment Pathfinder reached its record altitude, maintaining its height for about 13 minutes. Even at its peak distance from Earth, a span of more than 67 nautical miles (124 kilometers), AeroVironment Pathfinder transmitted data without any problems. The successful test was part of the Environmental Research Aircraft and Sensor Technology Program at NASA's Dryden Flight Research Center.[280]

*9 July*
NASA reported receiving 265 million hits on its Web site during the four days following Mars Pathfinder's landing. In response to what some observers called the "biggest event in computer network history," NASA buttressed its own computers with donated equipment from corporate sponsors, so that the government site could handle the high volume of viewers. The Internet statistics validated the widespread popularity of the mission to Mars. The CNN network dedicated nearly half of its programming to the Pathfinder story on the day of the Mars landing, doubling its network ratings. Entrepreneurs sold a variety of Mars- and Pathfinder-themed products, and visitors jammed the National Air and Space Museum in Washington, DC. *USA*

---

[278] Michael R. Gordan, "Russia Launches Supply Craft To Repair the Damaged Mir Space Station," *New York Times*, 6 July 1997.
[279] John Noble Wilford, "On the Surface of Mars, Sojourner Rolls to Work," *New York Times*, 7 July 1997; Knight-Ridder News Service, "Data from Mars Indicate It Could Have Supported Life," 7 July 1997; Peter N. Spotts, "Tiny Land Rover Begins Relaying Detailed Data About Surface of Red Planet," *Christian Science Monitor*, 7 July 1997.
[280] Michael A. Dornheim, "Pathfinder Aircraft Hits 71,500 Ft.," *Aviation Week and Space Technology* 147, no. 3 (21 July 1997): 40; NASA, "Solar-Powered Pathfinder Sets New Record; Prepares To Monitor Deforestation of Hawaiian Island," news release 97-153, 14 July 1997.

*Today* summed up the success of the Mars Pathfinder Mission: "NASA's successful landing on the Red Planet has set off a Marsfest on the blue one."[281]

NASA announced the retirement of veteran astronaut Jeffrey A. Hoffman, who had flown on five Shuttle missions. Hoffman planned to continue his NASA service, becoming NASA's European representative in Paris. Hoffman was one of the four astronauts to take part in the spacewalking mission to service the Hubble Space Telescope (HST) in 1993. In total, Hoffman had spent more than 1,200 hours in space and traveled more than 21.5 million miles (35 million kilometers).[282]

The crew aboard Space Shuttle *Columbia* lit more than 200 small fires, testing the flammability of dozens of gases and materials. The experiments took place under the direction of Paul D. Ronney, a professor of mechanical engineering at the University of Southern California. Ronney explained that the experiments had as much application for Earth as for space. According to Ronney, the weaker the flame in space's weightless environment, the cleaner its fuel would burn on Earth. The astronauts found that, without the effect of gravity, the flames often formed the shape of a ball and heat did not rise. The experiments created some attention for Mission STS-94, largely overshadowed by the success of the Pathfinder mission to Mars.[283]

*14 July*
Routine medical testing of the crew aboard *Mir* revealed that the Russian commander of the space station, Vasili V. Tsibliyev, had an irregular heartbeat and, according to Russian Mission Control, a problem with his "cardiovascular system." The tests, indicating that Tsibliyev was overtired and suffering from stress, did not surprise space officials, considering his traumatic experience of late June, when a cargo craft had crashed into *Mir*. Tsibliyev reported that he had felt "irregularities" while he was exercising and when he was trying to sleep. The news of crew health problems complicated further the arduous repair plan. A day after learning of Tsibliyev's test results, Russian officials proposed that U.S. astronaut C. Michael Foale take Tsibliyev's place in the repair spacewalk. NASA subsequently deliberated over the request before granting Foale permission to participate. Ultimately, however, Russian Mission Control decided not to allow *Mir*'s exhausted crew to perform the tasks. Instead, Mission Control planned for the replacement crew scheduled to arrive in August 1997 to make the repairs. The response to the latest of *Mir*'s challenges demonstrated an increasingly dependent and trusting partnership between the Russian and American space agencies. During the Soviet-led race for space, few could have imagined Russian officials asking for the aid of an American crew member in a time of crisis.[284]

---

[281] NASA, "Pathfinder Gets Hit Hard on the Internet," news release I97-8, 9 July 1997; Martha T. Moore, "Mars Mania Invades Earth," *USA Today*, 9 July 1997; Associated Press, "'Mars Live' Draws Millions to Web Site," 9 July 1997; United Press International, "Mars Tops Internet Hit List," 8 July 1997.
[282] NASA, "Jeff Hoffman Retires from Astronaut Corps," new release 97-151, 9 July 1997.
[283] Associated Press, "Shuttle Tests of Tiny Fires Spur Delight," 9 July 1997; Wayne Tompkins, "Columbia Crew Ready for Hot Time in Space," *Florida Today* (Brevard, FL), 4 July 1997.
[284] Paul Hoversten, "Latest Mir Trouble: Commander's Heart," *USA Today*, 15 July 1997; Michael Gordon, "Russians Suggest American Take On Mir Repair Task," *New York Times*, 16 July 1997; Associated Press, "Commander of Russian Space Station Shows Heart Irregularities," 14 July 1997; Reuters, "Tests on Mir Commander Show Cardiovascular Problem," 14 July 1997; Associated Press, "Backs to the Wall, Russians Ask for Help from American Astronaut," 15 July 1997; Reuters, "NASA Approves Mir Spacewalk Training for Foale," 17 July 1997; *New York Times*, "A Rescue Assignment for NASA," editorial, 17 July 1997; Associated Press, "Mir Crew Gets Disappointing News; Next Crew To Do Key Repairs," 19 July 1997.

*15 July*
Vice President Albert A. Gore Jr., U.S. Department of Transportation Secretary Rodney E. Slater, and NASA Administrator Daniel S. Goldin joined in leading an aviation safety event at Dulles International Airport in Virginia. NASA and the Federal Aviation Administration (FAA) had invited media representatives and interested spectators to the event, to observe demonstrations of the new technologies that the agencies had developed to improve the safety of flying aboard commercial, private, and government aircraft. NASA made available one of its Boeing B-757 research aircraft, providing hands-on demonstrations of the new techniques and technologies produced by NASA's and the FAA's Task Force on Aviation Safety.[285]

*16 July*
As Mars Pathfinder and its surveyor Sojourner triumphantly explored the surface of Mars, NASA named as Software of the Year the Dynamics Algorithms for Real-Time Simulation (DARTS) software—a program used on the Pathfinder mission. Abhinandan Jain, Guillermo Rodriguez, and Guy K. Man of JPL had created the software to provide real-time simulations to test other software and hardware. DARTS allowed NASA mission managers to test procedures in a cost-effective manner.[286]

*17 July*
Space Shuttle *Columbia* landed safely at KSC in Cape Canaveral, Florida, just after sunrise, returning its seven-person crew to Earth. The landing marked the conclusion of a successful, 16-day mission, during which the astronauts completed 33 laboratory experiments. The low-cost mission was the first reflight in NASA's history, concluding the research agenda planned for the April 1997 mission that NASA had terminated as a safety precaution. Although the press coverage of events on Mars and *Mir* had overshadowed the *Columbia* crew's efforts, NASA Mission Manager Teresa B. Vanhooser stated confidently that scientific journals would give the mission due respect.[287]

*25 July*
Two Yemeni men filed a lawsuit against NASA, claiming that NASA had committed trespassing violations when it landed Pathfinder on Mars. The men claimed that their ancient ancestors had deeded the planet to them as an inheritance, requesting that Yemen's prosecutor-general bring U.S. Ambassador David G. Newton to court to stop the violation. Although no legal restitution was expected, the lawsuit highlighted the excitement surrounding NASA's Mars mission and the widespread interest in the Red Planet's exploration.[288]

*30 July*
Astronomers using the HST and the W. M. Keck Observatory in Hawaii announced the discovery of the most distant object ever observed from Earth. The discovery team comprised

---

[285] NASA, "Vice President To Showcase Aviation Safety Technologies Developed by NASA and FAA," news release N97-47, 14 July 1997.
[286] NASA, "Software Used on Mars Pathfinder Wins NASA's Software of the Year Award," news release 97-154, 16 July 1997.
[287] Associated Press, "Shuttle Returns Full of Research Information," 17 July 1997.
[288] Reuters, "Get Off Our Planet, Yemeni Men Tell U.S.," 25 July 1997.

researchers from the University of Groningen in the Netherlands and the University of California at Santa Cruz. The scientists' calculations indicated that the distant light they had discovered was an infant galaxy, approximately 13 billion light-years from Earth. Formally named Red Arc in CL1358-62, the discovery gave astronomers a glimpse into the universe as it had existed when only 7 percent of its present age. Lead researcher Garth D. Illingworth described the images as "a pathfinder for deciphering what is happening in young galaxies."[289]

## AUGUST 1997

*1 August*
Frank L. Culbertson Jr., NASA's Manager for the Space Shuttle–*Mir* research program, announced that in the future NASA would likely require astronauts to have training in spacewalking. The announcement coincided with reports in the *New York Times* and other U.S. newspapers that David A. Wolf had replaced astronaut Wendy B. Lawrence as the next NASA representative aboard *Mir*. NASA had replaced Lawrence because of her small stature. At 5 feet, 3 inches (1.6 meters) tall, she was too small to fit into the bulky Russian Orlan spacesuits worn on *Mir*. In addition, she had received no training in spacewalking. Previously, NASA had planned for Wolf to replace Lawrence aboard *Mir* following her research stint. NASA and the Russian Space Agency addressed the sensitive issue of replacing Lawrence because of her size, explaining the need for flexibility aboard *Mir* during significant repairs to the station. Lawrence reacted professionally, despite the disappointment of her replacement less than two months before her planned once-in-a-career mission.[290]

NASA Administrator Daniel S. Goldin addressed the Experimental Aircraft Association fly-in convention, outlining NASA's vision to "revitalize general aviation." Goldin boldly predicted that the U.S. aviation industry would be able to deliver 10,000 aircraft annually within 10 years, and 20,000 annually within 20 years. The fly-in convention, featuring flyovers by NASA's SR-71, had the theme, "Boomers Turn 50," celebrating both the 50th anniversary of the U.S. Air Force and that of the first breaking of the sound barrier. The Association announced the winner of the National General Aviation Design Competition, sponsored by NASA and the Federal Aviation Administration (FAA), a competition encouraging university engineering students to focus on the aviation industry. The conference also featured the unveiling of the new FAA flight-training curricula, as well as a national aviation safety initiative.[291]

*2 August*
The Rasmussen Research Group reported that seven out of 10 Americans had seen at least one image of Mars obtained through the Mars Pathfinder Mission, denoting a remarkable level of public awareness of NASA's mission to the Red Planet. The researchers also reported that nearly half of all Americans believed that an American would land on Mars within 10 years. Despite the

---

[289] Malcolm W. Browne, "Astronomers Discover Most Distant Object," *New York Times*, 31 July 1997.
[290] Warren E. Leary, "Future Astronauts Must Walk Before They Fly," *New York Times*, 1 August 1997; Todd Halverson, "Mir Prerequisite: Spacewalking," *Florida Today* (Brevard, FL), 1 August 1997; United Press International, "U.S., Russia Cite Logic of Replacing Female Astronaut," 1 August 1997; Associated Press, "NASA: Decision To Replace Astronaut Difficult but Necessary," 31 July 1997.
[291] NASA, "NASA To Showcase New Activities at Upcoming Aircraft Fly-In Convention," news release N97-51, 25 July 1997.

widespread interest and optimism regarding the project, only 41 percent of those polled saw merit in the mission.[292]

*5 August*

The Russian *Soyuz* spacecraft lifted off from Kazakhstan carrying cosmonauts Anatoly Y. Solovyev and Pavel V. Vinogradov on a mission to repair the disabled *Mir* space station. Russian space officials tasked the two men with restoring power to *Mir* after the 25 June collision with a cargo craft cut by half the orbiting research center's energy. Regarding the importance of the mission, some observers of Russian space activity went so far as to say that the country's piloted space program depended on Solovyev and Vinogradov's success.[293]

During the terminal countdown of a demonstration test of the Titan IV-B rocket designed for the Cassini mission to Saturn, engineers noticed liquid hydrogen and liquid oxygen leaking from the Centaur stage of the rocket. NASA had scheduled Cassini's launch for early October 1997. NASA officials delayed the launch until they could complete further tests, offering no immediate estimate of how long that might take. Interested scientists and engineers, however, predicted that the problem could postpone the mission by months or even years. Under the mission's original timetable, NASA had expected Cassini to launch in October 1997, reaching Saturn in July 2004. Because of the length of the journey from Earth to Saturn, the scientific community expected that NASA would find it difficult to reschedule the flight. Moreover, experts warned that the alignment of planets providing an ideal launch date, such as that in October 1997, would not occur again for 12 to 14 years. At US$3.2 billion, Cassini was one of the last of NASA's high-stake, multibillion-dollar missions, drawing considerable attention during the climate of steady cost reductions that characterized the 1990s.[294]

*7 August*

A relief crew of two Russian cosmonauts, Anatoly Y. Solovyev and Pavel V. Vinogradov, arrived at *Mir* to begin repair work on the damaged space station. On 25 June 1997, *Mir* had suffered its worst accident, when a robotic supply craft attempting to dock pierced the exterior of one of its research modules. The arriving crew, carrying tools and equipment, also had difficulty docking. As *Soyuz* approached *Mir*, its autonomic docking system failed, forcing Solovyev to override the system and pull *Soyuz* back. Solovyev then successfully docked the relief vehicle using manual controls. The crew planned to repair *Mir*'s failing oxygen-generating system, before reconnecting the damaged research module to the rest of the station. Both NASA and the Russian Space Agency stressed that *Mir* had sufficient oxygen reserves in the event that the oxygen generator failed altogether. After repairing the oxygen system and the research module, the new crew planned to make as many as six spacewalks to repair the external damage caused by the June collision. Cosmonaut Solovyev brought significant experience to the mission; he had resided on *Mir* during four earlier missions, spending a total of 456 days in space. NASA

---

[292] Mark Carreau, "Mars Trip a Monster Summer Hit," *Houston Chronicle*, 2 August 1997.
[293] Associated Press, "2 Russians Blast Off on Mir Repair Mission," 6 August 1997.
[294] NASA, "Countdown Test Reveals Fuel Leaks on Cassini Mission Centaur Upper Stage," 7 August 1997; Associated Press, "Rocket Leak Could Postpone Landing on Saturn by Years," 8 August 1997.

planned for British-born U.S. astronaut C. Michael Foale to remain on *Mir*, although the Russian Space Agency was replacing its crew aboard the space station.[295]

Shuttle *Discovery* Mission STS-85 launched, carrying Commander Curtis L. Brown Jr., Pilot Kent V. Rominger, and astronauts N. Jan Davis, Robert L. Curbeam Jr., Stephen K. Robinson, and Canadian Bjarni V. Tryggvason, who planned to conduct studies on Earth's ozone layer and to test a newly engineered robotic arm. The test on the small robotic arm, a prototype of the one planned for the International Space Station (ISS), continued NASA's trend of using Shuttle missions to prepare for the much-anticipated space station. *Discovery* also carried a 7,000-pound (3,200-kilogram) German satellite, which would orbit freely for most of the 10-day mission, gathering data on the declining ozone layer. Brown emphasized the international flavor of the crew and cargo saying, "one word for our mission would be diversity." With this mission, its 23$^{rd}$ flight, *Discovery* tied with *Columbia* for the most missions by a single vehicle.[296]

*8 August*
After exceeding even the most optimistic predictions made during its primary mission, the Mars Pathfinder spacecraft concluded its initial research maneuvers and began extended operations. The initial mission had returned to Earth 1.2 gigabits of data and more than 9,000 images of the Martian landscape. Project Manager of the Mars Pathfinder Mission Brian Muirhead said that the mission had not only had gathered an impressive amount of data, but also validated NASA's new approach to exploring space. "This mission demonstrated a reliable and low-cost system for placing science payloads on the surface of Mars," he stated. "We've validated NASA's commitment to low-cost planetary exploration." After completing the planned 30-day mission and traveling 171 feet (52 meters) across Mars's landscape, the rover Sojourner remained in excellent condition. Operators gave Sojourner a two-day rest period to recharge its battery, before it began an extended exploration of the planet.[297]

*9 August*
As the two cosmonauts worked to repair the failing *Mir* space station, President Boris N. Yeltsin allowed the Russian government to borrow US$100 million from foreign banks, giving the Russian Space Agency an infusion of capital. Yeltsin announced the decision after touring the construction facilities where Russia planned to build its contributions to the ISS. The Russian Space Agency's cash-flow problems had resulted in construction delays and failure to meet deadlines, postponing the completion of the entire ISS.[298]

After widespread speculation concerning the possibility of an indefinite delay of the Cassini Mission to Saturn, NASA announced that it would be able to repair a fuel leak in the spacecraft

---

[295] *Baltimore Sun* (MD), "Relief Team Reaches Crippled Mir Space Station," 8 August 1997; Reuters, "Two Cosmonauts Blast Off on Mission To Repair Mir," 6 August 1997; Associated Press, "Russians Rocket To Repair Station," 6 August 1997; Reuters, "New Crew Faces Hydra of Woes on Mir Space Station," 5 August 1997.
[296] NASA, "August 7 Selected for STS-85 Space Shuttle Mission Launch," news release 97-159, 24 July 1997; Marcia Dunn, "Shuttle Blasts Off on Ozone Study," *USA Today*, 8 August 1997; Mark Carreau, "Discovery Mission Plans Tests with International Flair," *Houston Chronicle*, 4 August 1997.
[297] NASA, "Mars Pathfinder Results Generating New Picture of Mars as Mission Moves into Extended Operation," news release 97-174, 8 August 1997; *Los Angeles Times*, "Mars Mission '100 Percent Success', NASA Says," *Los Angeles Times*, 9 August 1997.
[298] Associated Press, "Government Will Borrow $100 Million for Space Agency," 9 August 1997.

in time to meet an October 1997 launch date. NASA had enlisted U.S. Air Force engineers to make the repairs. The spacecraft's need for repairs increased the tension surrounding the ambitious and expensive mission. Lieutenant Ken Hoffman of the U.S. Air Force, who supervised the repairs, attempted to reassure the public that the problem was "a correctable condition." Nevertheless, opponents of the mission—particularly those opposed to the spacecraft's use of radioactive plutonium—worried that the rocket's launch might release hazardous materials. Michio Kaku, a City University of New York physics professor, said that, despite NASA's announcement that it could quickly correct the problem, the leak was "just one more indication that NASA does not have things under control."[299]

*10 August*
Astronauts aboard Shuttle *Discovery* successfully tested a new robotic arm, considerably smaller than *Discovery*'s massive 50-foot (15-meter) crane, intended for tasks of higher precision. *Discovery*'s crew used the arm to unlock a cargo bay door and lift a small box. NASA had engineered the Japanese-made robotic arm for US$100 million, expecting that it would eventually serve astronauts aboard the ISS. Of the robotic arm's test performance, astronaut Stephen K. Robinson stated simply, "it operated beautifully."[300]

*13 August*
NASA and the Boeing Company unveiled a new stitching machine that experts believed would change the method of fabricating aircraft wing structures. The demonstration took place at Boeing Stitched Composites Development Center in Huntington Beach, California. The project team of engineers had replaced the large metal structures used in airplanes with strong, but light, composite materials. By reducing the weight of an airplane, engineers hoped to decrease fuel usage, with the eventual result of cost savings for consumers. Rather than using rivets to hold together an airplane's wings, the NASA-developed stitching machine sewed together precut fabric layers in the shape of a wing. Next, the machine added braided stiffener materials to the wing, before setting the wing skeleton with resin. The process created a wing as strong as one built with metal structures, but at a fraction of the weight.[301]

*14 August*
Two Russian Space Agency cosmonauts, Vasili V. Tsibliyev and Alexander I. Lazutkin, returned to Earth after spending more than six months aboard the *Mir* space station. The *New York Times* echoed the feelings of many in the space community, calling the six months the Russians had spent in space, "six of the least glorious months in the history of manned space flight." During their departure from *Mir*, Tsibliyev and Lazutkin continued to experience the same type of problem that had characterized their tenure on board the space station. The two cosmonauts, scheduled to disengage from *Mir* in the *Soyuz* spacecraft and then to circle the space station, cancelled the exercise because of concern that the two spacecraft would collide. After the cosmonauts landed in the desert of Kazakhstan, Tsibliyev recounted to an international press corps the difficult mission that he and Lazutkin had endured aboard *Mir*, remarking that the two cosmonauts should not become the scapegoats for Russia's problems in space. The collision

---

[299] Robyn Suriano, "Cassini Won't Be Delayed," *Florida Today* (Brevard, FL), 9 August 1997.
[300] Associated Press, "Astronauts Aboard Discovery Test New Japanese Robot Arm," 9 August 1997; Associated Press, "Shuttle Tests a Robot Built To Spacewalk," 11 August 1997.
[301] NASA, "New Manufacturing Method Could Lower Air Travel Costs," news release 97-176, 13 August 1997.

between the Progress cargo craft and *Mir* had occurred during the two men's tenure, and shortly thereafter, doctors had diagnosed Tsibliyev's irregular heartbeat. Space officials planned extensive debriefings for the two men. Meanwhile, President Boris N. Yeltsin congratulated the cosmonauts for their "persistence, courage, and heroism," in spite of his having claimed only days earlier that the recent problems with *Mir* were solely the result of human error.[302]

Abe Silverstein received the Guggenheim Medal, honoring his significant contributions to the advancement of flight. Silverstein's 40-year career at NASA had included a period as the Director of Lewis Research Center. The citation praised him for his leadership in the Mercury and Gemini programs and for "advancing technology of aircraft and propulsion performance," as well as crediting him with proposing the name "Apollo" for the lunar landing mission. Previous winners of the Guggenheim Medal included Orville Wright, William E. Boeing, and Charles A. Lindbergh.[303]

*18 August*
*Mir*'s main computer failed during a cargo craft docking, setting the Russian space station adrift. The malfunction postponed a planned spacewalk to begin repairs to *Mir*, forcing the crew to turn its attention to reorienting the space station toward the Sun. The crew shut down most of *Mir*'s operating systems to conserve power.[304]

*19 August*
Shuttle *Discovery* returned to Earth, landing at Kennedy Space Center in Cape Canaveral, Florida, after a 12-day mission devoted to environmental research. The crew hoped that the atmospheric data gathered during the mission would help scientists understand the ozone damage and climate change affecting Earth.[305]

*25 August*
NASA launched an unpiloted Delta rocket carrying the Advanced Composition Explorer (ACE), a US$100 million solar-observatory satellite bound on a 1 million-mile (1.6 million-kilometer) journey. The launch took place one day later than expected because, on the original launch date, a commercial fishing boat had come within range of the launch danger zone. Scientists launched ACE toward an imaginary point 1 million miles (1.6 million kilometers) from Earth and 92 million miles (148 million kilometers) from the Sun, a point at which scientists believe that the gravity of the Earth and Sun balance each other. NASA planned for ACE to orbit for as long as five years, analyzing particles streaming from the Sun. In keeping with NASA's focus on improved efficiency, NASA engineers collaborated with the Boeing Company to build ACE and launch the mission, at a cost of US$200 million—more than US$30 million under budget.[306]

---

[302] Daniel Williams, "Cosmonauts Return Safely from Space," *Washington Post*, 15 August 1997; Michael Spencer, "Troubled Mir's Crew Leaves, Landing Safely in Kazakhstan," *New York Times*, 15 August 1997; Associated Press, "Mir Crew Returns to Earth Safely, but to Many Questions," 14 August 1997.
[303] NASA, "Former NASA Official Receives Guggenheim Medal," news release 97-177, 14 August 1997.
[304] Shannon Tangonan, "Computer Shutdown Sends Mir Adrift," *USA Today*, 19 August 1997; Michael Specter, "Computer Failure on the Mir Sends It Out of Control," *New York Times*, 19 August 1997.
[305] Mark Carreau, "We're Very, Very Happy, Shuttle Crew Says of Flight," *Houston Chronicle*, 20 August 1997.

# SEPTEMBER 1997

*2 September*

Astronaut F. Story Musgrave retired from NASA after 30 years of service and six Shuttle flights. During his more than 1,200 hours in space, Musgrave had flown on the first voyage of *Challenger* and had been a member of the first crew to service the Hubble Space Telescope (HST). Flying aboard STS-80 in 1996, Musgrave became the oldest person to fly in space. NASA Director of Flight Crew Operations David C. Leestma said of Musgrave, "throughout the Shuttle program, from its earliest stages to the present, Story has been instrumental in developing the techniques crew members use to perform spacewalks."[307]

*3 September*

The embattled effort to launch the Cassini spacecraft suffered a setback when NASA engineers discovered ripped insulation in its Huygens probe, while the craft was sitting on the launchpad in Cape Canaveral, Florida. Investigators concluded that a ground-air-conditioner malfunction had caused the tears. The possibility of a launch delay, even for only a few days, concerned scientists, because Saturn was moving farther and farther from Earth. A longer journey to Saturn would require the spacecraft to use more fuel for transport, leaving less for scientific exploration. The Cassini Mission to Saturn had received significant public criticism because 72 pounds (33 kilograms) of plutonium—the largest amount of radioactive material ever rocketed into space—powered the probe. Although protesters had organized marches and petitioned government officials to stop the launch, NASA spokesperson Douglas Isbell had pointed out NASA's perfect record of using nuclear fuel safely. Scientists from the European Space Agency and NASA's Jet Propulsion Laboratory (JPL) had planned and engineered the building of the 23rd nuclear-powered Cassini for the 11-year mission. Experts from both sides of the Atlantic were optimistic that engineers would repair the insulation damage quickly.[308]

*4 September*

Using the HST, astronomers from a number of universities and other institutions reported in *Science* that they had discovered a massive crater on the asteroid Vesta. Scientists had long suspected that the asteroid might have a large crater, because Vesta is the "parent body" of many smaller asteroids, suggesting that a significant collision once occurred on Vesta. However, even with hints of such a history, the sheer size of Vesta's crater surprised the observers. Measuring 285 miles (469 kilometers) across, the crater was nearly equal to Vesta's 330-mile (531-kilometer) diameter. Scientists had waited for the asteroid to move closer to Earth, so that they could examine its surface closely. In May 1996, Vesta had moved within 110 million miles (177 million kilometers) of Earth, its closest approach in a decade. The science team stated further

---

[306] Associated Press, "After 1-Day Delay, NASA Launches Bargain-Priced Solar Observatory," 25 August 1997; Associated Press, "Launch Delayed by Errant Shrimpers," 25 August 1997.
[307] NASA, "Astronaut Story Musgrave Retires from NASA," news release 97-188, 2 September 1997.
[308] John Noble Wilford, "Delay Expected for Spacecraft Going to Saturn," *New York Times*, 4 September 1997; John Kennedy and Seth Borenstein, "Nuclear Launch Draws Protest," *Orlando Sentinel* (FL), 30 August 1997; Sharon K. Spry, "Danger of Cassini Accident Too Great," *Florida Today* (Brevard, FL), 31 August 1997; Associated Press, "Mishap May Delay Mission to Saturn," 4 September 1997.

that unveiling Vesta's complexion and history had yielded new insights into the effects of "a large impact on a small object."[309]

*6 September*
British-born U.S. astronaut C. Michael Foale and Russian cosmonaut Anatoly Y. Solovyev conducted a 6-hour spacewalk outside the *Mir* space station, to investigate further the damage caused by the June 1997 collision with a robotic supply craft. However, they were unable to find the puncture that space officials believed *Mir* had sustained. The exercise was only the second joint spacewalk of a U.S. astronaut and a Russian cosmonaut. During the excursion, Foale used a crane to hoist Solovyev toward the damaged region of the craft. The two crew members endured temperatures ranging from 200°F (93°C), when *Mir* was in the sunlight, to -150°F (-101°C), when it passed into the darkness. Although Foale had spent many hours preparing for the spacewalk, learning how to maneuver in the heavy Russian spacesuit, he still exercised extreme caution to avoid snagging the suit on of *Mir*'s jagged edges. NASA had approved Foale's participation in the spacewalk only shortly before it occurred.[310]

*9 September*
Mars Global Surveyor passed a critical test on its journey toward the Red Planet, when it successfully opened valves allowing high-pressure rocket propellant to enter its fuel lines. The same task had thwarted the 1993 Mars Observer at a similar juncture during its trip to Mars. Mars Global Surveyor had traveled for 10 months and for more than 435 million miles (700 million kilometers) to reach Mars. According to Glenn E. Cunningham of NASA's JPL, the process of transforming Global Surveyor from a high-speed-travel spacecraft into a slow-moving craft subject to Mars's gravitational pull was "a nail-biter." Without the powerful propellant, the transition could not occur, which would have ended any hope of the mission's gathering data. With the opening of the valves complete, NASA scientists were optimistic that the US$250 million probe would descend slowly into a tight orbit of Mars, yielding new information about the planet.[311]

*10 September*
Software magnate James W. Benson announced the formation of SpaceDev, a new company focused on launching a robotic craft to an asteroid. Benson argued that space "is a place, not a government project," claiming that his company would gather new scientific data for a fraction of the money NASA spent annually on space exploration. Specifically, Benson stated that his Near Earth Asteroid Prospector would land on an asteroid, gathering images and scientific readings. He estimated that his company could complete such a feat for under US$50 million, about US$200 million less than NASA's budget for a comparable mission. Benson and SpaceDev entered the realm of space exploration with overtly capitalist intentions. Benson planned to sell the data the company gathered to NASA and to universities for profit.

---

[309] Peter C. Thomas et al., "Impact Excavation on Asteroid 4 Vesta: Hubble Space Telescope Results," *Science* 277, no. 5331 (5 September 1997): 1492–1495; NASA, "Hubble Reveals Huge Crater on the Surface of the Asteroid Vesta," news release 97-191, 4 September 1997.

[310] Michael R. Gordon, "Astronauts Fail To Locate a Puncture in Crippled Mir," *New York Times*, 7 September 1997; Michael Gordon, "Russian and American Astronauts Begin Space Walk," *New York Times*, 6 September 1997; Associated Press, "U.S. Astronaut Gets NASA's Approval for Space Walk Outside Mir," 5 September 1997.

[311] Paul Hoversten, "Mars Surveyor Passes Key Test," *USA Today*, 10 September 1997; Associated Press, "Craft Bound for Mars Clears Crucial Hurdle," 10 September 1997.

Additionally, SpaceDev consultants estimated that asteroids might contain significant deposits of gold and other precious metals, worth US$1–4 trillion, which would become the property of whoever succeeded in accessing it. Of his overall theory on space and science, Benson clarified: "When we deliver science, we expect to get paid. We expect a profit. We offer more science for the dollar."[312]

*11 September*
NASA, the U.S. Department of Defense, and the Federal Aviation Administration signed a joint memorandum of agreement providing guidelines for commercial spaceports. The agreement was the culmination of a long debate over how the federal government could encourage private companies and state governments to use available federal launch sites. The White House Office of Science and Technology initiative had proposed the agreement, which sought to limit the regulatory steps a private company would have to take to become a launch-site operator. The measure continued the federal government's trend of encouraging private industry to explore space.[313]

Mars Global Surveyor fired its engine enabling Mars's gravitational pull to capture it. Researchers from NASA's JPL declared that the spacecraft had behaved exactly as planned during the crucial entry to Mars's gravitational field. Experts predicted that the mission would produce 700 million bits of data, more than the amount acquired in all previous Mars missions combined. The mapping mission continued a new era of Mars exploration for NASA, involving dozens of probes surveying the planet's surface and atmosphere.[314]

*12 September*
For the first time, NASA turned over control of a satellite project to a group of non-NASA scientists, placing the Advanced X-ray Astrophysics Facility (AXAF) under the control of the Smithsonian Astrophysical Observatory at Harvard University. Harvard scientists had designed and built the satellite, but in previous NASA-university collaborations, NASA had always maintained control over implementation projects that it had funded. NASA Administrator Daniel S. Goldin said that NASA hoped the move would embolden non-NASA scientists: "What we wanted to do is get the scientists closer to their spacecraft and close to their sources of data." NASA intended the AXAF, which had cost US$1.4 billion to develop, to study black holes and collisions between stars and galaxies.[315]

NASA announced that, in one week, engineers had completed repairs to the damaged insulation of the Huygens probe, a part of the Cassini Mission to Saturn. Some scientists had feared that the tears to the insulation would set back the ambitious mission indefinitely. NASA engineers, however, had worked rapidly to disassemble the probe, completely inspect the damage, and make the appropriate repairs. Additionally, during the repair period, the project had received a

---

[312] Associated Press, "Company Promotes First Private Launch to Asteroid," 10 September 1997.
[313] NASA, "NASA, DOD, and FAA Sign Joint Agreement on Spaceports Launch Guidance," news release 97-194, 11 September 1997.
[314] Kathy Sawyer, "New Visitor from Earth Poised for Mars Orbit," *Washington Post*, 11 September 1997; Associated Press, "Everything 'Perfect' as Surveyor Craft in Countdown to Mars Orbit," 11 September 1997; Associated Press, "Mars Global Surveyor," 12 September 1997.
[315] Jules Crittenden, "X-ray Tour of the Universe To Begin in Cambridge Lab," *Boston Herald*, 13 September 1997; Peter J. Howe, "Putting the Universe into Focus," *Boston Globe*, 13 September 1997.

welcome voice of support from the National Space Society, an independent space advocacy organization. The Society had issued a statement expressing its full confidence in NASA and in the mission, offsetting the protests against the mission's use of nuclear fuel. The National Space Society declared, "saying 'no' to Cassini would be saying 'no' to knowledge," defending the use of nuclear power as the only safe and viable option for such a lengthy mission: "Cassini's Radioisotope Thermoelectric Generators (RTGs) . . . have proven their safety and capability in 23 prior missions, including human missions. RTGs are the only realistic option for sending probes great distances from the Sun and will certainly play a part in future human missions."[316]

*15 September*
The *Mir* space station nearly collided with a U.S. military satellite, forcing *Mir*'s crew to prepare hastily for an emergency exit. When the satellite came within 500 to 1,000 yards (457 to 914 meters) of *Mir*, U.S. astronaut C. Michael Foale and the two resident Russian cosmonauts sealed themselves in the *Soyuz* spacecraft so that they would be able to leave if the satellite hit the station. According to NASA, problems with space debris and close encounters between satellites are common, but usually miles, rather than yards, separated the rapidly traveling spacecraft. Russian space officials had taken a different approach to the situation than their American counterparts would have chosen. Rather than ordering *Mir* to adjust its position, as NASA had instructed Shuttle crews to do when debris approach, Russian officials had ordered *Mir*'s crew back into *Soyuz*, prepared to take off immediately. The near miss of *Mir* and the satellite was not the first time that *Mir* had received a collision alert. NASA had begun monitoring *Mir*'s path when American astronauts joined the permanent crew and reported that space debris passed within a few kilometers of *Mir* about once every six weeks.[317]

Data gathered by the U.S.-French satellite *TOPEX/Poseidon* confirmed theories that a "full-blown El Niño condition" existed in the Pacific. Scientists believe that an El Niño forms when steady, westward-blowing trade winds weaken or reverse direction, altering typical, atmospheric jet-stream patterns around the world. *TOPEX/Poseidon* measured sea-surface heights and atmospheric water vapor to confirm the National Oceanic and Atmospheric Administration's earlier prediction. Based on previous El Niño events, scientists predicted heavy rainfall for the western United States, mild winters in the east, and extreme droughts in Australia, Africa, and Indonesia.[318]

*18 September*
U.S. House Committee on Science and Technology Chairperson F. James Sensenbrenner Jr. (R-WI) issued a recommendation that NASA cancel its plan to replace C. Michael Foale aboard *Mir* with another astronaut. Sensenbrenner issued the bipartisan statement with U.S. Representative George F. Brown Jr. (D-CA), calling on NASA Administrator Daniel S. Goldin to alter the plan for the next Shuttle mission, retrieving Foale without delivering another American to the station. "I don't think the increased risk is worth the benefit," Sensenbrenner told reporters after a 4-hour

---

[316] NASA, "Repair Work on Cassini Huygens Probe Completed Successfully," news release 97-198, 12 September 1997; Associated Press, "National Space Society Adds to Support for Saturn Mission," 9 September 1997.
[317] Edmund L. Andrews, "Satellite Gives Mir One More Close Call," *New York Times*, 17 September 1997; Shannon Tangonan, "Too-Close Satellite Gives Mir a Scare," *USA Today*, 17 September 1997.
[318] NASA, "Independent NASA Satellite Measurements Confirm El Niño Is Back and Strong," news release 97-200, 15 September 1997.

hearing on *Mir*'s safety issues. Goldin testified that, despite some setbacks, NASA had no reason to suspect that sending an American astronaut to *Mir* was more dangerous than any other space exploration. NASA's Shuttle–*Mir* Program Director Frank L. Culbertson Jr. responded to Sensenbrenner's comments, "I take the safety of my friends [the astronauts] very seriously and would not send anyone on something that I would not do myself." The hearing was the first time that NASA had formally defended the safety of *Mir* to Congress.[319]

*22 September*
Vice President Albert A. Gore Jr. began talks with officials in Russia, focused primarily on the International Space Station (ISS), economic cooperation, and controlling nuclear proliferation. The talks occurred in the context of a torrent of problems aboard Russia's *Mir* space station and in Russia's failures to meet its commitments to help build the ISS. NASA Administrator Daniel S. Goldin traveled with the Vice President to participate in the space-related discussions.[320]

*25 September*
NASA's Shuttle *Atlantis* lifted off from Kennedy Space Center in Cape Canaveral, Florida, for Mission STS-86. The launch took place only hours after an independent task force chaired by astronaut Lieutenant General Thomas P. Stafford II reported that the *Mir* mission was sufficiently safe for the American astronauts. Specifically, the panel had determined that sending astronaut David A. Wolf to replace C. Michael Foale aboard the Russian space station would be reasonable, as well as important to continuing scientific research. NASA Administrator Daniel S. Goldin made the last-minute decision for the mission to go forward after weighing the panel's findings against NASA's risk reports and the opinions of the House Committee on Science and Technology. Committee Chairperson F. James Sensenbrenner Jr. (R-WI) had been particularly aggressive in calling for the end of the program to place astronauts aboard *Mir* for significant periods. James D. Wetherbee served as Commander of Space Shuttle *Atlantis*, embarking on his fourth flight to space. The Shuttle also carried Jean-Loup Chrétien of the French Space Agency, Vladimir G. Titov of the Russian Space Agency, Pilot Michael J. Bloomfield, and Mission Specialists Scott E. Parazynski, Wendy B. Lawrence, and David A. Wolf. The Shuttle carried the SPACEHAB module in its payload, with supplies for *Mir* research and logistical equipment aboard.[321]

## OCTOBER 1997

*1 October*
Cosmonaut Vladimir G. Titov and Mission Specialist Scott E. Parazynski made a 5-hour spacewalk outside *Mir* to test jetpack systems and to retrieve experiments for return to Earth. The two men, who left *Atlantis* while the Shuttle remained docked with *Mir*, became the first Russian-American team ever to conduct a spacewalk from a NASA Shuttle. The joint spacewalk

---
[319] Larry Wheeler, "Panel to NASA: Stay Off Mir," *Florida Today* (Brevard, FL), 19 September 1997; Knight-Ridder News Service, "Top Congressmen Say U.S. Astronaut Shouldn't Board Mir," 19 September 1997.
[320] Reuters, "Al Gore in Moscow for Talks on Space, Energy," 21 September 1997; Associated Press, "Gore Calls for Reforms To Boost Investment in Russia," 22 September 1997.
[321] NASA, "Panels Give Astronaut a 'Go' for Launch to Mir," news release 97-214, 25 September 1997; NASA, "Astronaut Wolf Continuing American Presence on Mir, Joint US-Russian Spacewalk Highlights STS-86 Mission," news release J97-27, September 1997; Traci Watson, "Amid Controversy, Atlantis Lifts Off," *USA Today*, 26 September 1997; Associated Press, "Astronaut Leaves for Tour on Mir," 26 September 1997.

was also the first in which *Mir* crew members used NASA spacesuits rather than those provided by the Russian Space Agency. On this busy day of operations, two of the Russian crew also successfully installed a new guidance computer, which *Atlantis* had transported to the station.[322]

NASA announced that the construction schedule for the International Space Station (ISS) remained on target to begin delivery launches in December 1998. All 15 nations participating in the building of the ISS had traveled to Johnson Space Center (JSC) to finalize the station's construction sequence and compare progress updates. The international consortium agreed on a 45-flight sequence for delivering the already-assembled portions of the station from Earth to their orbiting home. To assemble the colossal station completely, NASA and its partners planned five years of steady launch-and-construct missions, expecting to send the U.S.-constructed Node 1 into space sometime during June 1998.[323]

*3 October*
The White House Office of Science and Technology Policy formally approved the mid-October launch plan of the controversial plutonium-powered Cassini Mission. Critics had subjected to intense scrutiny the exploratory mission to Saturn and its moon Titan, because it used nuclear power. The Florida Coalition for Peace and Justice, among other organizations, had protested the mission and attempted to derail it, but had failed to spark widespread public outrage. Activists worried that the rocket might release plutonium if the launch failed, despite NASA's more than a dozen safe launches using the hazardous fuel. The mission required nuclear power because Saturn has extremely dim sunlight—not enough to provide adequate solar power to prevent the Cassini spacecraft's instruments from freezing. NASA's policy required that the White House give final approval of the launch, because the mission used nuclear fuel. NASA had begun work on the Cassini Mission in 1989, with cooperation from the European Space Agency and the Italian Space Agency. Many regarded the mission—the first time a space probe had attempted to land on the moon of another planet—as NASA's boldest and most ambitious endeavor. This aspect of the mission especially interested scientists, who theorized that Titan has an Earth-like atmosphere. NASA Administrator Daniel S. Goldin said of the mission: "I fully expect that it will return spectacular images and scientific data about Saturn."[324]

*4 October*
The international scientific community observed the 40th anniversary of *Sputnik*'s successful launch. On 4 October 1957, the Soviet Union had sent into orbit *Sputnik I*, a small sphere-shaped satellite, setting off a space race that continued throughout the Cold War. Although the United States had initially responded to the Soviet Union's achievement with fear and apprehension, NASA commemorated the event 40 years later with a celebratory symposium. The United States had launched NASA itself less than one year after the USSR's *Sputnik* success, partly because of the American public's distress over the Soviets beating the United States into space. The son of

---

[322] Robyn Suriano, "Spacewalk," *USA Today*, 1 October 1997; Mark Carreau, "Atlantis, Mir Crews Prepare for Spacewalk, Computer Task," *Houston Chronicle*, 1 October 1997.
[323] NASA, "Control Board Reports International Space Station Launch on Target, Finalizes Assembly Sequence," news release 97-222, 1 October 1997.
[324] NASA, "Cassini To Survey Worlds of Saturn and Titan; Sends First Probe to Moon of Another Planet," news release 97-190, 3 September 1997; NASA, "NASA Receives Approval To Launch Cassini Mission," news release 97-225, 3 October 1997; Maurice Tamman, "Protestors Set Stage," *Florida Today* (Brevard, FL), 29 September 1997.

Soviet Premier Nikita Khrushchev, Sergei N. Khrushchev, a prominent professor of International Studies at Brown University in Rhode Island, spoke at NASA's *Sputnik* symposium. With the fall of the Soviet Union, Russia had unveiled to the world's scientists many of the developmental plans of the *Sputnik* program. However, the end of communism had also precipitated enormous budget shortfalls for the Russian Space Agency. Kazakhstan's President Nursultan Nazarbayev took the opportunity of *Sputnik*'s widely celebrated anniversary to complain publicly that Russia owed more than US$400 million in rent for Baikonur Cosmodrome—the site of *Sputnik*'s launch and of ongoing Russian space activity. When the Soviet confederation broke up, Russia had agreed to pay US$100 million annually for the right to continue using the launch facilities in Kazakhstan.[325]

*6 October*
Shuttle *Atlantis* touched down at Kennedy Space Center in Cape Canaveral, Florida, safely returning British-born U.S. astronaut C. Michael Foale to Earth after his 145-day stay aboard *Mir*. Foale's tenure in space was among the longest in NASA's history, surpassed only by Shannon W. Lucid's 188-day mission. Although Foale was determined to return to Earth's gravity in good physical condition, NASA doctors planned to keep the astronaut under medical supervision for several days.[326]

*7 October*
NASA announced that astronomers using the Hubble Space Telescope (HST) had identified the most luminous star ever charted. The "celestial mammoth," large enough to fill the diameter of the Earth's orbit, reportedly releases up to 10 million times the power of the Sun. The team from the University of California at Los Angeles estimated that the star, possibly as old as 3 million years, weighs up to 200 times the mass of the Sun. The team nicknamed it the Pistol Star because it has a pistol-shaped nebula. Despite the star's mass and brightness, it was not visible to the naked eye because of the interstellar dust clouds between Earth and the center of the Milky Way. Therefore, astronomers had used infrared technology to observe the star. Team researcher Mark R. Morris argued that the discovery of the massive star near the center of the galaxy would force scientists to rethink their conceptions of how stars formed in the first place.[327]

*8 October*
After more than nine days without sending a signal, Mars Pathfinder's main transmitter resiliently delivered a message to Earth. To anxious project managers, the message was a welcome relief. The breakdown in the spacecraft's communication to NASA controllers had necessitated NASA's activating a contingency plan for Sojourner. The rover had stopped gathering data and reverted to slowly circling its parent spacecraft, waiting to receive directions, a pattern that some scientists feared would be the long-term fate of the rover. The spacecraft's aging batteries had caused the breakdown in communication of data. However, in spite of the malfunction, NASA officials reiterated that the mission had fulfilled its objectives, emphasizing

---

[325] NASA, "Reconsidering Sputnik: Forty Years Since the Soviet Satellite," news release N97-64, 5 September 1997; Associated Press, "Plea for Rent Marks Sputnik Anniversary," 5 October 1997.
[326] Mark Carreau, "Astronaut Ends Marathon Ride; Foale Lands on Shuttle After 145 Days Aboard Russian Space Station," *Houston Chronicle*, 7 October 1997.
[327] Associated Press, "Space Monster," 8 October 1997; NASA, "Hubble Identifies What May Be the Most Luminous Star Known," news release 97-227, 7 October 1997.

that solving the problem of communicating with the craft while it was resting on the Red Planet would benefit future explorations of Mars.[328]

Mars Pathfinder scientists announced that Mars appeared to have a crust, a mantle, and an iron core—attributes similar to those of Earth. The researchers suggested that Mars may once have been warm and wet, although they remained unable to determine whether Mars had a molten core, such as that of Earth or Mercury, or a "dead" core, such as that of the Earth's Moon. The findings contributed to growing evidence that Mars might be more than a solid ball of rock and that the planet might once have had water and life. Signs of possible erosion and weathering encouraged such hypotheses, as well, although scientists clarified that proof of such theories was still probably years away.[329]

A Russian cargo spacecraft docked without incident at the *Mir* space station, using the craft's automatic pilot to complete the maneuver. It was the first successful docking after two previous failed attempts, one of which had resulted in the crash that seriously damaged *Mir*. To address the earlier problem, the Russian Space Agency had reconfigured the mathematical formula guiding the procedure. The Progress M-36 cargo craft carried a backup computer for the space station. *Mir*'s crew planned to unload the more than 6,000 pounds (2,700 kilograms) of cargo in the days following the delightfully uneventful docking.[330]

*10 October*
NASA's Marshall Space Flight Center (MSFC) selected Scientific and Commercial Systems Corporation of Falls Church, Virginia, to provide support services for its Huntsville, Alabama, facilities. The US$66 million contract included transportation and equipment services, as well as publishing, mail, and food support. MSFC arranged the contract as a cost-plus-incentive-fee, performance-based contract, to provide NASA with cost savings, as well as with flexibility to reorganize services as necessary.[331]

*14 October*
NASA announced that new smart software, with more realistic landing simulations for Shuttle pilots, would become an integral part of astronaut flight training. Hamid Berenji, Ping-Wei Chang, and Steven R. Swanson of JSC engineered the software, which used "adaptive fuzzy logic," allowing the program to learn from experience and to function in a manner closer to human thinking than that of previous software. The software team predicted that the program would help not only Shuttle pilots, but also commercial pilots and even people learning to use machinery unrelated to aviation. Berenji predicted that, because the new training system would

---

[328] Associated Press, "Lacking Signal, Mars Vehicle Is Left To Be Roving in Circles," 7 October 1997; Associated Press, "Pathfinder Checks in After Weeks of Silence," 8 October 1997.
[329] K. C. Cole, "Internal Structure of Mars, Earth Similar, Scientists Say," *Houston Chronicle*, 9 October 1997; Associated Press, "Mars Once Warm and Wet, Data Suggests," 9 October 1997; Associated Press, "Three Mars Layers Suggests Life on Planet Was Possible," 9 October 1997.
[330] *Washington Post*, "Cargo Ship Docks with Space Station," 9 October 1997; Associated Press, "Successful Docking Gives Mir a Life in Morale," 8 October 1997; Reuters, "Mir Crew To Unload Cargo After Text-Book Docking," 8 October 1997.
[331] NASA, "Virginia Firm Chosen To Provide NASA Institutional Services," news release C97–r, 10 October 1997.

be easier to use than previous programs, new pilots would feel much more confident than before.[332]

In celebration of the 50th anniversary of his most historic achievement, Charles E. "Chuck" Yeager broke the speed of sound once again, flying an F-15 fighter jet in California's Mojave Desert. NASA aired the recreation on NASA TV as part of its celebration at Edwards Air Force Base. Some historians call Yeager's 1947 feat "the greatest achievement since the first successful flight of the Wright Brothers." The U.S. Postal Service observed the anniversary by unveiling a commemorative stamp, the "50th Anniversary of Supersonic Flight." The ceremonial jaunt was the 74-year old Yeager's last official U.S. Air Force flight.[333]

*15 October*
The Cassini spacecraft lifted off from Cape Canaveral, Florida, beginning a seven-year, 2.2 billion-mile (3.5 billion-kilometer) journey to and around Saturn. After several delays, including a two-day delay immediately before the launch because of high winds, the takeoff was flawless. Before the launch, activists had protested Cassini's fuel source, 72 pounds (33 kilograms) of radioactive plutonium; to calm public anxiety, NASA planned to sample air quality at 25 stations surrounding the launchpad, including several samples taken as far as 10 miles (16 kilometers) away. Additionally, a U.S. Department of Energy aircraft flew through the trailing gases of the rocket booster to make sure that no radioactive material had leaked. The US$3.4 billion mission set out to survey the planet Saturn, its rings, and its moon Titan. Engineers had programmed the Cassini spacecraft to conduct a number of "gravity-assist swingbys," flying twice past Venus and once past Earth and Jupiter, to build up the speed necessary to make the long journey to Saturn. About seven years into the mission, NASA planned for Cassini to release a disc-shaped Huygens probe toward Titan. Throughout the mission, Cassini was to make more than 40 close flybys of Titan, gathering far more vivid images of the moon than ever before. NASA intended Cassini's camera to take more than 300,000 colored pictures of Saturn and its moons. Scientists hoped that the mission would expand considerably the knowledge of the planet gained through the earlier Voyager missions.[334]

President William J. Clinton used his newly bestowed line-item veto power to cut funding for a "reusable space plane." The US$10 million appropriation was a priority of the U.S. Air Force Space Command and space officials. In total, Clinton's line-item vetoes trimmed about US$144 million from the 1998 U.S. Department of Defense appropriations bill, less than one-tenth of 0.1 percent of the total bill.[335]

---

[332] H. R. Berenji et al., "Refining the Shuttle Training Aircraft Controller," *Fuzzy Systems, 1997: Proceedings of the Sixth IEEE International Conference* 2 (1–5 July 1997): pp. 677–682; NASA, "Shuttle Landing Simulations To Improve with Smart Software," news release 97-229, 14 October 1997.
[333] Jay Levine, "Yeager Bids Farewell to 800,000-Plus," *Antelope Valley Press* (Palmdale, CA), 21 October 1997; NASA, "Breaking the Sound Barrier Reenactment on NASA TV," news release N97-72, 9 October 1997.
[334] NASA, "Cassini Launch," news release, October 1997; Earl Lane, "Plutonium-Bearing Craft Set for Saturn Trip," *Birmingham News* (AL), 5 October 1997; Paul Hoversten, "Nuclear Powered Craft Fuels Fury Before Launch," *USA Today*, 13 October 1997; Karl Grossman, "The Risk of Cassini Probe Plutonium," *Christian Science Monitor*, 10 October 1997; Associated Press, "Cassini Spacecraft Lifts Off for Saturn," 16 October 1997.
[335] John M. Broder, "Clinton Gently Vetoes $144 Million in Military Budget Items," *New York Times*, 15 October 1997; David L. Chandler, "Clinton Veto Delays Space Plane Plans," *Antelope Valley Press* (Palmdale, CA), 21 October 1997.

*20 October*
Russian cosmonauts Anatoly Y. Solovyev and Pavel V. Vinogradov conducted an "internal" spacewalk, entering the damaged *Spektr* module of *Mir*. Once inside the darkened module damaged by the June 1997 cargo craft collision, the men found floating space debris. Seven bags and a refrigerator door were among the items banging around inside the *Spektr* module. After securing the debris, the cosmonauts went to work installing new solar-panel cables to restore full power to *Mir*. The cosmonauts, who were wearing bulky spacesuits, performed the 6-hour spacewalk inside a cramped module. Although the men became exhausted before they could finish the repair work, the arduous exercise had boosted the station's power supply by 30 percent. Russian Space Agency officials planned for the cosmonauts to conduct another repair operation during the following days.[336]

*21 October*
The Pentagon announced that it had successfully tested its Mid-Infrared Advanced Chemical Laser (MIRACL), at White Sands Missile Range in New Mexico. The laser, constructed under President Ronald Reagan's Strategic Defense Initiative, generated a beam that traveled more than 260 miles (418 kilometers) from Earth, hitting an orbiting satellite. According to Pentagon representative U.S. Air Force Lieutenant Colonel Robert Potter, if turned to its highest setting, MIRACL would have destroyed the satellite target. The test, which took place two years after the expiration of a 1985 congressional ban on tests of this type, was the first time the United States had tested a laser aimed at an orbiting satellite. Representatives of defense-related industries and scientists around the world debated the acceptability of such a test and the ethics of deploying the instrument. However, whether or not the test was justifiable, the successful hit had demonstrated the vulnerability of satellites to Earth-based weapons.[337]

*23 October*
Representatives of NASA and the U.S. Department of Health and Human Services signed an agreement to enlist NASA technologies in the fight against breast cancer and other women's illnesses, creating a cooperative framework for the exchange of research information between the two agencies. According to Carolyn A. Krebs of NASA's Goddard Space Flight Center, the technology allowing HST to map distant galaxies also had the potential to scan human tissue for malignant masses.[338]

*28 October*
NASA announced the retirement of J. Wayne Littles, Director of MSFC. Littles had served as Director from 1996 to 1997, the capstone to his 30-year career at the facility. Wayne Littles had twice received the Presidential Rank of Distinguished Executive.[339]

---

[336] David Hoffman, "A Partial Repair Gives Mir a Power Boost," *Washington Post*, 21 October 1997; Associated Press, "A Spacewalk of Surprises Aboard Mir," 21 October 1997.
[337] Associated Press, "Laser Test-Fired into Space Strikes Orbiting U.S. Satellite," 21 October 1997.
[338] NASA, "NASA/Heath and Human Services To Join Forces To Fight Women's Illnesses," news release N97-74, 20 October 1997.
[339] NASA, "Marshall Center Director Dr. Wayne Littles To Retire," news release 97-247, 28 October 1997.

# NOVEMBER 1997

*3 November*
*Mir* cosmonauts Anatoly Y. Solovyev and Pavel V. Vinogradov engaged in a 6-hour spacewalk, successfully dismantling a damaged solar panel so that *Mir*'s crew could install a replacement panel. The Russian Space Agency nearly called off the spacewalk when Solovyev discovered a problem with the radio-monitoring system on his spacesuit, just as the men prepared to leave the station. Although the crew's attempts to repair the monitoring system were unsuccessful, Russian Mission Control eventually decided to proceed with the spacewalk. The cosmonauts worked effectively, communicating with Russian space officials only periodically. This was Solovyev's fourth spacewalk during his tenure on *Mir*, confirming his status as the world's most experienced spacewalker, with more than a dozen spacewalks to his credit. While the cosmonauts worked outside *Mir*, U.S. astronaut David A. Wolf monitored their progress from inside.[340]

*4 November*
After a four-month mission, NASA officials announced that Mission Control had halted communications with Mars Pathfinder and its rover Sojourner. Sojourner's batteries were nearly drained, and many of the rover's instruments had frozen in Mars's harsh climate. Mission controllers expected the rover simply to come to a halt on Mars's vast plains, ceasing signal transmissions altogether. Nevertheless, NASA considered the US$266 million mission largely successful. The mission had lasted much longer than planned, transmitting, from the surface of the Red Planet to Earth, nearly 10,000 pictures of the Martian landscape, including images showing evidence of erosion, as well as more than 1.2 billion bits of data. Sojourner's remarkable service, even in the midst of harsh dust storms and temperatures reaching -50°F (-6°C), had excited scientists and public alike. Mars Pathfinder Web sites had received more than 565 million hits during the mission's tenure. Using the treasure trove of information Pathfinder had gathered, scientists had concluded that Mars's surface had many similarities to that of Earth. Perhaps the most compelling was the evidence compiled during the mission in support of the theory that, at one time, Mars had water on its surface. In addition, the mission's research team concluded that water and wind had played crucial roles in shaping the planet's surface. Mission controllers spoke of their gratitude for the data gathered, which would keep scientists busy for years, as well as about their excitement at the possibility of building upon Mars Pathfinder's success in future missions.[341]

*5 November*
A team of scientists from the University of California at Riverside announced that NASA's orbiting Compton Gamma Ray Observatory had enabled them to discover a "vast halo" of gamma rays surrounding the Milky Way Galaxy. The astronomers clarified that the discovery of the halo did not make clear how or why the gamma rays encircled the galaxy. NASA's orbiting Compton Gamma Ray Observatory uses gamma rays to expose highly energetic particles that would otherwise be invisible because of their absorption in the atmosphere. The scientific community greeted the discovery with interest and questions.[342]

---

[340] Associated Press, "Mir," 3 November 1997.
[341] Paul Hoversten, "Mars Pathfinder Comes to Quiet End Today," *USA Today*, 4 November 1997; Reuters, "NASA Announces End of Mars Pathfinder Mission," 5 November 1997.
[342] K. C. Cole, "Halo Around Milky Way Reported," *Los Angeles Times*, 6 November 1997; Reuters, "Gamma Rays Found Surrounding Milky Way," 5 November 1997.

Russian Space Agency spokesperson Sergei Gorbunov announced the agency's support for a film company to shoot a Hollywood-style movie aboard *Mir*. Russian director Yuri Kara had submitted a proposal to bring actors and film crews to *Mir* to shoot the film, entitled "Space Flight Has a Price." Although the Russian Space Agency had not officially approved the project, the cash-strapped agency welcomed the unique opportunity to earn money. The biggest obstacle to the project seemed to be finding financial backers willing to pay to send Kara's crew into space. The prospect of filming a movie aboard *Mir* raised some concerns in the space community, but Kara noted that Russian cosmonauts already had acting experience, having filmed commercials aboard the space station for products, including Pepsi-Cola, Omega watches, and Israeli milk.[343]

*6 November*
A group of astronomers announced that they had used NASA's Rossi X-ray Timing Explorer to observe a black hole "literally dragging space and time around itself as it rotates." The discovery confirmed the existence of *frame dragging*, an extrapolation of Albert Einstein's general theory of relativity. According to the findings, the black hole had such a massive gravitational field that nothing nearby, not even light, could escape its pull. To make the discovery, the team of researchers brought together academic scientists from Italy and the United States, as well as NASA personnel. By studying x-ray emissions, the astronomers were able to find a pattern of forces that clearly revealed the strength of the black hole's gravitational pull. Alan N. Bunner, Director of the Structure and Evolution of the Universe Program at NASA Headquarters, called the discovery "exciting work," encouraging scientists to test the findings. Before the discovery, frame dragging persisted as one of the last unverifiable theories of relativity, which scientists had suspected but had been unable to observe. Popular press reports touted the breakthrough as another victory for Einstein.[344]

NASA announced that lighting technology developed to grow plants in space might help treat cancerous brain tumors in children. According to a series of trials, some tumor-fighting drugs performed more proficiently when illuminated with the light-emitting diodes (LEDs) created for NASA's Shuttle plant experiments. To gauge further the effectiveness of LEDs in assisting cancer-fighting drugs, Dr. Harry Whelan of the Medical College of Milwaukee obtained permission from the Food and Drug Administration to use the LEDs on a limited number of children with cancer. Whelan's method involved injecting his patient with cancer-fighting drugs and then placing an LED probe near the affected area. Whelan believed that the light focused the drugs on the dangerous tumors only, leaving the rest of the patient's brain unaffected. NASA's Small Business Innovation Research program, allowing small businesses to adapt NASA's technologies for extended public uses, initiated the trial of LEDs to fight cancer.[345]

*7 November*

---

[343] Associated Press, "Mir—the Movie," 5 November 1997.
[344] NASA, "First Observation of Space-Time Distortion by Black Holes," news release 97-258, 6 November 1997; K.C. Cole, "Finding Backs Einstein's Theory of Gravity," *Los Angeles Times*, 7 November 1997; Kathy Sawyer, "Score Another One for Einstein," *Washington Post*, 7 November 1997.
[345] NASA, "Space Research Shines a Light on Tumors To Save Lives," news release 97-259, 6 November 1997.

Following on the heels of Mars Pathfinder's successful mission to map the surface of Mars, two researchers used sonar information and satellite images to create a highly detailed map of the global ocean floor, another largely unexplored area. David T. Sandwell of the Scripps Institution of Oceanography and Walter H. F. Smith of the National Oceanic and Atmospheric Administration improved by 30 times the precision of the world's most exact ocean maps. Their new map, revealing ridges and canyons never before delineated, helped explain ocean currents and tides across the world. The new mapping technology interested oil companies, looking for clues to the location of fossil fuel deposits, and fishermen seeking likely habitats of fish.[346]

*10 November*

Geochemists from the University of Michigan and the University of Tennessee announced that they had arrived at the most accurate and precise estimation of the age and origin of the Moon yet calculated. They suggested that a collision between the Earth and another planet, 4.5 to 4.52 billion years ago, resulted in the creation of the Moon. To reach their conclusion, the researchers used a special spectrometry technique to examine 21 lunar samples. The scientists also concluded that the collision had set Moon's orbit.[347]

*13 November*

After altering the spacecraft's path to move it farther away from Mars's violent winds, NASA launched Mars Global Surveyor from its safe elliptical orbit of the Red Planet, sending the spacecraft into the planet's harsh atmosphere. NASA officials indicated that, because of the change in flight plans, Mars Global Surveyor would begin its mapping mission of Mars in 1999, one year later than originally planned. The change of plan highlighted both the difficulty of exploring Mars and NASA's ability to adapt to difficult conditions. When Mars's atmosphere had proven more hostile than previously anticipated, mission controllers had proven able to readjust Surveyor's orbit. In effect, the move had placed Surveyor in a holding pattern, keeping it from harm. Once conditions had improved, however, NASA would not be able simply to proceed as planned. Because Surveyor ran on solar power, the mission planners must wait for Mars and the Sun to realign, to provide energy for the craft. However, although NASA would have to postpone the mission for one year, the delay would provide an added benefit, allowing NASA's researchers to set Surveyor on a circular orbit that would actually bring Surveyor closer to the planet than originally projected.[348]

NASA announced the retirement of Wilbur C. Trafton, who had guided NASA's International Space Station (ISS) planning for nearly four years. Trafton resigned from his position as Director of the Space Station Program one week after a contentious hearing, in which NASA officials informed Congress that cost overruns on the ISS had reached nearly US$1 billion. However, at Trafton's departure, NASA Administrator Daniel S. Goldin praised him, stating, "Will Trafton has met or exceeded all of the challenges I have placed before him." During Trafton's tenure,

---

[346] Robert Lee Hotz, "Map Dramatically Deepens Learning About Sea Floor," *Los Angeles Times*, 7 November 1997.
[347] United Press International, "Researchers Narrow Moon's Origins, Age," 10 November 1997.
[348] John Noble Wilford, "Mars Map Will Come a Year Late," *New York Times*, 11 November 1997; Mark Carreau, "Mars Map Still on Tap," *Houston Chronicle*, 11 November 1997; A. J. Hostetler, "Surveyor Has New Course," *Times-Dispatch* (Richmond, VA), 13 November 1997.

Russia had officially joined the ISS effort as a full partner, and the ISS had moved from planning stage to construction.[349]

*17 November*
Reaffirming their commitment to the international consortium's expensive and increasingly complicated task of constructing *Mir*'s replacement, NASA and the Russian Space Agency announced appointments to the first four crews that would live and work aboard the ISS. NASA chose astronaut William M. Shepherd as the first expedition commander. Russia selected cosmonauts Yuri P. Gidzenko and Sergei K. Krikalev to accompany Shepherd on the first increment crew. The second crew would have a Russian commander, Yury V. Usachev, accompanied by a crew of American astronauts, James S. Voss and Susan J. Helms. Each member of the first two crews possessed significant space experience. Moreover, the selection of the crews represented careful diplomatic choices meant to support the strong, but still strained, space partnership of the United States and Russia.[350]

*18 November*
The Assignation Records Review Board made public more than 1,500 pages of classified records, including a memorandum detailing a plan to blame Prime Minister Fidel Castro and the government of the Republic of Cuba if astronaut John H. Glenn Jr.'s attempt to orbit Earth in the *Mercury* capsule was unsuccessful. The 2 February 1962 document stated, "by manufacturing various pieces of evidence which would prove electronic interference on the part of the Cubans," the United States could shift the blame to Castro. Glenn's safe return to Earth rendered any such plans unnecessary.[351]

*19 November*
Space Shuttle *Columbia* lifted off from Earth on Mission STS-87 with great fanfare, executing an unusual 180° roll 6 minutes into the 8½-minute flight, a maneuver planned to put the Shuttle in radio contact with communication satellites. The launch itself took place at the "exact moment on the exact day as planned" for the sixth consecutive mission. The six-person crew included Takao Doi, a Japanese astronaut who planned to make his country's first spacewalk; Kalpana Chawla, the first Indian-born woman to travel into space; Leonid K. Kadenyuk, the first Ukrainian to fly on a U.S. Shuttle; and U.S. astronauts Kevin R. Kregel, Steven W. Lindsey, and Winston E. Scott. The crew planned to conduct a spacewalk to test instruments and procedures for the ISS. The Shuttle also carried the U.S. Microgravity Payload-4, which the astronauts planned to use to test the responses to conditions in space of a variety of materials and liquids.[352]

*24 November*

---

[349] NASA Trafton Announces His Departure from NASA," news release 97-264, 13 November 1997; *Orlando Sentinel* (FL), "NASA's Space Station Chief Resigns," *Washington Post*, 14 November 1997.
[350] NASA, "U.S., Russia Name International Space Station Crews," news release 97-269, 17 November 1997.
[351] George Lardner Jr. and Walter Pincus, "Military Had Plan To Blame Cuba if Glenn's Space Mission Failed," *Washington Post*, 19 November 1997.
[352] NASA, "Solar and Microgravity Research Highlight Final Shuttle Mission of 1997," news release J97037, November 1997; William Harwood, "Columbia Corkscrews into Orbit," *Washington Post*, 20 November 1997; Reuters, "U.S. Shuttle Blasts Off with Multinational Crew," 19 November 1997; Associated Press, "Shuttle Executes Flip on Its Way into Orbit," 16 November 1997.

Two members of *Columbia*'s crew, U.S. astronaut Winston E. Scott and Japanese astronaut Takao Doi, conducted a spacewalk, successfully recapturing the unresponsive *Spartan* satellite released on 21 November. After the 3,000-pound (1,400-kilogram) satellite had failed to turn toward the Sun, *Columbia*'s robotic arm had bumped it away from the Shuttle. NASA officials had cancelled the mission's originally planned spacewalk to rescue the US$10 million spacecraft. Astronauts Scott and Doi later received a telephone call from President William J. Clinton congratulating them on grabbing the satellite.[353]

## DECEMBER 1997

*1 December*
The U.S. General Accounting Office (GAO) released a report on the threat of space debris to the planned International Space Station (ISS). According to the report, *Space Surveillance: DOD and NASA Need Consolidated Requirements and a Coordinated Plan*, NASA's surveillance capabilities, relying on the U.S. Department of Defense (DOD) network of radar and optical sensors for information about possible collisions, had not provided enough information to prevent collisions in space. GAO surmised that DOD had focused most of its attention on military objects, missing many of the smaller pieces of debris that could have caused significant damage to NASA's Shuttles or the ISS. GAO recommended that NASA, DOD, and the CIA collaborate to upgrade the space surveillance system. According to estimates, more than 110,000 pieces of debris, most measuring about 0.5 inches (1.3 centimeters) in diameter, were floating in space. With a closing velocity of 35,000 miles (56,000 kilometers) per hour, even such a small object would do significant damage to the ISS, when the station was in a prograde orbit. GAO's report noted that the fully completed ISS would be 10 times larger than any Shuttle flying in space, and that it would be in orbit for at least 10 years, increasing the chance of a collision occurring. Although they acknowledged the possibility of a collision, NASA scientists argued that the odds favored the ISS's avoiding such an occurrence. As NASA's expert, Chief Scientist for Orbital Debris, Nicholas L. Johnson, explained, "The odds are certainly in your favor that nothing will happen at all in those 10 years [of planned space station operation]."[354]

*2 December*
In a once-in-a-century occurrence, eight planets lined up in the sky "like pearls on a string" from west to east. Astronomers could view, in one line of sight, Pluto, followed by Mercury, Mars, Venus, Neptune, Uranus, Jupiter, and Saturn—with a crescent Moon visible as well. Jack Horkheimer, host of public television's *Star Hustler* described the event as "naked-eye astronomy at its best," a boon for amateurs. According to computer calculations, the orbits of the planets would not line up in such a pleasing manner for observers on the Earth for at least another century.[355]

---

[353] William Harwood, "Satellite Designed To Study Sun Fails," *Washington Post*, 22 November 1997; Associated Press, "Columbia Crew Prepares Plan To Recover Spinning Satellite," 24 November 1997; William Harwood, "Space Walkers Grab Satellite, Wrestle It Back onto Shuttle," 25 November 1997; Reuters, "Spacewalking Astronauts Grab Errant Satellite," 25 November 1997; Associated Press, "Clinton Calls Shuttle Crew with Thanks," 28 November 1997.

[354] U.S. General Accounting Office, "Space Surveillance: DOD and NASA Need Consolidated Requirements and a Coordinated Plan" (report no. NSIAD 98-42, Washington, DC, 1 December 1997); Sean Holton, "Worries Rise over Junk in Space," *Orlando Sentinel* (FL), 2 December 1997.

[355] Associated Press, "Planets in Once-in-Century Lineup," 2 December 1997.

*3 December*

Crew aboard Shuttle *Columbia* successfully released and operated the Autonomous Extra-vehicular Robotic Camera Sprint (AERCam Sprint). The 35-pound (16-kilogram), spherical camera propelled itself around the Shuttle after astronaut Winston E. Scott had released it during a spacewalk. The camera used periodic jet thrusts to maneuver, responding to control commands made from inside *Columbia*. The US$3 million camera transmitted images of *Columbia* and its crew back to Earth, providing unique vantage points for observation. The astronauts, who retrieved the small robotic device without any complications, concluded that it would be a helpful technology for future Shuttle missions and the ISS.[356]

*8 December*

The Astronauts Memorial Foundation recognized Major Robert H. Lawrence Jr. as the first black astronaut in NASA history, exactly 30 years after his death during a U.S. Air Force training mission. A ceremony adding Lawrence's name to the Astronauts Memorial Foundation's Space Mirror at Kennedy Space Center (KSC) ended a long bureaucratic debate over whether Lawrence qualified as an astronaut. Lawrence had never met the Air Force requisite that an astronaut must have flown at an altitude of least 50 miles (80 kilometers) above Earth, because, although he had received his appointment to the U.S. Air Force's Manned Orbiting Laboratory Program, he had died before flying in space. Therefore, the Astronauts Memorial Foundation had repeatedly refused to include his name on the Space Mirror at KSC. However, in 1996 U.S. Representative Bobby L. Rush (D-IL) had persuaded the Air Force to verify Lawrence's status as an astronaut. Thereafter, the Astronauts Memorial Foundation's board of directors voted, immediately and unanimously, to include Lawrence's name on the Space Mirror.[357]

At a meeting of the American Geophysical Union, scientists unveiled evidence of another cause, besides the Sun, of "killer electrons," which travel at high velocity. When electrons travel from Jupiter's massive magnetic field, their speed approaches the speed of light. Showers of high-velocity electrons cause many problems, including blackouts on Earth and damage to orbiting satellites and global positioning systems. Scientists had previously been unable to determine the source of the electrons causing the damage. Presenting the new information on the phenomenon of high-velocity electrons, the researchers declared that the ability to forecast "space weather," including the patterns of high-velocity electrons, would become increasingly important during the coming age of the ISS.[358]

*9 December*

U.S. Senators John McCain (R-AZ) and Bill Frist (R-TN) sent a harshly worded letter to NASA Administrator Daniel S. Goldin, decrying NASA's chronic cost overruns and promising investigative hearings before the Senate Committee on Commerce, Science, and Transportation. Although McCain and Frist both expressed support for NASA, and for space exploration in

---

[356] Seth Borenstein, "Robot Set for 1st Space Walk in Space," *Orlando Sentinel* (FL), 3 December 1997; Associated Press, "Free-Flying Camera Flits High Above Columbia, Sends Down Video," 3 December 1997.
[357] Associated Press, "Crash Victim Recognized as 1st Black Astronaut," 8 December 1997.
[358] Tim Friend, "Mystery Solved: Jupiter Is Source of Killer Electrons," *USA Today*, 9 December 1997.

general, they wrote that "continual, unabated overruns undermine the public faith" in the space program.[359]

*10 December*
NASA dedicated a plaque in memory of Ellison S. Onizuka, celebrating his record of achieving the highest altitude for propeller-driven aircraft. Onizuka had been Hawaii's first astronaut and a member of the Shuttle crew aboard *Challenger* when it exploded in 1986. NASA's recognition of Onizuka's achievement, more than a decade later, occurred during the same year that the remotely controlled AeroVironment Pathfinder broke a distance record for unpiloted space travel, topping 71,000 miles (114,000 kilometers) during its 7 July 1997 flight. NASA honored Onizuka for his part in "reaching for the heavens and striving for excellence," noting that his achievement had contributed to that of Pathfinder. NASA dedicated the plaque honoring Onizuka at the U.S. Navy's Pacific Missile Range Facility on Kauai, where the record-setting Pathfinder plane had set flight on its historic mission.[360]

U.S. Senator D. Robert "Bob" Graham (D-FL) announced a plan to lease room aboard NASA Space Shuttles and the ISS to private entities, to offset the costs of space exploration. Graham announced his plan during a news conference at a new Cape Canaveral, Florida, air station dedicated to commercial launches. Congress had banned the use of the Shuttle for commercial purposes after the *Challenger* accident. Some space experts believed that Graham's plan would take advantage of a strong commercial market for the retrieval of satellites from space. Graham's plan joined a list of other initiatives of the mid-1990s, designed to merge government and private space interests to ease costs for the federal government.[361]

*16 December*
The Galileo spacecraft made the closest ever flyby of Jupiter's moon Europa and earned a place in space history by becoming the first spacecraft to orbit an outer planet. Galileo came within 124 miles (200 kilometers) of Europa, gathering images of what looked liked an icy ocean beneath the moon's frozen crust. The prospect of water on the moon's surface excited scientists, who had identified Europa's climate as unique among the moons and planets of the galaxy. The images of the possible ocean seemed to show signs of "slushy material" pushing up against the surface ice, a sign of a heated core. The flyby was the first encounter to occur during Galileo's two-year extension to its original mission. According to press accounts, Galileo, launched in 1989, had already gathered enough data to force scientists to "rewrite textbooks."[362]

*17 December*
NASA released images from the Hubble Space Telescope (HST) revealing more clearly than ever the "final blaze of glory" of aging Sun-like stars. Scientists had long hypothesized that stars gradually dimmed and died out as they cast off a shell of glowing gas. However, the HST revealed instead that dying stars create in space an intricate series of glowing

---

[359] Sean Holton, "Senators Blast NASA for Program Overruns," *Orlando Sentinel* (FL), 10 December 1997.
[360] NASA, "NASA Solar-Powered Altitude Record To Be Dedicated to Challenger Astronaut," news release N97-087, 3 December 1997.
[361] Seth Borenstein, "Graham: Let's Make Money in Space," *Orlando Sentinel* (FL), 11 December 1997.
[362] NASA, "Closest Europa Flyby Marks Start of Galileo Mission 'Part II'," news release 97-286, 16 December 1997; K. C. Cole, "Galileo Takes Close-Ups of Icy Europa," *Los Angeles Times*, 17 December 1997; Associated Press, "Galileo Flies By Europe, Sees Indication of Ocean Under Ice," 17 December 1997.

patterns—"pinwheels, lawn sprinkler style jets, elegant goblet shapes, and even some that look like a rocket engine's exhaust." Astronomers speculated that the new images would lead them to reanalyze "stellar evolution." According to Howard E. Bond of the Space Telescope Science Institute of Baltimore, Maryland, the images made old ideas about the death of stars seem simplistic. Hubble scientist David S. Leckrone of NASA Headquarters, referring to the eventual burning out of the Sun, went so far as to say, "In a very real way, these images show us our own destiny." Scientists believed that the images portrayed a complicated but very organized sequence of events, which experts might understand after further study.[363]

*Mir*'s crew had to abandon a German-built robotic camera called the Inspector, which Russian space officials had hoped would capture precise images of the holes in the *Spektr* module, made when the a cargo craft crashed into *Mir* in June 1997. The camera had failed immediately after its launch, diverting from its planned course. After attempting for several hours to redirect the Inspector, the crew made plans to change *Mir*'s orbit, to avoid a collision with the unruly robot. The malfunctioning Inspector was the latest in the long string of accidents and malfunctions that had forced the three Russian-American crew members aboard *Mir* to spend nearly as much time on maintenance as on scientific research.[364]

*18 December*
President William J. Clinton awarded Congressional Space Medals of Honor to Edward H. White II and Roger B. Chafee, who died in the *Apollo 1* fire. Clinton presented the medals to members of the men's families in a ceremony in the Oval Office. White and Chafee were part of the three-man crew involved in the tragedy, which had occurred more than 30 years earlier. On 27 January 1967, the *Apollo 1* space capsule had caught fire on the launchpad, killing the three men on board the spacecraft. The third member of the crew, Virgil I. "Gus" Grissom, had received the Congressional Space Medal of Honor in 1978. The accident had caused NASA to institute aggressive safety procedures, placing a premium on the well-being of its astronauts. Clinton noted the contributions of the fallen astronauts, "though they never got there, astronauts Chaffee, White, and Grissom's footprints are on the moon." Besides *Apollo 1*'s crew, only nine other astronauts had received the Congressional Space Medal of Honor.[365]

*21 December*
NASA announced that, because of a US$100 million budget shortfall, it would lay off 600 workers from its 6,000-person workforce, including 500 workers from KSC. United Space Alliance, the joint venture of Lockheed Martin and the Boeing Company, which NASA had contracted to take over Shuttle operations, planned to solicit volunteers for early retirement

---

[363] NASA, "Hubble Witnesses the Final Blaze of Glory of Sun-Like Stars," news release 97-287, 17 December 1997; Paul Hoversten, "Hubble Observes Stars' Swan Songs," *USA Today*, 18 December 1997; John Noble Wilford, "Hubble Takes Gaudy Photos of Dying Stars," *New York Times*, 18 December 1997; Associated Press, "Hubble Previews Death of Earth in 6 Billion Years," 18 December 1997.
[364] Reuters, "Mir To Abandon Robot Space Camera After Failure," 17 December 1997; Associated Press, "Robot Fails, Floats Free Near Mir," 18 December 1997.
[365] Associated Press, "Apollo 1 Astronauts Honored: 3 Killed in 1967 Fire on Launch Pad," 18 December 1997; Associated Press, "Clinton Gives Congressional Space Medal," 17 December 1997.

before resorting to "pink slips." To ensure the maintenance of safety standards, NASA was to maintain the right of final approval of all personnel decisions.[366]

*24 December*
EarthWatch Inc. of Longmont, Colorado, launched into orbit *Early Bird 1*, the first private spy satellite. The launch, which took place in Russia, ended government domination of space spying. According to EarthWatch, *Early Bird 1* had the capability to track objects on Earth as small as 10-feet (3-meters) long. The company also reported that "customers were lining up to buy images." Customers for *Early Bird 1* services included public interest groups, planning to track military movement and arms control measures, and foreign governments, without the resources to launch their own satellites. Because *Early Bird 1* was an American-produced satellite, the U.S. federal government retained the right to switch it off during wartime, as well as the right to filter images sold to foreign customers. Experts predicted that the security and privacy questions surrounding the first private spy satellite would take years to sort out. However, they also surmised that *Early Bird 1* was only the first in the wave of satellites that private industry would launch during the coming decade.[367]

*25 December*
The Russian Space Agency paid cosmonauts Vasili V. Tsibliyev and Alexander I. Lazutkin belated bonuses for their six-month *Mir* mission, payments that Russia had withheld until investigators had determined the two men's roles during the troubled mission, which included the crash of a robotic cargo craft into *Mir*. S. P. Korolev Rocket and Space Corporation Energia (RSC Energia), the company that built *Mir*, had blamed Tsibliyev and Lazutkin, suggesting that Russia deny the two cosmonauts their salaries. However, once the Russian government had determined that human error had been only one among many causes of the crash, RSC Energia's President Yuri P. Semyonov had retracted the company's demand that Russia punish the cosmonauts and had supported their right to receive payment. Tsibliyev received about US$100,000 and Lazutkin about US$80,000. Both men had expressed dismay at the negative reception they had received upon returning to Earth after the difficult mission.[368]

## JANUARY 1998

*1 January*
NASA's Langley Research Center in Hampton, Virginia, opened a new Collaborative Engineering Center equipped with virtual reality, teleconferencing, and cutting-edge technology. Using a combination of a joystick and computerized voice commands, engineers could control the three-dimensional, virtual space station, which included a docking Space Shuttle orbiter.[369]

*4 January*

---

[366] Dan Klepal, "NASA Confirms 600 Jobs Will Be Cut," *Florida Today* (Brevard, FL), 21 December 1997; Associated Press, "Space Center Layoffs," 23 December 1997.
[367] William J. Broad, "First Civilian Spy Satellite Soars into Space, Launched in Russia by a U.S. Company," *New York Times*, 25 December 1997; Reuters, "1st Civilian Spy Satellite Launched in Russia," 26 December 1997.
[368] Associated Press, "Mir's Pay Day," 25 December 1997.
[369] Akweli Parker, "NASA Reshaping the Way It Works," *Virginian-Pilot* (Norfolk, VA), 16 January 1998.

The crew of the *Mir* space station replaced the malfunctioning computer block, which had failed on 2 January, and restarted the station's orientation system. This common type of malfunction, which had occurred seven times in 1997, did not create life-threatening conditions, but it did reduce the power supply to the space station, thereby crippling the gyroscope that oriented the station's solar panels. Using the *Soyuz* escape module engines to control its orientation, *Mir* turned off all nonessential systems, to conserve energy until the crew had completed the repair.[370]

Steve Fossett's *Solo Spirit* balloon, carrying NASA's prototype Aerobot Science Payload, landed prematurely when it experienced equipment problems. NASA had been testing the Aerobot Science Payload prototype for use on balloons in outer space, to collect information from other planets' atmospheres. On this trip, Aerobot returned accurate data on position, balloon velocity, vertical wind velocity, temperature, and humidity, but experienced problems with a pressure sensor and a satellite communications link.[371]

*6 January*
An Athena II rocket carrying Lunar Prospector launched from the renovated U.S. Navy test-launch site, Launch Complex 46, marking the first time that the government-based space transportation agency, Spaceport Florida Authority, had managed a launch from the complex. The rocket carried aboard some of the ashes of geologist Eugene M. Shoemaker, who had died in a car crash in July 1997. Among his many professional accomplishments, the sixty-nine-year-old Shoemaker had selected lunar landing sites and trained Apollo astronauts. On 5 January, NASA had cancelled Athena II's first scheduled launch for safety reasons, after U.S. Air Force radar necessary to track the rocket failed. The unpiloted rocket launched at 9:28 p.m. (EST) and, after 1 hour, propelled the 4-foot (1.2-meter), 650-pound (291-kilogram) Lunar Prospector out of low-Earth orbit toward the Moon.[372]

*8 January*
Astronomers announced that pictures from a new camera on the Hubble Space Telescope (HST) supported the theory that a new planet was forming around Beta Pictoris. Images focusing on the disc (the halo of dust and gas rotating around the star) revealed that, instead of a smooth flow, one quadrant has a large bulge, indicating the orbital path of a planet that could be many times larger than Jupiter. No actual planet was visible in the images, only the effects of its gravitational pull. Nevertheless, this was the strongest evidence to date in visible light, of the presence of a possible planet. Beta Pictoris is younger (20 million to 100 million years old compared to the Sun's 4.5 to 5 billion years) and about eight times brighter than the Sun. The formation of planets around a young star suggested that planets form very early in the life of a solar system.[373]

---

[370] *Washington Post*, "Mir Crew Replaces Computer Block," 4 January 1998; *Chicago Tribune*, "New Year, Old Woe—Mir Computer Glitch," 3 January 1998.
[371] *Florida Today* (Brevard, FL), "Balloon Trip Helps Martian Studies," 11 January 1998.
[372] Lynne Bumpus-Hooper, "Prospector Off on Historic Trip to Moon," *Orlando Sentinel* (FL), 7 January 1998; Jim Erickson, "Late Flagstaff Astronomer's Ashes Are Headed for Moon in Fitting Tribute," *Arizona Daily Star* (Tucson, AZ), 7 January 1998; *Washington Times*, "NASA Launches Mission to Moon Seeking Water for Future Settlers," 7 January 1998.
[373] *Washington Post*, "Images Support Theory of New Planet: Hubble Photos Show Bulge in Star's Halo Dust—63 Light-Years from Sun," 9 January 1998.

NASA named current Director of Goddard Space Flight Center (GSFC) Joseph H. Rothenberg to head its Office of Space Flight, which oversees the Human Exploration and Development of Space Enterprise, which includes the Shuttle program and the planned International Space Station (ISS). Rothenberg replaced Wilbur C. Trafton, who had announced his resignation in November 1997. The Office of Space Flight accounted for about 40 percent of NASA's US$13.6 billion annual budget. Rothenberg was the first Director of the Office of Space Flight who had no experience in military aviation, the astronaut corps, the Space Shuttle, or management of the space station program. However, he had a strong background in engineering and astronomy. Rothenberg was closely involved in planning and using the HST and had worked as a space engineer at Kennedy Space Center (KSC). NASA named Alphonso V. Diaz to succeed Rothenberg at GSFC.[374]

*9 January*
The two Russian crew members of the *Mir* space station completed a record-breaking spacewalk. Commander Anatoly Y. Solovyev and Pavel V. Vinogradov dismantled and retrieved in record time the *Kristall* module's optical-monitoring equipment made by the United States. The scheduled 6-hour spacewalk took only 3 hours and 8 minutes. However, it was unclear whether the cosmonauts had been able to repair a leaky hatch. The problem was not with the rubber sealing, as Russian ground control had suspected, but rather with a malfunctioning lock that was preventing the hatch from sealing hermetically. The cosmonauts activated 10 reserve locks and pumped air into the airlock, planning to monitor the hatch for leaks over the next 24 hours. The spacewalk was Commander Solovyev's 16$^{th}$ in five flights over 10 years. He had spent more than 70 hours in free space, more than any other person had accrued.[375]

*11 January*
Russian Mission Control monitored *Mir*'s airlock pressure following the repair work of 9 January. On Friday, 8 January, when the crew entered the airlock without spacesuits to retrieve some equipment, the repair was holding, but the pressure had fallen slightly. Russian ground control near Moscow, which was analyzing the data, planned to decide on Monday whether *Mir*'s *Kvant-2* module was safe.[376]

At 7:15 a.m. (EST), Lunar Prospector executed the first of three engine bursts needed to settle into orbit. In a flawlessly executed maneuver, Prospector fired its engines for 30 minutes to slow down, permitting the Moon's gravity to capture it. The small robotic spacecraft was NASA's first visitor to the Moon since the astronauts from *Apollo 17* had walked on its surface in 1972. The Lunar Prospector project had cost about US$63 million.[377]

*12 January*

---

[374] Sean Holton, "NASA Picks an Outsider To Run Shuttles, Space Station," *Orlando Sentinel* (FL), 9 January 1998; *Washington Post*, "Rothenberg To Head Office of Space Flight," 9 January 1998; NASA, "Diaz Named Goddard Space Flight Director," news release 98-4, 8 January 1998.
[375] *Washington Post*, "Mir Commander Walks into Spaceflight History," 10 January 1998; *Washington Times*, "Cosmonauts Possibly Fix Leaky Hatch, Do Spacewalk in Record-Short Time," 10 January 1998.
[376] Reuters, "Mir's Airlock Pressure Down a Bit—Russian Control," 11 January 1998.
[377] *Washington Times*, "After 25 Years, NASA Shoots for the Moon," 12 January 1998; Robyn Suriano, "Prospector Slides into Lunar Orbit," *Florida Today* (Brevard, FL), 12 January 1998.

Pressure in *Mir*'s *Kvant-2* module had been dropping steadily, to about three-quarters of the normal atmospheric pressure of Earth, indicating that the hatch was still not closing properly. Despite these problems, the team delayed further repairs, scheduling them to take place after other planned events, including a spacewalk scheduled for later in the week, the docking with Space Shuttle *Endeavour* planned for 25 January, and the arrival of a Russian-French team on 30 January.[378]

*13 January*
Lunar Prospector settled into its orbit after three flawless engine burns. The five scientific instruments (magnetometer, electron reflectometer, gamma-ray spectrometer, neutron spectrometer, alpha particle spectrometer, and Doppler gravity experiment using an S-band antenna) began transmitting streams of data. The neutron spectrometer scanned the lunar surface for evidence of hydrogen; the gamma-ray spectrometer determined concentrations of elements such as uranium and iron; and the alpha-particle spectrometer searched for indications of volcanic and tectonic activity.[379]

The contractor operating the Space Shuttle program, United Space Alliance (USA), announced that it would cut 500 jobs at KSC by 30 January, to reduce total workforce by 10 percent. The announcement raised concerns in Congress about Shuttle safety.[380]

Florida state leaders began forming a task force to attract Lockheed Martin's VentureStar™ Reusable Launch Vehicle to the coast's spaceports. U.S. Representative David Weldon Jr. (R-FL) confirmed verbal commitments from the Spaceport Florida Authority and from Enterprise Florida to help fund the task force. The Florida VentureStar™ Capture Team, the proposed name for the task force, estimated that they needed between US$200,000 and US$1 million to carry out a successful bid for the VentureStar™ business.[381]

*15 January*
American astronaut David A. Wolf made his first spacewalk, accompanied by veteran spacewalker Anatoly Y. Solovyev, Commander of the *Mir* space station. Solovyev, making his 18th spacewalk, set the world's record for the highest number of spacewalks by one individual.[382] The spacewalk began 20 minutes later than scheduled, because of problems opening locks on the malfunctioning outer hatch. The scheduled task had been to assess wear and tear on the exterior of the station using a photoreflectometer designed in the United States. David A. Wolf worked for 2 hours trying to verify whether the device was working properly, but only succeeded in taking readings on the shell of the nine-year-old *Kvant-2* module, one of the station's oldest

---

[378] Associated Press, "Mir," 12 January 1998; Reuters, "Mir Airlock Continues To Leak After Repair," 12 January 1998.
[379] Reuters, "Corrected—U.S. Space Probe Begins Hunt for Water on the Moon," 14 January 1998; NASA, "Solar System Exploration: Lunar Prospector," *http://solarsystem.nasa.gov/missions/profile.cfm?MCode=LunarPr&Display=ReadMore* (accessed 30 July 2007).
[380] Robyn Suriano, "500 KSC Layoffs Jan. 30 Expected: Budget Shortfall Will Reduce Work Force 10 Percent," *Florida Today* (Brevard, FL), 14 January 1998; *Florida Today* (Brevard, FL), "Latest Round of KSC Layoffs Needs Congressional Scrutiny," 15 January 1998.
[381] Todd Halvorson, "Task Force To Focus on Attracting Lockheed Martin's VentureStar to State," *Florida Today* (Brevard, FL), 14 January 1998.
[382] Seth Borenstein, "American Gets Ready for Spacewalk Today," *Orlando Sentinel* (FL), 14 January 1998.

compartments. Engineers planned to use the data to help make selections for materials and protective coatings for the new ISS. Flight Director Vladimir A. Solovyov explained in a news conference at Mission Control outside Moscow, "the device proved to be far from mature, so it was decided to skip the examination of the core module." Instead, the two men studied the holes in the *Spektr* module, which had suffered punctures during a collision with a cargo craft on 25 June 1997.[383]

*16 January*
*Mir* Commander Anatoly Y. Solovyev celebrated his 50$^{th}$ birthday—in space, enjoying a "semi-officially" sanctioned one-gulp-per-person dose of cognac and a birthday dinner chosen from the space station's supplies. Solovyev had logged an impressive 80 hours of spacewalking during 17 walks. As of this *Mir* mission, Solovyev's fifth, the Russian cosmonaut had spent a cumulative total of approximately 20 months on the space station.[384]

NASA officially named U.S. Senator John H. Glenn Jr. (D-OH), veteran astronaut of the Mercury program, as a crew member aboard Space Shuttle *Discovery*'s Mission STS-95 planned for October 1998. At age 77, Glenn would become the oldest person to fly in orbit. In 1962 he had been the first American to orbit Earth. NASA described the research planned during Glenn's spaceflight, a joint effort between NASA and the National Institute of Aging of the National Institutes of Health. Scientists intended to supplement peer-reviewed research on sleep disorders, muscle atrophy, balance, and clinical evaluations of blood and heart function with data collected during *Discovery*'s planned October mission. Senator Glenn had undergone medical tests with NASA physicians, as well as with independent consultants, all of whom had determined that Glenn was medically qualified for spaceflight, with an excellent fitness level. NASA Administrator Daniel S. Goldin remarked that Glenn "brings a unique blend of experience to NASA. He has flight, operational, and policy experience."[385]

*18 January*
The Russian Space Agency announced that the first module of the ISS was completed and ready to ship to the launchpad. Under a contract with Boeing Company, Khrunichev's Moscow factory had designed and built this first module to hold cargo. The new module was similar to two existing modules in the *Mir* space station. Russia had originally scheduled the module for launch in November 1997, but fell behind in the production of the second unit. Despite the delay, Director General of the Russian Space Agency Yuri N. Koptev emphasized the importance of the day during a news conference: "This event symbolizes the fact that Russia was, is, and shall be a space nation, capable of realizing large-scale projects penetrating into space." The ISS project, a joint effort of Russia, the United States, the European Space Agency (ESA), Canada, and Japan, was especially significant to Russia since the ISS would replace *Mir* as the only piloted element of Russia's space program.[386]

---

[383] *USA Today*, "American Looks at Mir from Outside," 15 January 1998; Oleg Shchedrov for Reuters, "Hitches Marred Mir Crew Space Walk, Says Official," 15 January 1998; *Houston Chronicle*, "Astronaut, Cosmonaut Take Short Spacewalk: Excursion Marked by a Few Small Difficulties," 15 January 1998.
[384] Associated Press, "Mir-50-Year-Old Cosmonaut," 16 January 1998.
[385] NASA, "Sen. Glenn Gets a 'Go' for Space Shuttle Mission," news release 98-8, 16 January 1998.
[386] Associated Press, "Russia-Space Station," 18 January 1998.

*20 January*
Following a meeting of the Defence Council, President Boris N. Yeltsin issued a decree giving overall control of Russia's space industry to the civilian Russian Space Agency. Yeltsin's intention was "to extract maximum economic benefit from one of the few manufacturing industries in which Russia is a world leader." With this decree, Yeltsin required the government to create a plan for a restructured space industry, giving priority to national security, to economic and scientific missions, and to strengthening Russia's position in the world marketplace.[387]

*21 January*
A NASA Space Shuttle program manager announced that a special safety study would not prevent USA, the prime Space Shuttle contractor, from cutting up to 10 percent of its workforce at KSC. The safety report, not yet released, called for some changes in the layoff plan, but would still permit the private company to cut up to 600 jobs.[388]

NASA modified its contract with Orbital Sciences Corporation in order to produce a second flight vehicle for the X-34 program. The modified contract expanded test objectives, allowing US$7.7 million in additional funds for the purchase of hardware and another US$2 million for additional testing and analysis, wind tunnel testing, and a second leading-edge thermal protection system.[389]

NASA announced the change in name of one of its four strategic enterprises. NASA's Mission to Planet Earth Strategic Enterprise was renamed Earth Science Enterprise. Mission to Planet Earth Strategic Enterprise had originated 10 years ago, based on the concept that NASA should observe the Earth in the same way that it studies other planets. NASA said that the new name reflected the enterprise's reformulated goals. NASA's Earth Science Enterprise would seek to answer key questions about land-surface cover, near-term and long-term climate change, natural hazards, and atmospheric ozone.[390]

The Russian Space Agency nominated veteran cosmonaut Valery V. Ryumin to fly as a mission specialist aboard *Discovery* during Mission STS-91 in May. In his last spaceflight in 1980, Ryumin was a member of the Soyuz 35 Mission. During that mission, he had spent 185 days in space from April through October. Ryumin had spent a total of 362 days in space in the course of his three missions.[391]

*22 January*
Space Shuttle *Endeavour* Mission STS-89 successfully launched from KSC in Cape Canaveral, Florida, despite forecasted thunderstorms. The crew of STS-89 included Commander Terrence W. Wilcutt, Pilot Joe F. Edwards Jr., Payload Commander Bonnie J. Dunbar, and Mission Specialists Michael P. Anderson, James F. Reilly, Salizhan S. Sharipov, and Andrew S. W. Thomas, who would replace David A. Wolf on *Mir*. This was the eighth Shuttle–*Mir* docking,

---
[387] Reuters, "Russia Puts Civilian Agency in Charge of Space," 20 January 1998.
[388] Seth Borenstein for Knight-Ridder Tribune Business News, "Kennedy Space Center in Florida To Cut 600 Private-Company Shuttle Jobs," 22 January 1998.
[389] NASA, "NASA Commits to Second Vehicle for X-34 Program," news release 98-11, 21 January 1998.
[390] NASA, "Mission to Planet Earth Enterprise Name Changed to Earth Science," news release 98-12, 21 January 1998.
[391] NASA, "Veteran Cosmonaut Nominated To Fly on Final Shuttle/Mir Mission," news release, 21 January 1998.

but the first for an orbiter other than *Atlantis* and the first overseen by Launch Director David A. King. NASA had appointed two new rotational launch directors, David A. King and Ralph R. Roe Jr., following the retirement of veteran Launch Director James F. Harrington III.[392]

NASA Launch Integration Manager Donald R. McMonagle announced that Space Shuttle *Columbia*'s tune-up would take place at the Boeing Company's facility in Palmdale, California, rather than at KSC in Cape Canaveral, Florida. NASA Launch Integration and Shuttle operator USA advised Boeing officials "to get ready to map out a plan for *Columbia*'s work in California." The announcement, following the recent notification about significant job cuts, was more bad news for KSC officials. NASA officials had proposed in 1996 that KSC handle much of the Shuttle's maintenance and modification but had decided that the facility was "too busy with launches to handle [S]huttle renovation." Alan Buis, a Boeing representative, suggested that, because *Columbia* would be getting "a complicated internal glass cockpit, Palmdale is 'a logical choice'."[393]

*24 January*
Shuttle *Endeavour* docked with *Mir* at an altitude of 214 nautical miles (2,467 miles or 3,970 kilometers), bringing the seventh American astronaut to work on board *Mir* since the program began in 1995. Andrew S. W. Thomas, a mechanical engineer, replaced David A. Wolf, a physician, who had been on board *Mir* since September. The hatches opened at 5:25 p.m. (EST), and Thomas's transfer occurred at 6:35 p.m. (EST).[394]

*25 January*
Astronaut Andrew S. W. Thomas was unable to fit into his Russian-made emergency spacesuit before boarding *Mir*, but after initial orders that Thomas spend the first night in the Space Shuttle, Mission Control granted him permission to board. The suit was necessary only in case of emergency evacuation aboard *Soyuz*. Once aboard *Mir*, Thomas altered his spacesuit, adjusting and removing some straps in the armpit and torso areas. After the alterations, the suit fit exactly as it should, ensuring that Thomas would not have to return to Earth with *Endeavour* on Thursday.[395]

During the Shuttle's docking at *Mir*, a sensor aboard *Endeavour* failed, requiring *Mir* to take control of *Endeavour*. *Mir* was to keep the Shuttle pointed in the right direction until Mission Control could transmit software to fix the sensor.[396] Crews transferred about half of the 7,400 pounds (3,356 kilograms) of supplies and equipment aboard the Shuttle to the space station.[397]

---

[392] *USA Today*, "Shuttle Lifts," 23 January 1998; Seth Borenstein, "Weather May Keep Endeavour on Ground," *Huntsville Times* (AL), 21 January 1998; NASA, "Mission Archives: STS-89," http://www.nasa.gov/mission_pages/shuttle/shuttlemissions/archives/sts-89.html (accessed 11 June 2007); NASA Kennedy Space Center, "KSC Launch Director Harrington To Retire; Successors Named," news release 223-97, 17 December 1997, http://www-pao.ksc.nasa.gov/kscpao/release/1997/223-97.htm (accessed 5 August 2008).
[393] Seth Borenstein, "California Plant Gets Shuttle Work Again," *Orlando Sentinel* (FL), 24 January 1998.
[394] *New York Times*, "Shuttle Docks at Space Station with Astronaut and Supplies," 25 January 1998; NASA, "STS-89."
[395] Steve Marshall, "Mir Spacesuit, Astronaut Prove To Be a Poor Fit," *USA Today*, 26 January 1998; *Washington Post*, "Astronaut Moves into Mir Despite Spacesuit Problem," 26 January 1998; *Washington Times*, "Thomas Finally Fits in on Mir: Crucial Alterations Made to Space Suit," 27 January 1998.
[396] Pauline Arrillaga for Associated Press, "Space Shuttle," 26 January 1998.
[397] Brad Liston for Reuters, "Balky Suit and Other Problems Fixed on Mir," 26 January 1998.

*26 January*
A Russian Space Agency official insisted that there had been no problem with Andrew S. W. Thomas's spacesuit, describing the U.S. astronaut as bad tempered and expressing concern about the implications of Thomas's attitude for the remainder of the mission. Deputy Flight Commander Viktor Blagov indicated that, since major work would take place on *Mir*, "any fuss about the space suit is out of place." Blagov explained that Thomas would keep both David Wolf's spacesuit and his own.[398]

*Endeavour* lost power to its small forward-maneuvering jets for 30 minutes. Because the jets, which deliver 25 pounds (11 kilograms) of thrust, keep the complex stabilized as it orbits Earth, the loss of power left the *Endeavour–Mir* complex adrift. During this malfunction, the crew continued its main task, to transfer supplies and equipment from the Shuttle to the space station. NASA officials explained that the crew had never been in danger. However, if the Shuttle's small maneuvering jets had not regained power, *Endeavour* would have had to rely on its primary jets, which deliver 875 pounds (397 kilograms) of thrust. Using the primary jets would have risked damage to *Mir* upon separation of the two spacecraft.[399]

NASA announced the discovery of a fast-spinning pulsar, providing evidence of an evolutionary link between strong-field, slower-spinning, energetic pulsars and weak-field, millisecond pulsars. Frank Marshall, William W. Zhang, and Eric V. Gotthelf of GSFC found the pulsar by examining x-ray emissions that NASA's Rossi X-ray Timing Explorer spacecraft had recorded in 1996. Observations of the Advanced Satellite for Cosmology and Astrophysics spacecraft, a satellite jointly owned by Japan and the United States, confirmed the discovery. The scientists reported that the pulsar was spinning at a rate of 60 times per second, and that it could have been spinning as fast as 150 times per second when it formed 4,000 years ago. An astrophysicist at the U.S. Department of Energy's Los Alamos National Laboratory remarked, "this is the fastest high-energy pulsar of its type we know about." The research team reported that the pulsar is "likely associated with the remnant of a supernova (N157B) that exploded in the Large Magellanic Cloud about 4,000 years ago." Comparing the pulsar found in N157B to the high-energy pulsar in the Crab Nebula, the scientists reported that the central source of x-ray light from N157B is this fast-spinning pulsar associated with a supernova remnant, proving the hypothesis that the weaker the magnetic field, the faster the pulsar spins at birth.[400]

*27 January*
Engineers at KSC made corrections to *Endeavour*'s software, completely repairing the Shuttle's jets overnight. After the transmittal of the corrected software to the computers on board the Shuttle, the problem was resolved. NASA believed that a malfunctioning relay latch, an electrical connection on the spacecraft, had caused the problem.[401]

---

[398] Reuters, "Balky Suit."
[399] Brad Liston for Reuters, "Faulty Jets Are Fixed, and Space Complex Is Stabilized," *New York Times*, 27 January 1998.
[400] NASA, "Fast-Spinning Pulsar Discovery Provides Evolutionary Link," news release 98-14, 26 January 1998.
[401] Reuters, "Faulty Jets Are Fixed."

*28 January*
The crew closed the hatches used to transfer equipment between *Endeavour* and *Mir* at 5:34 p.m. (EST) in preparation for *Endeavour*'s return flight to Earth. The closure of the hatches concluded five days of *Endeavour*'s continuous link-up with *Mir*, during which time the astronauts and cosmonauts had transferred 9,600 pounds (4,350 kilograms) of materials. In addition to equipment exchanges, the crew exchanged American astronaut David A. Wolf for Andrew S. W. Thomas, a U.S. astronaut born in Australia, who was to remain aboard *Mir* for a planned four-and-one-half-month mission.[402]

NASA and the National Oceanic and Atmospheric Administration awarded a US$423 million contract to Hughes Space and Communications of El Segundo, California, to manufacture and launch up to four weather-monitoring Geostationary Operational Environmental Satellites (GOES). The basic contract covered the manufacture and launch of two spacecraft. Separate fixed-price options for two additional spacecraft would cost US$190 million and US$185 million. The spacecraft would carry instruments to provide regular measurements of the Earth's atmosphere, cloud cover, and land surfaces. Two spacecraft would also carry the Solar X-ray Imager and Space Environment Monitor instruments. These four spacecraft, known as GOES-N, -O, -P, and -Q were to continue, enhancing the services of the current GOES-I through -M series, a "mainstay of modern weather forecasting." The GOES series provided meteorologists and hydrologists with "visible and infrared images of weather systems, and precise atmospheric soundings." The positions of the spacecraft in orbit would allow scientists to monitor storms when they were first forming in the Atlantic and Pacific oceans.[403]

*29 January*
An Atlas rocket launched from Cape Canaveral Air Force Station in Florida at 1:37 p.m. (EST), after NASA had scrubbed the flight three days in a row. The Lockheed Martin rocket, which carried a classified spacecraft for the Pentagon's National Reconnaissance Office, separated from its passenger 1½ hours after launching, with the satellite successfully placed in orbit. The satellite, viewing two-thirds of the globe, was to circle the Earth in an egg-shaped orbit, taking it from 24,000 miles (39,000 kilometers) above the northern latitudes to just 200 miles (322 kilometers) above Antarctica.[404]

*Endeavour* separated from *Mir* to head home at 11:57 a.m. (EST), returning David A. Wolf to Earth and leaving Andrew S. W. Thomas at the *Mir* station for a four-and-one-half-month stay. Wolf was returning to Earth aboard *Endeavour* after completing a 119-day stay on *Mir*. The new crew of *Mir*, two cosmonauts, Kazakh Talgat A. Musabayev and Russian Nikolai M. Budarin, and a French astronaut Léopold Eyharts, lifted off from Baikonur Cosmodrome in Kazakhstan on *Soyuz TM-27*, 23 minutes before *Endeavour* undocked from *Mir*.[405]

---

[402] Pauline Arrillaga for Associated Press, "Space Shuttle," 30 January 1998; NASA, "STS-89."
[403] NASA, "Hughes Selected To Build Weather Satellites," contract announcement release C98-B, 28 January 1998.
[404] Robyn Suriano, "Atlas Launches on Fourth Try: Classified Craft Likely To Serve as Orbital Switchboard," *Florida Today* (Brevard, FL), 30 January 1998.
[405] *New York Times*, "Shuttle Departs, Leaving Last American on Mir," 30 January 1998; Paul Hoversten, "Endeavour Separates from Mir," *USA Today*, 30 January 1998; NASA, "STS-89."

Fourteen nations and space agencies participated in a ceremony, signing final accords to build and manage the future ISS. Participating countries were Belgium, Canada, Denmark, France, Germany, Italy, Japan, the Netherlands, Norway, Russia, Spain, Sweden, Switzerland, the United Kingdom, and the United States. Acting Secretary of State Strobe Talbott signed, on behalf of the United States, the 1998 Intergovernmental Agreement on Space Station Cooperation, and NASA Administrator Daniel S. Goldin signed separate bilateral memoranda of understanding with heads of the Russian Space Agency, ESA, and the Canadian Space Agency. The new agreements superseded previous agreements concerning the ISS, which Canada, Europe, Japan, and the United States had signed in 1988.[406]

*30 January*
Raytheon Systems Company's training unit, formerly known as Hughes Training, appointed former astronaut Henry W. Hartsfield Jr. as Managing Director for the company's NASA support business. In this position, Hartsfield was to manage the development of simulators for the ISS program.[407]

*31 January*
Just hours before *Endeavour* touched down in Florida, a new crew from Russia docked at *Mir* to begin a three-week transition from *Mir 24* to *Mir 25*. *Mir 25* Commander Talgat A. Musabayev replaced Anatoly Y. Solovyev, Flight Engineer Nikolai M. Budarin replaced Pavel V. Vinogradov, and Researcher Léopold Eyharts, a 40-year old French air force pilot on his first spaceflight, joined the crew.[408]

The U.S. Air Force Space and Missile Museum in Cape Canaveral, Florida, celebrated the 40th anniversary of the successful launch of the modified Redstone rocket, which had carried into orbit *Explorer I*, the United States' first satellite. The launch occurred four months after the Soviet Union had placed *Sputnik I* in orbit. In addition to achieving orbit, the 31-pound (14-kilogram) *Explorer I* had discovered the Van Allen Radiation Belt. The museum at the Cape Canaveral Air Force Station held an open house and a space fair and launched a model rocket to celebrate the anniversary of this milestone and to educate the public about the United States' 40 years in space.[409]

## FEBRUARY 1998

*1 February*
One thousand Puerto Ricans protested against NASA's plans to launch, from a nature preserve in Puerto Rico, suborbital rockets filled with chemicals. The Coqui Dos program planned the launch to measure high-level winds and turbulence in the upper atmosphere. Coqui Dos continued the El Coqui program of 1992, part of NASA's Suborbital Sounding Rocket program, which launched 25 sounding rockets annually from locations worldwide. Because the rockets

---

[406] *New York Times*, "Space Station Accord Signed," 30 January 1998; Associated Press, "Agreements Signed for Launching of Space Station," 29 January 1998; NASA, "Space Station Agreements To Be Signed in Washington," news release 98-17, 29 January 1998.
[407] David Welch for Knight-Ridder Tribune Business News, "Raytheon Systems Names Chief of NASA Deals," 30 January 1998.
[408] NASA, "STS-89"; *Washington Post*, "2 Russians, Frenchman Blast Off for Mir," 30 January 1998.
[409] *Florida Today* (Brevard, FL), "Explorer 1 Anniversary Event Will Bring History to Life," 31 January 1998.

would disperse aluminum trimethol before crashing into the sea, the protesters demanded that NASA produce studies proving that the launch would not damage the environment. The mayor of Vega Baja led the protest outside the entrance to the Tortuguero Recreation Area, the section of the nature preserve where the rocket launches would take place, threatening to sue NASA to force it to release environmental impact studies from the 1992 launch.[410]

*2 February*
Administrator Daniel S. Goldin announced that NASA planned to reserve an additional US$750 million of federal funds over the next two years, enabling Lockheed Martin to continue work on a replacement vehicle for the Space Shuttle. Lockheed Martin and NASA had signed an agreement in 1996 to develop the X-33 suborbital test craft. At that time, the federal government had pledged to dedicate US$900 million to develop the craft and fund 15 suborbital test flights, beginning in the spring of 1999. Goldin explained that if the test program exceeded the original ceiling of US$900 million, the US$750 million funding pool would be available, beginning in the year 2000, to develop the X-33 replacement vehicle.[411]

NASA Administrator Daniel S. Goldin announced the selection of Ghassem R. Asrar as the new Associate Administrator for Earth Science Enterprise. Before this appointment, Asrar had served as Chief Scientist for the Earth Observing System in the Office of Earth Science Enterprise at NASA Headquarters. He led an international team in developing scientific priorities and measurements "to be obtained from a series of advanced Earth-orbiting satellites that promise fundamental new insights into the connections between Earth's land, oceans, atmosphere, ice, and life."[412]

*4 February*
A West European Ariane 4LP rocket blasted off from French Guiana, carrying two communications satellites, *Brasilsat-3* and *Inmarsat-3*. Twenty-one minutes after takeoff, the first satellite, belonging to Brazil, separated from the rocket. Five minutes later, the second satellite, a 2-ton (1,800-kilogram or 1.8-tonne) worldwide mobile communications satellite, separated from the rocket. The launch successfully placed both satellites in orbit.[413]

*6 February*
Shuttle operator United Space Alliance (USA) cut 363 jobs at Kennedy Space Center (KSC), one of the largest cutbacks in decades. Combining the layoffs with 194 voluntary resignations, USA eliminated 557 jobs at KSC. In implementing the cutbacks, the company hoped to "trim $100 million from the current [S]huttle budget," thereby reducing costs, improving efficiency, and attracting new payload customers without compromising safety.[414]

*10 February*

---

[410] NASA, "NASA Studying Space Weather from Puerto Rico," news release 98-23, 6 February 1998; Mario Maldonado for Associated Press, "Puerto Rico-NASA," 1 February 1998; Lance Oliver, "11 Launches Will Attract Protesters: Many in Puerto Rico Fear Damage by Research Rockets," *Orlando Sentinel* (FL), 2 February 1998.
[411] Frank Sietzen Jr. for United Press International, "More NASA Money for Shuttle Replacement," 3 February 1998.
[412] NASA, "Asrar Named Associate Administrator for Earth Science," news release 98-19, 2 February 1998.
[413] *Orlando Sentinel* (FL), "European Ariane Rocket Puts 2 Satellites into Orbit," 5 February 1998.
[414] Richard Burnett and Seth Borenstein, "363 Lose Jobs at Space Center," *Orlando Sentinel* (FL), 7 February 1998.

At a news conference at NASA in Washington, DC, Robert P. Kirshner, astronomer at the Harvard-Smithsonian Center for Astrophysics, and other astronomers discussed new Hubble Space Telescope (HST) images, showing a "shock wave lighting a knot of gas some 100 billion miles [161 billion kilometers] wide." In 1987, when a telescope in Chile detected the first light of an exploding star, the supernova designated SN1987A, astronomers had observed a glowing ring of gas around the remnant of the star. Although the gas ring had cooled and faded, the new HST images showed the shock wave of energy colliding with and lighting up the edges of the ring. The astronomers had been studying the images closely to learn more about how stars evolve and become supernovas. Anne L. Kinney of the Space Telescope Science Institute explained that studying stellar evolution and the inner workings of supernovas is important, because the explosions that create supernovas "fertilize the galaxies" with the enriched material of heavy elements.[415]

The National Research Council released a report entitled "Space Technology for the New Century," finalizing two years of independent study by an expert panel, conducted at the request of NASA. The report's key finding stated that, because of its "faster, better, cheaper" short-term goals, NASA "may be neglecting key technologies needed for longer-term advances in space." The panel identified six technologies that NASA ought to "spur with annual investments of $3 million to $5 million each." The technologies, all with broad-based applications, include: 1) high-volume, planet-to-planet video and data communications based on laser technologies instead of radio frequencies; 2) precisely controlled space structures (space telescopes with more sophisticated and sensitive steering and pointing); 3) micro-machines; 4) safer and more efficient nuclear power systems; 5) radiation-resistant computers and electronics; and 6) advanced space mining and manufacturing technologies.[416]

*11 February*
The Commercial Aviation Safety Strategy Team (CASST) announced a campaign to focus attention on "the most pressing aviation safety problems." CASST was composed of the Air Transport Association, Boeing Company, Airbus Industrie, the Air Line Pilots Association, and the Aerospace Industries Association, as well as major jet engine manufacturers, General Electric Company, Pratt & Whitney of United Technologies Corporation, and Rolls-Royce. CASST identified the top safety issues needing improvement worldwide, including controlled-flight-into-terrain accidents, engine failures, runway incursions, and maintenance errors. CASST recommended working to reduce these safety hazards, as well as implementing technologies to detect in-flight turbulence.[417]

*12 February*
Elizabeth H. Moore, Head of the Art and Archaeology Department of the School of Oriental and African Studies, the University of London, and Anthony Freeman, a radar scientist at NASA's Jet Propulsion Laboratory (JPL), discussed new radar imagery of Angkor, Cambodia, during a

---

[415] NASA, "Shock Wave Sheds New Light on Fading Supernova," news release, 10 February 1998; John Noble Wilford, "Ring of Fire Could Hold Clues to Events Leading to Supernova," *New York Times*, 11 February 1998; Associated Press, "Bright, Exploding Star of 1987 Readies a Reprise of Fireworks," *Washington Post*, 11 February 1998.
[416] Sean Holton, "Report: NASA Should Work More on New Technology," *Orlando Sentinel* (FL), 11 February 1998.
[417] Tim Dobbyn for Reuters, "Aviation Industry Unveils Safety Priorities," 11 February 1998.

press briefing at JPL. The researchers announced that the Airborne Synthetic Aperture Radar (AIRSAR), developed by JPL to produce radar maps, had helped them to detect and locate a distribution of prehistoric mounds and undocumented temples to the northwest of Angkor. The three-dimensional maps, created using radar interferometry, revealed evidence of temples and earlier civilization that were absent or incorrectly marked on modern topographic maps and early twentieth-century archaeological reports. These discoveries brought into question long-standing ideas about the urban evolution of Angkor. The 1996 AIRSAR Pacific Rim Deployment on board a NASA DC-8 jet was a follow-up to a 1994 study of Angkor that used Spaceborne Imaging Radar-C/X-band Synthetic Aperture Radar (SIR-C/X-SAR) on board Space Shuttle *Endeavour* during Mission STS-59, producing the radar images used to create the three-dimensional maps. Unlike SIR-C/X-SAR, AIRSAR had P-band with a longer wavelength, able to penetrate the forest canopy. AIRSAR also flew in a mode that allowed it to measure topography and create three-dimensional images of the surface.[418]

As part of the Small Business Innovation Research Program, NASA officials chose more than 300 research proposals to share US$23 million in funding. Out of 28 award categories, 21 focused on space-based applications, such as achieving routine space travel, exploring and settling the solar system, and developing new technologies in astrobiology. The range of topics indicated that research associated with high-profile space transportation and exploration efforts, curtailed because of cuts in NASA's 1999 budget, would continue on a smaller scale.[419]

The Voyager 1 spacecraft became the most distant human-made object in space, at 6.5 billion miles (11 billion kilometers) from Earth. At 2:10 p.m. (PST), the Voyager 1 spacecraft surpassed the distance of the Pioneer 10 spacecraft, which was traveling in the opposite direction. Voyager's signal took 9.5 hours to reach Earth. The spacecraft had to rely on radioisotope thermal electric generators, because at the far edge of the solar system, very little solar power was able to reach it—at that distance, the Sun is only $1/5,000^{th}$ as bright as it is on Earth. Edward C. Stone, a Voyager scientist and Director of NASA's JPL, remarked, "the fact that the spacecraft is still returning data is a remarkable technical achievement." NASA had launched Voyager 1 from Cape Canaveral, Florida, on 5 September 1977.[420]

Wesley T. Huntress Jr. announced that he would resign from NASA in the near future. As NASA's Associate Administrator for Space Science, Huntress had been responsible for programs in astrophysics, planetary exploration, and space physics. NASA Administrator Daniel S. Goldin credited him with the revitalization of NASA's Space Science Enterprise, in particular, with overseeing critical components of the program such as the HST and Mars Pathfinder. Huntress had joined NASA's JPL in the 1960s, a research scientist specializing in ion chemistry and planetary atmospheres, and NASA had appointed Huntress to head the Office of Space Science in March 1993. In his retirement announcement, Huntress stated, "I have served in this position for more than five years now and it is simply time to move on."[421]

---

[418] NASA, "New Insights into Ancient Angkor by NASA Radar To Be Subject of Press Briefing," news release N98-11, 5 February 1998; NASA, "NASA Radar Reveals Hidden Remains at Ancient Angkor," news release 98-28, 12 February 1998; Jane E. Allen for Associated Press, "Unseen Angkor," 12 February 1998.
[419] Anne Eisele, "Small Business Grants Span Range of NASA Goals," *Space News*, 23 February–1 March 1998.
[420] NASA, "Voyager 1 Now Most Distant Human-Made Object in Space," news release 98-30, 13 February 1998; *Los Angeles Times*, "Voyager 1 Becomes Mankind's Farthest Object from Earth," 18 February 1998.
[421] NASA, "Huntress Announces His Departure from NASA," news release 98-31, 18 February 1998.

*19 February*

The *Soyuz* landing capsule, carrying two Russian cosmonauts and a French astronaut returning from *Mir*, landed safely in the snow and bitter cold (-22°F or -30°C) about 20 miles (32 kilometers) from the northern Kazakhstan village of Arkalyk. Only one of eight rescue helicopters was able to meet the crew because of wind, clouds, and thick snow, limiting visibility to less than 160 feet (49 meters). Anatoly Y. Solovyev and Pavel V. Vinogradov returned to Earth after a six-month mission on *Mir*. During their stay on *Mir*, the cosmonauts had focused on repairing the aging space station. Léopold Eyharts returned to Earth with six salamanders. The two cosmonauts who had replaced Solovyev and Vinogradov had flown the salamanders to *Mir*, carrying them on board the space station in January. While in space, Eyharts had conducted experiments on the salamanders, including monitoring the effects of antigravity on the reptiles' egg-laying capabilities.[422]

*20 February*

This date marked *Mir*'s 12th birthday. Russian Space Agency cosmonauts, Kazakh Talgat A. Musabayev and Russian Nikolai M. Budarin, and U.S. astronaut, Australian-born Andrew S. W. Thomas, took a 45-minute trip outside the space station, moving the escape capsule to the other side of *Mir* to make space for the next cargo craft to dock. Ground controllers briefly lost contact with the crew when *Mir* lost its connection to a communications satellite. For the first time, Thomas wore the Russian-made spacesuit that had been too small when he first arrived at the space station.[423]

*21 February*

After two delays, one because of mechanical problems, the other because of strong winds, a Japanese H-2 rocket blasted off from Tanegashima Space Center, carrying a US$370 million communications satellite intended to conduct a three-year series of communications and broadcast tests. However, an engine breakdown prevented the satellite from entering a geostationary orbit; instead, the launch left the satellite floating in space, useless for its intended mission. Engineers were looking for ways to rescue the mission.[424]

The Alabama Engineering Hall of Fame inducted into its roll of honors James B. Odom, an engineer at NASA's Marshall Space Flight Center (MSFC) who had served NASA for 33 years in the International Space Station, Space Shuttle, and HST programs. The Hall of Fame also honored Marshall's X-ray Calibration Facility, which had tested and calibrated x-ray telescopes since the 1970s. MSFC's X-ray Calibration Facility supported the Advanced X-ray Astrophysics Facility in achieving a resolving power 10 times greater than any previous x-ray telescope.[425]

*24 February*

---

[422] Shavkat Rakhmatullayev for Reuters, "Focus: Russia-France Space Crew Returns to Deep Cold," 20 February 1998.
[423] Associated Press, "BRF—Mir," 20 February 1998; Adam Tanner for Reuters, "Focus: Mir Marks 12th Birthday with Brief Excursion," 20 February 1998.
[424] Mari Yamaguchi for Associated Press, "Japan-Satellite," 21 February 1998.
[425] *Huntsville Times* (AL), "MSFC Facility, Ex-official To Be Lauded," 21 February 1998.

A US$35,000 study commissioned by the Florida Space Business Roundtable revealed that commercial satellite companies did not favor Cape Canaveral as a launch site. The Florida Space Business Roundtable presented its findings at the Florida Space Launch Symposium in Melbourne, Florida. In the survey, executives from satellite manufacturing companies rated spaceports, evaluating them in 11 categories, such as their launch base facilities, flight scheduling, governmental red tape, and on-site technical services. The executives' favorite launch site was Vandenberg Air Force Base in central California, and they selected Xichang Satellite Launch Center in China as the worst place to launch. Florida's spaceports did well in two categories, but business executives considered red tape at those facilities a problem.[426]

*25 February*
NASA partially cancelled the *Clark* satellite program, keeping only the US$13 million launch vehicle service portion of the contract with Lockheed Martin. As reasons for the cancellation, NASA cited cost overruns, an uncertain launch schedule, and questions concerning the effectiveness of *Clark*'s instruments in orbit, because its primary instrument had been "sitting on the ground two years longer than planned." The intended mission of the program was to take black-and-white stereo images, showing detail as small as 10 feet (3 meters) across, for NASA's Office of Earth Science Enterprise (formerly the Office of Mission to Planet Earth), as well as for commercial customers. NASA had paired the *Clark* satellite with the *Lewis* satellite. However, after its August 1997 launch, the *Lewis* satellite had gone into a spin, which controllers were unable to repair. Consequently, the *Lewis* satellite plunged into the atmosphere where it burned up.[427]

*26 February*
A valve in an air purifier on *Mir* overheated, spreading smoke aboard the station. Although the incident caused elevated carbon monoxide, the levels continued to drop steadily. The crew was not in danger, but the incident was a reminder of the terrifying fire that had occurred last year, when an oxygen generator burned for 14 minutes, filling *Mir* with smoke and nearly causing the crew to abandon the station.[428]

**MARCH 1998**

*3 March*
Russian space officials cancelled a spacewalk after cosmonauts broke three wrenches on the last of 10 locks, while trying to open a hatch. The cosmonauts had intended to repair a damaged solar panel. NASA rescheduled the spacewalk for April. The incident further increased concerns about the safety of *Mir*.[429]

*5 March*

---

[426] Todd Halvorson, "Survey: Brevard Launch Sites Fail To Measure Up—Space Coast Falls Short for Commercial Liftoffs," *Florida Today* (Brevard, FL), 25 February 1998.
[427] NASA, "NASA Terminates Clark Earth Science Mission," news release, 25 February 1998; Justin Ray, "NASA Cans Clark, an Imaging Satellite," *Florida Today* (Brevard, FL), 26 February 1998.
[428] Associated Press, "BRF—Smoke on Mir," 27 February 1998.
[429] Vladimir Isachenkov for Associated Press, "Mir," 3 March 1998.

At a news conference at Ames Research Center (ARC), NASA scientists announced that the Lunar Prospector spacecraft had found "evidence of ice crystals sprinkled throughout the Moon's shadowy north and south poles." Alan B. Binder, a Lunar Prospector scientist, stated that the team believed they had detected anywhere from 10 million to several 100 million tons of water over areas of 3,600 to 18,000 square miles (9,300 to 47,000 square kilometers) at the northern pole and 1,800 to 7,200 square miles (4,700 to 19,000 square kilometers) at the southern pole. However, the ice, detected using a neutron spectrometer that scans the lunar surface for signs of hydrogen, was not in the form of ice sheets, but "present in small crystals mixed in with the moon's rocky soil, scattered across thousands of square miles in lunar polar regions." Although evidence of ice was quite strong, the water signal was relatively weak. Because the Moon is very dry, scientists theorized that comets or asteroids colliding with the Moon's surface had carried the ice to the Moon.[430]

NASA named astronaut Eileen M. Collins to command Space Shuttle *Columbia* on the upcoming Mission STS-93 in December 2008. Collins was the first woman to command a Space Shuttle mission. NASA had selected her as an astronaut candidate in 1990, and she had become an astronaut in July 1991. She flew her first mission aboard Shuttle *Discovery* in February 1995. On that mission, *Discovery* came within 30 feet (9 meters) of *Mir* during a practice run for docking with the space station. Collins was also the first woman ever to pilot a Space Shuttle, flying Shuttle *Atlantis* in May 1997, to deliver British-born U.S. astronaut C. Michael Foale to *Mir* and to return astronaut Jerry M. Linenger to Earth.[431]

Russian cosmonauts aboard *Mir* located a lost wrench among 12 years' worth of accumulated spare parts. They used the wrench to loosen the bolt that had broken three wrenches earlier in the week, preventing the two cosmonauts from undertaking a planned spacewalk.[432]

*10 March*
NASA officially declared Mars Pathfinder "dead" six months after its mission ended, after getting no response to one last set of commands. Pathfinder's mission had ended on 27 September 1997, when NASA lost communication with the craft. Pathfinder and its rover Sojourner had landed on Mars on 4 July 1997, and NASA had expected the two craft to "live" on Mars for only 30 days. However, Pathfinder had exceeded its mission by eight weeks, and Sojourner had roamed the Martian terrain 10 weeks longer than the originally planned one-week operation.[433]

*12 March*
The X-38 atmospheric vehicle completed its first unpiloted test flight successfully. Project Manager John F. Muratore commented that the successful test was "the culmination of two years of hard work by a team from the Johnson Space Center and the Dryden Flight Research Center."

---

[430] NASA, "Lunar Prospector Finds Evidence of Ice at Moon's Poles," news release 98-38, 5 March 1998; Ruth Larson, "Probe Finds Signs of Water on Moon: Prospect of Ice in Soil at Poles Tantalizes NASA," *Washington Times*, 6 March 1998; Kathy Sawyer, "Spacecraft Finds Solid Evidence of Water at Moon's Poles," *Washington Post*, 6 March 1998.
[431] NASA, "Collins Named First Female Shuttle Commander," news release 98-37, 5 March 1998; Reuters, "Woman Air Force Officer To Command Shuttle CNN," 4 March 1998.
[432] Reuters, "Mir Cosmonauts Solve Mystery of Missing Wrench," 5 March 1998.
[433] Dan Whitcomb for Reuters, "Scientists Declare Pathfinder Officially 'Dead'," 10 March 1998.

NASA had created the X-38, which could hold up to seven passengers, as an escape vehicle for the International Space Station (ISS). Described as a "whale-shaped, wingless vehicle with fins that look as if they were inspired by a 1956 Cadillac," the X-38 was the first new spacecraft in more than 20 years to reach the flight-testing stage. Engineers had designed a fully automated craft, "so that even badly injured crew members could climb in, [the vehicle would] separate from the station, and the vehicle would then serve as an ambulance, using satellite-based navigation aids" to carry passengers directly back to a designated landing field on Earth.[434]

NASA Administrator Daniel S. Goldin publicly acknowledged for the first time that the ISS would cost more than originally proposed, with a revised cost of US$21.3 billion, and would take longer to assemble than was initially planned, with a revised completion date of December 2003. Since 1993, NASA had consistently reported that the cost of the ISS would be US$17.4 billion, and had targeted a 2002 completion date. The revised cost was 200 percent more than the original US$8 billion estimate in President Ronald Reagan's 1984 proposal.[435]

*13 March*
New data from NASA's Mars Global Surveyor Mission provided insights into the planet's deeply layered terrain, its strong, localized magnetic fields, and the genesis and evolution of a Martian dust storm. Scientists analyzed data collected in October and November 1997, publishing the first set of formal results in the 13 March 1998 issue of *Science*. Mars Global Surveyor had used a variety of instruments to collect data. The thermal emission spectrometer had mapped the temperature and opacity of the Martian atmosphere during the sandstorm, and a camera had returned visual images of the storm's effects. The spectrometer had also obtained some infrared emission spectra of Mars's surface, indicating "the presence of pyroxene and plagioclase, minerals which are common in volcanic rocks, with a variable amount of dust component." Surveyor's magnetometer and electron reflectometer had measured strong localized magnetic fields on Mars. A laser altimeter had collected data from Mars's northern hemisphere, revealing a flat surface, becoming increasingly rough toward the equator. Scientists had interpreted a variety of landforms in the Mars images, including the northern polar cap and gigantic canyons, ridges, craters, and shield volcanoes.[436]

*14 March*
Progress M-38 launched on a Soyuz-V rocket from Baikonur Cosmodrome in Kazakhstan, carrying a 900-kilogram (2,000-pound) propulsion unit. *Mir*'s crew planned to attach the unit to the *Quantum* module of the space station. The cargo craft also carried 1,500 kilograms (3,300 pounds) of repair tools, replacement parts, food, and water.[437]

Minnie, the last living astro-chimp, died of old age at forty-one-years-old. Minnie had lived out her retirement at Holloman Air Force Base under the care of the Coulston Foundation. Minnie, an understudy for chimps Ham and Enos, was the only female chimpanzee to train for the

---

[434] NASA, "X-38 Atmospheric Vehicle Completes First Unpiloted Flight Test," news release 98-44, 12 March 1998; David Colker, "Emergency Space Vehicle Model Takes To Test Flight," *Los Angeles Times*, 13 March 1998.
[435] Larry Wheeler, "Station Cost Overrun Now Pegged at $6 Billion," *Florida Today* (Brevard, FL), 24 March 1998.
[436] NASA, "New Global Surveyor Data Reveals Deeply Layered Terrain, Magnetic Features and Genesis of a Martian Dust Storm," news release 98-45, 13 March 1998.
[437] *Spacewarn Bulletin*, no. 533, 1 April 1998, *http://nssdc.gsfc.nasa.gov/spacewarn/spx533.html* (accessed 20 October 2008).

Mercury Project in the early 1960s, but she had never flown in space. Ham had become the first chimpanzee in space when he had made a 15-minute suborbital flight in 1961, before Alan B. Shepard Jr.'s flight that May. Enos's orbital flight had been a precursor to John H. Glenn Jr.'s flight in February 1962.[438]

*17 March*
At the request of the United States, Israeli officials arrested Ehud Tenenbaum, the eighteen-year-old computer hacker who called himself The Analyzer, along with two other Israeli teenagers. Tenenbaum was ringleader of a group called Israeli Internet Underground. Prosecutors accused the teenage hackers of breaking into U.S. government and military computer systems, including NASA's Jet Propulsion Laboratory (JPL), as well as into Israeli government computers and commercial and educational computer systems in the United States. The FBI said that all three had admitted their involvement, stating that they had worked in tandem with two juveniles in Cloverdale, California.[439]

A Progress cargo spacecraft delivering supplies to *Mir* veered off course while in automatic mode, forcing the space station's crew to perform a manual docking. Mission Control ordered Commander Talgat A. Musabayev to perform the docking when the two craft were only 65 feet (20 meters) apart. Musabayev steered the cargo craft smoothly and avoided repeating last June's collision.[440]

Director General of the Russian Space Agency Yuri N. Koptev responded to U.S. criticism of Russian delays in constructing components of the ISS, "NASA [is] trying to 'concentrate attention entirely on Russia's difficulties' even though [NASA is] 'three months behind' in constructing a laboratory module." Koptev stated that the Russian Space Agency had only received one-fifth of the US$100 million President Boris N. Yeltsin had allocated for construction of the service module, leading to a seven-month delay in the program. In an effort to improve the Russian Space Agency's financial situation, Vice President Albert A. Gore Jr. and Prime Minister Viktor S. Chernomyrdin discussed the possibility of Russia's launching more foreign commercial satellites, with the condition that Russia would "keep its promise to stop assisting Iran's weapons program."[441]

*18 March*
NASA announced the arrest of "the suspected leader of a group of computer hackers who broke into the network of a NASA laboratory." Law enforcement agents accused twenty-year-old Calldan Levi Coffman of Carson, Washington, suspected leader of a group called ViRii, of infiltrating the networks of various corporations, universities, and other government agencies, in addition to breaching the security of computer systems at NASA's JPL last June. NASA's Computer Crimes Division had worked with other government agencies, including the U.S. Naval Intelligence Service, the U.S. Air Force Office of Special Investigations, and the FBI, to

---

[438] *Los Angeles Times*, "Minnie, Last of the Early U.S. Space Program 'Astro-Chimps', Dies at 41," 28 March 1998.
[439] Mike Billington for United Press International, "Hacker 'Confederacy' Hits Pentagon," 20 March 1998; Michael J. Sniffen for Associated Press, "US-Israel-Analyzer," 20 March 1998; *Washington Post*, "Five Teens Suspected of Hacking into Pentagon Computers," 20 March 1998.
[440] Reuters, "Focus: Technical Failure Mars Russian Space Docking," 17 March 1998; Vladimir Isachenkov for Associated Press, "Mir," 17 March 1998.
[441] *Houston Chronicle*, "Russian Official Blames NASA for Delaying Space Station," 18 March 1998.

uncover a link between Coffman's group and Israeli hacker Ehud Tenenbaum, known as The Analyzer. Law enforcement officials suspected Tenenbaum of organizing a confederation of hacker groups, including his own group and Coffman's group.[442]

*19 March*
NASA's Office of Earth Science Enterprise selected the U.S. Naval Research Laboratory in Washington, DC, and the University of Colorado's Laboratory for Atmospheric and Space Physics to conduct "parallel six-month definition studies of a new small satellite to monitor variations in the amount of radiant solar energy that reaches Earth," as part of the preparations for the Total Solar Irradiance Mission.[443]

NASA officials appeared before a hearing of the U.S. House Committee on Science and Technology to explain the increasing price tag of the ISS. The anticipated cost of the ISS had increased to US$4 billion more than originally projected, drawing ire from supporters and opponents alike. NASA's new Associate Administrator for Space Flight Joseph H. Rothenberg led the lawmakers through the reasons for the increasing budget, explaining the series of revised out-year estimates. He attributed part of the problem to cost overruns related to "recurring schedule and budget problems with the Russian [S]pace [A]gency," and part to the fact that NASA's primary contractor, the Boeing Company, was already US$600 million to US$817 million over budget. However, Rothenberg stated, "only a fraction of these revised out-year estimates are tied to actual overruns."[444]

At a NASA Advisory Council meeting at Marshall Space Flight Center, Jay Chabrow, an aerospace industry consultant, presented findings of a study conducted to determine the true costs of assembling and maintaining the future ISS. U.S. Senator John McCain (R-AZ) and other concerned lawmakers had tasked Chabrow and his seven-member team of private and government analysts with undertaking this investigation, conducted over five months. The analysts' findings showed that the ISS could cost as much as US$6 billion more than NASA officials had estimated, reaching a total of US$24 billion, and that the outpost likely would not be ready to house a full crew until 2006. The report also provided estimates regarding the total cost of the space station over its entire lifetime. The analysts projected that Space Shuttle missions to assemble the station, resupply flights, and staff operating the Shuttle from the ground, would cost NASA US$94 billion. The study also found that planned caps on congressional spending were unrealistic, and that NASA and Boeing officials had been overly optimistic in their cost and schedule estimates. Factors driving costs higher included Russian nonperformance, software development and integration, construction of a crew return vehicle, and Boeing cost overruns. The report did not account for further cost increases in the event of a total Russian pullout.[445]

*20 March*

---

[442] Associated Press, "NASA Hacker," 18 March 1998; Billington for United Press International, "Hacker 'Confederacy'," 20 March 1998.
[443] NASA, "Two Studies Will Refine and Expand Solar Monitoring Task," news release 98-47, 19 March 1998.
[444] Brett Davis, "Space Station Costs Irk Congress," *Huntsville Times* (AL), 20 March 1998; Tamara Lytle, "U.S. Legislators Criticize Increasing Cost of NASA's Space Station," *Orlando Sentinel* (FL), 20 March 1998.
[445] Wheeler, "Station Cost Overrun."

Boeing Company announced plans to cut an additional 8,200 jobs by the end of the year 2000, to streamline facilities, to focus on manufacturing and assembly operations, and to eliminate redundant laboratories. The job cuts, affecting work sites in Arkansas, California, and Texas, included planned product phaseouts of the MD-80 and MD-90, and brought into question the future of the MD-11. As it shut down selected facilities, including an electronic manufacturing facility and the Site 9 modification center, Boeing planned to transfer thousands of employees involved with aircraft and missile systems, reusable-space-systems support, and satellites.[446]

*22 March*
NASA marked its second full year of "continuous human presence in Earth orbit" aboard Russia's *Mir* space station. Shuttle *Atlantis* had launched on 22 March 1996, with U.S. astronaut Shannon W. Lucid aboard. Lucid then spent 188 days in space, an American record. She was the second NASA astronaut to live aboard *Mir*. Following her tenure, an unbroken succession of astronauts, John E. Blaha, Jerry M. Linenger, C. Michael Foale, David A. Wolf, and Andrew S. W. Thomas, had spent time on *Mir*. Norman E. Thagard, the first American astronaut to live on *Mir*, had arrived in March 1995, spending 115 days on *Mir* before Lucid's stay. Nine months separated Thagard's mission and Lucid's. Frank L. Culbertson Jr., head of NASA's cooperative human spaceflight program with Russia, commented, "an unbroken presence has been extremely valuable to our having an awareness of what it takes to fly a space station, to maintain an outpost on the frontier."[447]

*25 March*
The Boeing Company appointed Michael T. Kennedy to a newly created position, Vice President for the Boeing Evolved Expendable Launch Vehicle and Delta IV Programs. Since 1996, Kennedy had been Vice President of Delta III and had shown "his ability to manage launch vehicle development effectively by bringing Delta III on line for its first launch this summer." As Kennedy moved into his new position, Boeing promoted Daniel J. Collins to Program Director for Delta III, and Jay L. Witzling became Vice President of Delta II and Titan Programs. Kennedy had joined Boeing in 1970, working on the Skylab, Saturn SIVB, Titan, Delta, SPACEHAB, and ISS programs.[448]

*26 March*
An international team of researchers announced that observations of U.S. satellites orbiting Earth had proven one of physicist Albert Einstein's theories correct. In his general theory of relativity, Einstein had predicted that a spinning body could curve space, because the "spin of a body must change the geometry of the universe by generating space-time curvature." Einstein had called the phenomenon frame dragging, but scientists came to refer to it as the Lense-Thirring effect, naming it for two Austrian physicists who wrote that rotating celestial bodies, such as the Sun, "create a force that pulls space toward them." Ignazio Ciufolini, a physicist at Sapienza University of Rome, and his colleagues used lasers to measure changes in the orbits of the Laser Geodynamics Satellite I, or LAGEOS, a NASA spacecraft, and LAGEOS II, a satellite belonging to NASA and the Italian Space Agency. The research team observed changes that they could not

---
[446] United Press International, "Boeing Cuts 8,200 Jobs," 20 March 1998.
[447] Mark Carreau, "NASA Marks Second Year in Orbit: Continuous Human Presence in Space Accomplished Aboard Mir," *Houston Chronicle*, 21 March 1998.
[448] PR Newswire, "Boeing Selects Vice President for EELV/Delta IV Program," 26 March 1998.

account for by the laws of gravity or tidal forces, concluding that the Lense-Thirring effect exists. The satellites, launched in 1976 and 1992, were passive satellites dedicated exclusively to *laser ranging*. In this process, scientists had send laser pulses to the satellite from ranging stations on Earth, measuring the round-trip travel time of the pulses. The research team had analyzed data collected over a four-year period beginning in 1993.[449]

*27 March*
NASA awarded a contract worth US$148 million to Raytheon STX Corporation. The five-year award included a one-year base period followed by four one-year options. Contract services for providing federal information processing services at NASA's ARC included software development and maintenance, engineering operations and maintenance, network services, analysis, quality assurance, and reviews.[450]

France's Aérospatiale, British Aerospace, Germany's Daimler-Benz Aerospace, and Spain's CASA, partners in the Airbus Industrie airplane-manufacturing consortium, announced their intention to merge into a "unified civilian and military aerospace conglomerate that would compete in the world market with U.S. aerospace giants Lockheed Martin Corp[oration] and Boeing Co[mpany]" The partners provided no details regarding how the merger would take place, but the public statement represented "another move toward a unified European aerospace industry," an idea that had been elusive "because of the competing national interests of the various countries."[451]

*30 March*
Astronomers at the University of Manchester in England and the Hubble Space Telescope (HST) announced that six radio telescopes, in concert with the HST, had "captured the first image ever recorded of an unbroken 'Einstein ring'," a "circle of light created by the gravitational warping of space between a very distant galaxy and the earth." Physicist Albert Einstein's general theory of relativity had predicted that "light from a distant object would travel a curving path if it passed close to a massive object along the line of sight from Earth." The radio image showed a fragmented group of bright arcs, but the HST showed that the galaxy's infrared emission region was perfectly in line with the gravitational lens and Earth, confirming the existence of the Einstein ring.[452]

## APRIL 1998

*1 April*
In a spacewalk lasting nearly 7 hours, *Mir* cosmonauts failed to stabilize the solar panel damaged in a collision in 1997. Russian space officials had postponed this spacewalk from 3 March, when the crew had been unable to unlock the hatch, breaking three wrenches in their attempts. Russian Space Agency cosmonauts, Kazakh Talgat A. Musabayev and Russian Nikolai M. Budarin, spent most of the spacewalk setting up a handrail, left outside the space station by the previous crew.

---

[449] NASA, "Earth Dragging Space and Time as It Rotates," news release 98-51, 27 March 1998; Reuters, "Einstein Proved Right Yet Again," 26 March 1998.
[450] NASA, "NASA Awards $148 Million Contract to Raytheon STX Corp," contract announcement, 27 March 1998.
[451] Tim Smart and Anne Swardson, "European Space Firms Agree To Merge," *Washington Post*, 28 March 1998.
[452] Malcolm W. Browne, "'Einstein Ring' Caused by Space Warping Is Found," *New York Times*, 31 March 1998.

After struggling to unfold and assemble the handrail, the pair only had enough time to install one of two footrests. The cosmonauts returned to *Mir* after the elapse of 6 hours and 40 minutes; their oxygen tanks only held approximately 7 hours of air. Russian Mission Control had intentionally turned off the audio system during most of the spacewalk. Officials said that Mission Control would no longer allow journalists to listen to radio traffic between Mission Control and spacewalkers, because reporters had become overly critical. Russian officials accused reporters of exaggerating problems on the space station, despite the fact that *Mir* had been operating relatively trouble free for several months before this spacewalk.[453]

An Orbital Sciences Corporation Pegasus XL rocket launched from Vandenberg Air Force Base in California, carrying a NASA satellite to study solar events that could endanger astronauts and spacecraft. The Small Explorer (SMEX) satellite program, under the management of NASA's Goddard Space Flight Center, had developed the Transition Region and Coronal Explorer (TRACE) satellite.[454]

*2 April*
Eberhard R. M. Rees, deputy and later successor to Wernher von Braun as "chief of American rocketry efforts," died at age 89, after suffering from Alzheimer's disease. Rees, born in Germany, had graduated from the Dresden Institute of Technology in 1934 with an advanced degree in mechanical engineering. He had worked as a technical plant manager with the German Guided Missile Center beginning in 1940, but after Germany's defeat in World War II, Rees became one of 118 scientists who surrendered to the West under the so-called Project Paperclip, becoming a U.S. citizen in 1954. In the United States, he had first worked at White Sands Proving Ground in New Mexico, "where he shared his expertise on the V-2, a German rocket-propelled bomb, and on guided missile projects." In 1950 Rees had moved to Huntsville, Alabama, to serve as Deputy Chief of the Guided Missile Development Division at Redstone Arsenal. Among the projects Rees had managed were the Hermes II Project, the Redstone and Jupiter missile programs, the Explorer satellite project, and the Saturn program. Between 1956 and 1960, Rees had been Deputy Director of the U.S. Army Ballistic Missile Agency's Development Operations Division. He had received a Distinguished Service Medal from NASA for his role in the Apollo 11 Mission. Rees had succeeded Wernher von Braun as Director of NASA's Marshall Space Flight Center in 1970, retiring in 1973.[455]

*3 April*
The Organization of American Historians presented the 1998 Richard W. Leopold Prize to Andrew J. Butrica for his book, *To See the Unseen: A History of Planetary Radar Astronomy*. The organization awarded the biennial Richard W. Leopold Prize "for the best book written by an historian connected with federal, state, or municipal government in the areas of foreign policy, military affairs, the historical activities of the federal government, or biography in one of these areas." Butrica's book, a comprehensive history of planetary radar astronomy, was a

---

[453] Vladimir Isachenkov for Associated Press, "Mir Crew Undertakes Six-Hour Spacewalk To Repair Solar Panel," 1 April 1998; Vladimir Isachenkov for Associated Press, "Mir," 2 April 1998; Vladimir Isachenkov for Associated Press, "Mir Crew Runs Out of Time, Fails To Stabilize Solar Panel," 1 April 1998.
[454] Associated Press, "BRF—Satellite Launch," 2 April 1998; PR Newswire, "Orbital's Pegasus Rocket Successfully Launches TRACE Spacecraft for NASA; Advanced SMEX Satellite Will Study the Sun; Spacecraft's First Signals Indicate Early Success," 2 April 1998.
[455] Ford Burkhart, "Eberhard Rees, Rocketry Pioneer, Dies at 89," *New York Times*, 4 April 1998.

publication of the NASA History Series produced by the NASA Headquarters History Division's Office of Policy and Plans.[456]

*6 April*
NASA released images recorded by the Mars Global Surveyor spacecraft, showing the Cydonia region of Mars in 10 times more detail than that provided in the images recorded more than 20 years ago by one of the U.S. Viking space probes. In the new images, the Sun was shining from the opposite direction than in the images recorded by the Viking probe. Viking's images had revealed a surface formation in the hilly desert area, Cydonia, which some members of the public believed to be a monumental sculpture of a humanoid face, in spite of NASA scientists' insistence that the "face" was only a common geological formation called a mesa. Some of those people who believed the structure to be a sculpture had accused NASA of manipulating data in a conspiracy to keep the origins of the face a secret. Surveyor's new images confirmed NASA's previous analysis of the formation. However, in an effort to counteract the conspiracy rumors, NASA posted on the Internet the raw data sent by Mars Global Surveyor and took no official position regarding what the images did or did not reveal. Several planetary scientists remarked that "the new image contained no surprises and no evidence of artificial origin."[457]

Cosmonauts Talgat A. Musabayev and Nikolai M. Budarin, trying for a third time, successfully stabilized a bent solar panel during a spacewalk of 4 hours and 15 minutes. The solar panel was only one of 10 nonfunctioning panels. The crew had not intended to repair the damaged panel fully, but only to stabilize it, so that it would not break loose and hit the space station. The cosmonauts cut short the spacewalk, scheduled to take 5½ hours, when the thruster engine aligning *Mir* with the Sun ran out of fuel. Forced to return to the space station early, the cosmonauts were unable to remove some scientific equipment attached to the outside of the station.[458]

*7 April*
The European Space Agency (ESA) announced that Europe's Infrared Space Observatory, which had been orbiting Earth for two and one-half years, had located water in unexpected places. The most intriguing of these was the water vapor detected in the atmosphere of Saturn's largest moon, Titan. ESA Director of Science Roger M. Bonnet remarked, "the moon may duplicate the conditions that led to the creation of life on Earth . . . and the only thing you need is a little heat to heat up Titan, maybe [the] birth of life may be seen." Chairperson of ESA's Astronomy Working Group Reinhard Genzel said that the Observatory had "enabled scientists to see water throughout the universe," and that the water was likely brought "to the icy outer planets by comets." He explained further that comets might have deposited the water on Earth during the early formation of the solar system, when many comets constantly collided with the Earth.[459]

*8 April*

---

[456] NASA, "NASA History Wins Prestigious Prize," news release 98-53, 3 April 1998.
[457] Robert Lee Hotz, "California and the West Alien 'Face' Recedes into Martian Myth Space: NASA Photo Appears To Show a Naturally Weathered Mesa," *Los Angeles Times*, 7 April 1998; Kathy Sawyer, "With Mars Images, NASA Says: Face It, It's a Mesa," *Washington Post*, 7 April 1998.
[458] *Florida Today* (Brevard, FL), "Mir Crew Shortens Spacewalk After Thruster Fuel Runs Out," 7 April 1998.
[459] Edith M. Lederer for Associated Press, "Europe-Space Water," 7 April 1998; *Washington Times*, "Europe's Space Surveyor Finds Water Vapor Near Stars, Planets," 8 April 1998.

NASA announced the selection of nine researchers to form the Science Team for the Mars Microprobes, "a technology validation mission that will hitchhike to the [R]ed [P]lanet aboard NASA's 1998 Mars Polar Lander mission." NASA intended the probes to collect data measuring thermal and physical properties of the soil while searching for the presence of water ice. The nine selected scientists were David C. Catling, Julio A. Magalhaes, Jeffrey E. Moersch, James R. Murphy, and Aaron Zent of NASA's Ames Research Center; Ralph D. Lorenz of the University of Arizona; Paul Morgan of Northern Arizona University; Bruce C. Murray of California Institute of Technology; and Marsha Presley of Arizona State University.[460]

*9 April*
U.S. astronomers studying star formation reported that a body resembling a pinkish star, at the center of the sword in the constellation Orion, was "a hotbed of star formation called the Orion Nebula," the site of a gas cloud generating enough water "to fill the Earth's oceans 60 times a day." The team of astronomers, led by Martin O. Harwit of Cornell University, used the Infrared Space Observatory to observe the Orion Nebula, detecting "large amounts of water vapor within clouds of gas on their way to becoming stars in the nebula." The discovery, which demonstrated for the first time "the vital role that water plays in star formation," may "provide an important clue about the source of water in the [s]olar [s]ystem." Michael Werner of NASA's Jet Propulsion Laboratory (JPL) emphasized the significance of the discovery, saying that the data confirmed, "that water is prevalent in space" and "plays a vital role in the chemistry and physics of interstellar clouds."[461]

*15 April*
At a Kennedy Space Center (KSC) news conference, Space Shuttle Program Manager Thomas W. Holloway announced that an astronaut's error had caused a US$10 million solar satellite to malfunction after Space Shuttle *Columbia* released the satellite in November 1997 during Mission STS-87. An investigation revealed that astronaut Kalpana Chawla had not sent the *Spartan* satellite a key signal to activate itself. Chawla's efforts to recapture *Spartan* with the Shuttle's robotic arm had sent the satellite spinning, and two astronauts had to undertake a spacewalk to retrieve it. The review panel, comprising both external experts and NASA staff, found that Chawla's crewmates "could have helped her verify the command" but did not do so "because such vigilance was not part of the plan." The panel made nine major recommendations intended to help both the Shuttle and the International Space Station (ISS) programs. Holloway announced that, in response to the board's findings, NASA would make numerous changes to training, computers, and procedures.[462]

*16 April*
James K. Yungel, a NASA scientist studying Earth remote sensing, discussed a joint project underway at Monterey Peninsula Airport, involving NASA, the U.S. Geological Survey, and the National Oceanic and Atmospheric Administration. The researchers had parked at the airport a

---

[460] NASA, "Science Team Chosen for Technology Validation Mission To Explore the Subsurface of Mars," news release 98-59, 8 April 1998.
[461] Elizabeth Manning for United Press International, "Massive Water Find Linked to Star Birth," 9 April 1998.
[462] Craig Covault, "Shuttle/Spartan Verdict Sparks Station Concerns," *Aviation Week and Space Technology* 148, no. 16 (20 April 1998): 26; Seth Borenstein, "Astronaut Gets Blame Lack of a Signal Causes Satellite To Fail," *Orlando Sentinel* (FL), 16 April 1998.

Twin Otter airplane, with a scanning laser system and a global-positioning-satellite receiver, to make precise measurements of the West Coast, mapping the changes El Niño had caused along the coastline. The goal of the survey was to create "the most highly detailed map yet of the West Coast," increasing scientists' ability to predict erosion over the next decade.[463]

Jeremy Lin of Taiwan's Academia Sinica Institute and colleagues at the National Radio Astronomy Observatory in Socorro, New Mexico, reported in the journal *Nature* that measurements taken with the world's largest radio telescopes—the Very Large Array (VLA)—had detected "peculiar bulges in the atmosphere of a giant star." Huge plumes of gas thrust from beneath the surface of the red supergiant star, Betelgeuse, had likely caused the bulges. Betelgeuse is located in the shoulder of the constellation Orion, about 430 light-years from Earth. The research team used the Y-shaped, 22-mile-wide (35-kilometer-wide) cluster of 27 huge antennas of the VLA telescopes to capture the most detailed radio image ever taken of a star other than the Sun. The team discovered that some of the gas in the star's atmosphere was much cooler than previously believed—about 5,750°F (3,200°C). The discovery of the lower temperatures enabled astronomers to understand "how huge amounts of dust are constantly blown away from the star." Before the lower temperatures were identified, scientists had not been able to explain how the stellar dust formed, because at "higher temperatures the dust could not condense from the hot gas expelled from the star's interior." Astronomers believe that dust created and expelled by stars like Betelgeuse "is distributed throughout the universe and provide[s] the raw materials that gave rise to life on Earth."[464]

*17 April*
After problems with one of two network-signal processors caused a 24-hour delay, Space Shuttle *Columbia* (STS-90) blasted off on the Shuttle program's final Neurolab mission, carrying 26 experiments for basic research in neurosciences to "expand understanding of how the nervous system develops and functions in space." Test subjects aboard the Shuttle included crew members, rats, mice, crickets, snails, and two kinds of fish. The Neurolab mission was the joint effort of NASA, several U.S. partners, the space agencies of Canada, France, and Germany, the ESA, and the National Space Development Agency of Japan. Crew members included Commander Richard A. Searfoss; Pilot Scott D. Altman; Mission Specialists Richard M. Linnehan, Dafydd Rhys "Dave" Williams, and Kathryn P. Hire; and Payload Specialists Jay C. Buckey Jr. and James A. Pawelczyk. Mission Specialist Hire was the first KSC employee that NASA had selected as an astronaut candidate. Mission STS-90 was *Columbia*'s 25th flight.[465]

*21 April*
Two scientific teams, independently studying HR 4796, a star in the southern constellation Centaurus, announced they had captured "the strongest evidence yet" of a new group of planets forming around the star. The teams used the 10-meter (33-foot) W. M. Keck II Telescope on top of the extinct Mauna Kea volcano in Hawaii, and the National Science Foundation's 4-meter

---

[463] Michael McCabe, "NASA Mapping West Coast in Wake of El Niño: Most Rigorous Survey Ever Should Help Predict Erosion," *San Francisco Chronicle*, 17 April 1998.
[464] Malcolm W. Browne, "Atmosphere Bulges on None-Too-Distant Star," *New York Times*, 21 April 1998.
[465] NASA, "Neurolab Shuttle Mission To Launch April 16," news release N98-25, 3 April 1998; NASA, "Mission Archives: STS-90," *http://www.nasa.gov/mission_pages/shuttle/shuttlemissions/archives/sts-90.html* (accessed 11 June 2007).

(13-foot) Blanco Telescope at the Cerro Tololo Inter-American Observatory in Chile. Using both of these sensitive new instruments, the research teams believed they could see clearly, through the glare of the star, a planet-forming disc of gas and dust. A third international team of British and American astronomers released the first images of the huge discs, which appeared to be forming around two other stars, Fomalhaut and Vega, the star made famous in the novel and movie *Contact*. The third team used the 15-meter (49-foot) James Clerk Maxwell Telescope, also atop Mauna Kea. Scientists considered the discoveries of the discs around the stars a missing link in the study of planetary system formation, offering them clues about the formation of rocky planets like Earth.[466]

*22 April*
An international group of computer hackers, calling itself Masters of Downloading or MOD, announced that it had stolen key software programs from NASA. The same group had previously broken into the Pentagon's Defense Information Systems Network, claiming to have stolen enough information to "'take control' of military satellites and other systems." Although U.S. Department of Defense officials said that MOD had downloaded an application for management and record keeping, not for "anything that could perform a control function," the group claimed that its cyber attack on NASA had "stripped the U.S. space agency of its chief defense against computer intrusion." MOD had sent the stolen samples of NASA software to a computer expert who maintained a Web site devoted to information-security issues. The group, which included at least two Russian members, asserted that it had broken into NASA's system through JPL to obtain the software samples and threatened to sell the information to international terrorist groups or foreign governments.[467]

*23 April*
The Cost Assessment and Validation Task Force, an independent task force convened at the request of NASA and charged with analyzing the cost and schedule of constructing and maintaining the ISS, released its official report. The task force concluded that "program size, complexity, and ambitious schedule goals were beyond that which could be reasonably achieved within the [US]$2.1 billion annual cap or [US]$17.4 billion total cap." The task force concluded that "the International Space Station will be delayed by up to three years and cost as much as [US]$250 million more per year relative to the FY 1999 budget submission," recommending that NASA "revise the current [s]tation plan so that it is achievable with the funds available." NASA Administrator Daniel S. Goldin, who had appointed Jay Chabrow to lead the task force, told members of the U.S. Senate Appropriations Subcommittee on Veterans Affairs, Housing and Urban Development, and Independent Agencies that he would not accept the task force's conclusions, because he had not yet fully reviewed the report. Goldin told the panel that everyone involved in the space station program was very concerned about cost overruns and schedule slips, but that if Congress imposed a cap on ISS funding, it could affect the schedule significantly, delaying the project up to two years. Goldin warned, "such a funding shortfall would hurt relations with the international partners." The report also placed much of the blame for the budget and schedule issues on Russia. In 1993 the Clinton administration and top officials at NASA had advised Congress that, were Russia not involved in the project, NASA would

---
[466] Kathy Sawyer, "Oh, Baby! Telescopes Show the Birth of a Solar System; New Imaging Sharpens Delivery Room View," *Washington Post*, 22 April 1998.
[467] *Orlando Sentinel* (FL), "NASA Becomes Latest Victim of Masters of Downloading," 23 April 1998.

"save an estimated [US]$2 billion, expand the station's capabilities and enable the research station to be completed sooner." However, the task force's report concluded that "those assumptions were faulty."[468]

*29 April*
Astronauts and physicians Jay C. Buckey Jr. and Dafydd Rhys "Dave" Williams performed the first surgeries ever conducted on living specimens while in orbit. The two doctors injected chemical markers into the hind legs of six of the surviving baby rats to determine the role of gravity in the development of muscles. The crew anesthetized the baby rats before injecting them with the pink florescent chemical, the first time scientists had performed such a procedure in zero-gravity conditions. In addition to studying how muscles and nerves develop in space, the crew studied whether anesthesia really works without the help of gravity, whether patients have more difficulty recovering from anesthesia in space, and how wounds heal in space. Two other astronauts successfully completed "the most difficult medical procedure ever attempted on someone in space." The astronauts guided tiny needles into the nerves behind their knees, in an experiment aimed at understanding and easing the blood pressure problems and dizziness that astronauts suffer after returning to Earth. An estimated 500,000 Americans experience the same symptoms every day when they stand up. The needle enabled the astronauts to "listen" to electrical impulses, as the brain sent messages to the nerve.[469]

## MAY 1998

*1 May*
At the 13th annual NASA Continual Improvement and Reinvention Conference on Quality Management, NASA Administrator Daniel S. Goldin presented five aerospace companies with the 1998 George M. Low Award, NASA's highest "quality and excellence" award for contractors and subcontractors. The companies, representing both large and small businesses, were ILC Dover Inc. of Frederica, Delaware; Allied Signal Technical Services Corporation of Lanham, Maryland; DynCorp, Johnson Support Division of Houston, Texas; BST Systems Inc. of Plainfield, Connecticut; and Advanced Technology Company of Pasadena, California. Administrator Goldin said, "these companies exemplify excellence and outstanding achievements that prove beneficial to NASA and the [n]ation's industry." Associate Administrator for Safety and Mission Assurance Frederick D. Gregory added, "each of these companies has definitely made a positive impact on NASA's performance goals."[470]

*2 May*
The Chinese Long March 2C rocket launched two U.S. satellites into orbit, from the Taiyuan Satellite Launch Center in Shanxi province in northern China. The satellites were part of a

---

[468] NASA Advisory Council, "Report on the Cost Assessment and Validation Task Force on the International Space Station," 21 April 1998, *http://history.nasa.gov/32999.pdf* (accessed 24 September 2007); Larry Wheeler, "Russia's Role in Station Has Project over Budget, Behind Schedule," *Florida Today* (Brevard, FL), 24 April 1998; *Aerospace Daily*, "Goldin Not Sold on Report Finding Space Station Overruns," 24 April 1998.
[469] Seth Borenstein, "Shuttle Surgeons Hope Rats Will Survive," *Orlando Sentinel* (FL), 29 April 1998; Steven Young for Reuters, "U.S. Shuttle Crew Performs First Space Surgery," 29 April 1998; Marcia Dunn for Associated Press, "Space Shuttle," 30 April 1998; Robyn Suriano, "Astronauts Listen to Brain-Nerve Chatter," *Florida Today* (Brevard, FL), 30 April 1998.
[470] NASA, "NASA Awards Five Firms the George M. Low Award," news release 98-74, 1 May 1998.

mobile telecommunications system that Motorola Corporation had developed for the Iridium World Communications network scheduled to begin commercial service on 23 September 1998. The Iridium contract had "boosted China's launching industry," which had been "struggling to recover from a string of launch disasters in 1995 and 1996."[471]

*3 May*
Returning from the Neurolab mission, Space Shuttle *Columbia* landed smoothly, despite a malfunctioning hydraulic unit. Most of *Columbia*'s 1-hour-long descent relied on two hydraulic units, because the cooling system for the third unit had failed. To prevent the third unit from overheating, the crew turned it on just minutes before landing. After the landing, Mission Control congratulated the Neurolab astronauts on "a historic mission that elevated neuroscience research to record heights." During the 16-day Neurolab flight, the team had accomplished a number of experiments never before performed in space, including the first direct nerve recordings, the first joint recording of sleep and breathing, the first embalming of animals, and the first surgery on animals. Because only the rodents aboard *Columbia* were accessible during the flight, astronauts and scientists did not immediately know how the other animals, including fish, snails, and crickets, had fared. A quick inspection after landing, however, "revealed that only 25 of 225 baby swordtail fish survived the mission, a much higher mortality rate than expected." The baby rats also had a higher mortality rate than expected; 52 of the 96 baby rats had died in orbit after their mothers refused to nurse them. NASA veterinarians had expected a dozen deaths from natural causes. Animal rights groups criticized the experiments.[472]

*6 May*
At a NASA news conference, astronomers discussed the discovery of the largest explosion ever sighted in space. The journal *Nature* published three papers about the event, detected on 14 December 1997. The explosion, a gamma-ray burst (GRB) designated GRB971214, had occurred 12 billion years ago and had "apparently outshone all the rest of the universe" for about 40 seconds. The Italian-Dutch orbiting gamma-ray observatory BeppoSAX had measured the sharp pulse of gamma rays and "pinpointed the position of the rays' source." The United States' Compton Gamma Ray Observatory satellite had also detected the gamma-ray pulse, and scientific institutions around the world had monitored the after-effects of the explosion. Twelve hours afterward, John R. Thorstensen of Dartmouth College had used a 94-inch-diameter (49-centimeter-diameter) telescope at Kitt Peak Observatory in Arizona to detect a visible afterglow, which lasted for about two weeks. After the glow had faded, the scientists had discovered a faint galaxy in the same location. A team of American, Italian, and Indian astronomers, led by two California Institute of Technology scientists, used the Keck II Telescope in Hawaii to measure the distance from Earth to the source of the burst. No theorists were able to explain the explosion, and no existing models explaining GRBs could explain the amount of energy emitted from GRB971214. The scientists suggested the possibility that a superdense neutron star had

---

[471] Associated Press, "China Puts 2 Satellites into Orbit," 3 May 1998.
[472] *USA Today*, "Scientists To Study Effects on Critters," 4 May 1998; Paul Hoversten, "Mission's Data Has Scientists Feeling Antsy: Neurolab Analysis Begins Today," *USA Today*, 4 May 1998.

ripped apart as it disappeared into a neighboring black hole, or that a violent merger of two black holes had caused the burst.[473]

*7 May*
The U.S. House Committee on Science and Technology held a hearing about the International Space Station (ISS) project, focusing on the independent Cost Assessment and Validation Task Force, headed by Jay Chabrow. NASA Administrator Daniel S. Goldin had selected Chabrow to lead the investigation into the cost of the ISS. The task force's findings concluded that the space station may cost US$7 billion more than the original estimate and may take two years longer than anticipated to complete. Congressional supporters and opponents of the program questioned NASA officials about the project's cost and slipping timetable, but officials indicated that they needed a few more weeks to study Chabrow's report and were unable to say whether they agreed with his team's figures. However, Goldin acknowledged that Russia's participation in the program was not working. He also promised to provide Congress with a new assessment of costs and with assembly schedules by 15 June, including estimates for continuing the project without critical Russian equipment. House Committee on Science and Technology Chairperson F. James Sensenbrenner Jr. (R-WI) seemed mollified after Goldin had spoken. However, Sensenbrenner "chastised the Clinton administration," remarking that its failure to send a representative from the Office of Management and Budget to the hearing reflected "the low priority" the administration placed on solving the problems facing the ISS program.[474]

A U.S. Air Force Titan IV rocket launched from Cape Canaveral, Florida, carrying a spy satellite. Civilian observers speculated that the spacecraft's name was *Orion*, and that it would "play an important role in listening in on foes of the United States." With a wire mesh antenna the size of a football field, the satellite was so sensitive that it could "detect transmissions from a radio the size of a wristwatch or a very small cell phone." The satellite could also monitor electronic transmissions between military bases. At the time of the launch, the Titan was the most powerful American booster in use.[475]

Launched from Baikonur Cosmodrome, a Proton-K Russian booster carried into orbit an American *EchoStar IV* communications satellite, to relay broadcast signals to the central and western United States. Lockheed Martin Corporation had built the satellite and, under a partnership agreement between Lockheed and the Russian Khrunichev company, International Launch Services launched the spacecraft.[476]

*10 May*
NASA dedicated as a U.S. Air Force conference center the newly renovated, historic control center, Complex 14 at Cape Canaveral Air Force Station in Florida, the site of the Mercury program launches. The domed blockhouse with 10-foot-thick (3-meter-thick), reinforced concrete walls had protected launch controllers from possible explosions of Atlas rockets

---

[473] Malcolm W. Browne, "Astronomers Detect Immense Explosion 2d Only to Big Bang," *New York Times*, 7 May 1998; NASA, "Most Powerful Explosion since the Big Bang Challenges Gamma Ray Burst Theories," news release 98-75, 6 May 1998.
[474] Warren E. Leary, "Space Plan's Costs Increase amid Delays," *New York Times*, 7 May 1998; Larry Wheeler, "Goldin Rips Russians, Promises New Station Schedule by June 15," *Florida Today* (Brevard, FL), 7 May 1998.
[475] Robyn Suriano, "Titan Rocket Sends Spy Satellite into Orbit," *Florida Today* (Brevard, FL), 9 May 1998.
[476] Associated Press, "U.S. Satellite Launched from Russian Rocket," 9 May 1998.

launched only 1,000 feet (305 meters) away. The renovation preserved the periscopes that controllers had used to safely view launches. The control center was at the launch site of the first piloted orbital flight of astronaut John H. Glenn Jr.[477]

*13 May*

The National Oceanic and Atmospheric Administration (NOAA) launched into orbit, from Vandenberg Air Force Base in California, a new weather satellite, the NOAA-K, aboard a 34-year old converted Titan II nuclear missile. The satellite was an Advanced Television Infrared Observation Satellite (TIROS-N) weather satellite, able to see inside clouds, a feature that scientists expected would greatly increase our knowledge of how hurricanes develop and move. The TIROS-N was the latest model of TIROS satellite in four generations. Lockheed Martin had last built a TIROS satellite for NASA and NOAA in 1960. After the launch of the TIROS-N, the satellite would undergo a series of tests by NASA, followed by a round of tests by NOAA, before becoming fully available in time for next year's hurricane season.[478]

*14 May*

At a NASA press conference, Ethan J. Schreier of the Space Telescope Science Institute in Baltimore and his colleagues discussed new Hubble Space Telescope (HST) images. The new images showed in greater clarity an old galaxy's black hole "feeding on a smaller, younger galaxy that collided with it." The scientists explained that the new images revealed young stars in the midst of an "obviously old galaxy," suggesting that a collision had occurred. The merged galaxy complex, known as Centaurus A, was about 10 million light-years away from the Milky Way. The images, taken with HST's newest infrared camera, also showed gas and dust feeding an apparent large black hole. An inner disc of gas, swirling around the black hole, was behaving in a manner opposite to scientists expectations, based upon the behavior of other galaxies. Hans-Walter Rix of the University of Arizona theorized that the collision could explain the difference, postulating that the HST "may have caught Centaurus A in the midst of trying to realign itself after its spiral companion plowed into it." Schreier offered an alternative explanation—that the mass of the merged galaxies was greater than the mass of the black hole, causing the gravitational pull to warp the inner disc.[479]

The Iridium World Communications system global, digital wireless communications network completed a series of deployment missions, with the launch from Vandenberg Air Force Base in California of five Iridium satellites aboard a Boeing Delta II launch vehicle. The successful launch had completed the entire Iridium constellation "in just twelve months." The world's first global, wireless telephone company scheduled its commercial service to begin four months following this last launch, once it had completed integrating and testing the Iridium system.[480]

*18 May*

---

[477] *USA Today*, "Cape Canaveral Restores Piece of Space Age History," 11 May 1998.

[478] Justin Ray, "New NOAA Weather Satellite Will Keep a Better Eye on Hurricanes," *Florida Today* (Brevard, FL), 14 May 1998; Business Wire, "Nation's Newest Advanced Weather Satellite Built by Lockheed Martin Launched Successfully," 14 May 1998.

[479] Elizabeth Manning for United Press International, "Hubble Reveals Feeding Black Hole," 14 May 1998; NASA, "Hubble Provides Multiple Views of How To Feed a Black Hole," news release 98-71, 14 May 1998.

[480] PR Newswire, "Boeing Delta II Rocket Places Iridium Satellites into Orbit," 17 May 1998.

NASA successfully tested a fuel tank made of a new metal mixture, aluminum lithium, never before used in major U.S. manufacturing. The tank made of the new alloy was 30 percent stronger and 5 percent lighter than the aluminum magnesium tanks in use since the days of the Apollo program. The tank showed no leaks, pressure bulges, or other problems, when loaded with 537,000 gallons (2,033,000 liters) of liquid oxygen and hydrogen fuel. Lockheed Martin had developed the new fuel tank to help ensure that NASA would be able to haul large sections of the ISS into orbit.[481]

*19 May*
NASA astronauts presented four employees of Boeing Reusable Space Systems with the Silver Snoopy, one of the most prestigious honors available to people working on NASA's Space Shuttle program. Mark Brewer of the orbiter electrical-avionics team, Jeffrey Lewis of the mid-module-major assembly team, Michael Argent of the external-tank, umbilical and payload latch-assembly programs, and Larry Echaves of the configuration-management team each received a letter of commendation, a poster, a certificate, and a Silver Snoopy pin that had flown in space.[482]

The PanAmSat-owned *Galaxy IV*, a five-year old communications satellite, "lost its bearings" so that its antennas were not directed toward Earth, causing the "worst outage in 37 years since communications satellites first entered service." Most of the 45 million pagers in the United States were unable to function during the outage, which "severed electronic links vital to thousands of retailers, news organizations, and broadcasters." PanAmSat officials said that an attitude-control mechanism failed for an unknown reason and a backup processor failed to turn on, causing the satellite to rotate. At that point, hundreds of thousands of satellite dishes in the United States lost contact with *Galaxy IV*.[483]

*21 May*
Chryssa Kouveliotou of the Universities Space Research Association led a team of astronomers at NASA's Marshall Space Flight Center in confirming the existence of *magnetars*, a special class of neutron stars with a magnetic field one thousand trillion times the strength of Earth's. The team calculated the strength of SGR1806-20, first discovered in 1979, by combining data from NASA's Rossi X-ray Timing Explorer satellite and the Advanced Satellite for Cosmology and Astrophysics.[484]

John S. Lewis, Codirector of the Space Engineering Research Center at the University of Arizona in Tucson, testified before the House Science Subcommittee on Space and Aeronautics concerning the potential danger of nearby asteroids. His testimony supported the idea that any asteroid colliding with Earth in the next 100 years could only be an asteroid not yet known to scientists. Lewis argued that, since an international scientific effort could identify such an

---

[481] Seth Borenstein, "Shuttle's New Fuel Tank Test a Success," *Orlando Sentinel* (FL), 19 May 1998.
[482] Romy Jacobson, "Hailing the Right Stuff; NASA Honors Boeing Workers," *Los Angeles Daily News*, 20 May 1998.
[483] Mike Mills, "Satellite Glitch Cuts Off Data Flow," *Washington Post*, 21 May 1998.
[484] *UniSci Science and Research News*, "'Magnetar' Generates Most Intense Magnetic Field in Universe," 21 May 1998; C. Kouveliotou et al., "An X-ray Pulsar with a Superstrong Magnetic Field in the Soft Gamma-Ray Repeater SGR1806-20," *Nature* 393, no. 6682 (21 May 1998): 235–237; NASA, "Strongest Stellar Magnetic Field Yet Observed Confirms Existence of Magnetars," news release 98-87, 20 May 1998.

asteroid easily, he advocated "a systematic, globally-coordinated search and characterization program that costs less than a single small space mission." Such a search, he argued, would "give us adequate warning of a threatened asteroid impact," so that we would "have ample time to design, build, test, and deploy an effective defense against the threat."[485]

The U.S. House of Representatives approved Representative David Weldon Jr.'s (R-FL) amendment to the US$271 billion defense authorization bill, to protect funding for launch ranges. In the past six years, the U.S. Department of Defense had diverted money, which would have modernized U.S. Air Force–controlled launch ranges, to cover costs associated with overseas deployments, such as ongoing operations in Bosnia. Weldon stated that diverting funds away from the launch ranges was a "national security issue as well as a critical issue for NASA and commercial space." Weldon's amendment reserved US$273.3 million for the Cape Canaveral Range in Florida and US$109.1 million for the Western Launch Range at Vandenberg Air Force Base in California.[486]

Scientists published a study in the journal *Science*, using images from the Galileo probe as evidence that an ocean beneath the surface of Europa, one of Jupiter's moons, was once warm and salty. James K. Crowley of the U.S. Geological Survey stated that Galileo data showed that "salt absorption patterns on Europa were similar to those on Earth." Gary B. Hansen, a University of Hawaii geophysicist, added the observation that "the size of the salt bands on Europa's surface are continuous and stretch across much of the moon," evidence of a salty ocean beneath the moon's icy surface. Jeffrey S. Kargel, also of the U.S. Geological Survey, commented on his colleague's study in *Science*, saying that salt is an important "piece in the puzzle," but in the absence of other evidence, does not prove that a salty ocean lies beneath Europa's surface. Scientists seek other evidence, such as shifting ice plates, cracks in the frozen surface, and a magnetic field.[487]

*27 May*
Images released at the American Geophysical Union meeting in Boston showed what scientists believed was the first evidence of ice on Mars outside of the planet's polar ice caps. The images showed a 30-mile-wide (48-kilometer-wide) crater containing a discolored area measuring about 12 to 18 miles (19 to 29 kilometers) across the bottom of the crater. Some scientists thought the discoloration indicated the presence of some sort of deposit, possibly frozen mud or sand, evidence that ice is present, or was present in the past. Others offered a different theory about the contents of the images: volcanic activity could explain the apparent deposit in the crater. The images, from Mars Global Surveyor, had 10 to 12 times better resolution than any previous image of the crater. Arizona State University researchers made another announcement regarding Surveyor's data; the Arizona researchers believed that a "concentration of a rust-colored mineral along the Mars equator indicates it once had boiling hydrothermal vents and perhaps huge lakes." The mineral, hematite, was the first clear evidence of widespread thermal activity on

---

[485] Federal Document Clearing House, "Asteroids: Perils and Opportunities," government news release of testimony of John S. Lewis before the U.S. House Science Subcommittee on Space and Aeronautics, Washington, DC, 21 May 1998.
[486] *Florida Today* (Brevard, FL), "Bill Protects Launch Range Money," 22 May 1998.
[487] Michael Kahn for Reuters, "Jupiter Moon Might Have Salt Ocean—U.S. Study," 22 May 1998; T. B. McCord et al., "Salts on Europa's Surface Detected by Galileo near Infrared Mapping Spectrometer," *Science* 280, no. 5367 (22 May 1998): 1242.

Mars. NASA officials added that the finding indicated that water "was once stable at or near the surface and that Mars had a thicker atmosphere in its early history, probably 4 billion to 6 billion years ago."[488]

*28 May*
At the American Geophysical Union meeting in Boston, astrophysicists Alexander G. Kosovichev of Stanford University and Valentina V. Zharkova of the University of Glasgow in Scotland presented evidence confirming their "solar-bomb theory." Using sequential images that the Solar and Heliospheric Observatory (SOHO) spacecraft had captured, the two scientists explained that, 20 minutes following the rupture of a solar flare from the surface of the Sun, "a seismic wave front surges across the [S]un's face like ripples from a stone thrown into a pond." The SOHO images proved that solar flares not only release energy into the Sun's atmosphere, but also "trigger huge waves of seismic energy that ripple through the [S]un like earthquakes."[489]

At a NASA press conference, a team of NASA and university astronomers presented an image captured with the HST, discussing their conclusion that the image provides the first direct evidence of a planet in another solar system. Previously, scientists had been able to infer the existence of such planets by observing their gravitational effect on suns, but were unable to see them with telescopes. Located 450 light-years away in the Taurus constellation, a binary star system, the planet appeared to be at least twice the size of Jupiter. Designated TMR-1C, the planet orbited 130 billion miles (210 billion kilometers) from its parent stars. Scientists considered TMR-1C, young by planetary standards, to be forming still.[490]

Director General of the Russian Space Agency Yuri N. Koptev announced that Russia had found the funds needed to continue participating in the ISS project. Although the Russian Cabinet had rejected Koptev's request for additional funds, a meeting with Finance Minister Mikhail M. Zadornov and Deputy Prime Minister Boris E. Nemtsov had led to an agreement that funding for the ISS would come from a "special budget section." However, the Russian Space Agency provided no details of the funding plan. Koptev also announced that the Russian Space Agency might scrap the *Mir* space station in the fall, a year ahead of schedule, to make more cash available for the ISS. Koptev suggested that the space agency would look for revenue sources outside of the government.[491]

The European Southern Observatory (ESO) successfully tested the first of four linked, identical telescopes of 8.2-meter (27-foot) diameters, comprising the Very Large Telescope (VLT). The VLT is the world's largest and most advanced Earth-based telescope, with a collective diameter equal to16 meters (52 feet). The telescope's "first light" images (its first astronomically useful images) met the design goals. Exposures lasting up to 10 minutes confirmed that the telescope's tracking was very accurate and stable. The ESO had built the new telescope in the Atacama

---

[488] Sharon L. Lynch for Associated Press, "Mars Discovery," 28 May 1998; *Washington Times*, "Minerals Suggest Life in Mars' Past: Scientists Eye Water Signs for Answers," 28 May 1998.
[489] Elizabeth Manning for United Press International, "Solar Flares Trigger Sunquakes," 27 May 1998; A. G. Kosovichev and V. V. Zharkova, "X-ray Flare Sparks Quake Inside Sun," *Nature* 383, no. 6683 (28 May 1998): 317.
[490] Martin Merzer, "Scientists: Distant Planet Discovered—Hubble Photo May Confirm Sighting Outside Solar System," *Miami Herald*, 28 May 1998.
[491] Associated Press, "Russia-Space," 29 May 1998.

Desert in northern Chile, where a dry atmosphere and stable climate guaranteed clear skies 350 nights of the year.[492]

*30 May*

A computer failure aboard *Mir* partially affected the space station's solar panels, causing the panel to fail to generate sufficient electricity and requiring the crew to use the *Soyuz* capsule thrusters for steering. To conserve energy, the crew turned off nonessential equipment, including the air conditioner and lights. Although they replaced the computer with a new one, the crew was unable to load the necessary software. The crew feared the failure might delay Space Shuttle *Discovery*'s launch to *Mir*, planned for 2 June.[493]

*31 May*

NASA's Transition Region and Coronal Explorer (TRACE) spacecraft captured a video of a "short-lived but extremely bright explosion in the atmosphere of the sun." The explosion was 55,000 miles (89,000 kilometers) long, 200 miles (322 kilometers) wide, and traveled at a rate of 2 million miles (3.2 million kilometers) per hour. The explosions overlapped each other and "snapped," releasing "vast amounts of energy" as solar flares. Solar flares cause magnetic disturbances that disrupt satellites and power supplies if directed at Earth. In April, NASA had launched the TRACE spacecraft, with a telescope 10 times more powerful than any other telescope focused on the Sun. The telescope studied the transitional region of the Sun's atmosphere, between the relatively cool surface of the lower atmosphere, with temperatures of 6,000°C (11,000°F), and the corona or upper atmosphere, where temperatures reach 1.7 million°C (35 million°F). The telescope was able to photograph ultraviolet light and to show temperature variations using false colors.[494]

## JUNE 1998

*1 June*

The 16 countries involved in the International Space Station (ISS) agreed, for the second year in a row, to delay construction, naming 2004 as the new target date for completion. The delay was primarily the result of Russia's inability to pay its share of the program's costs. NASA Associate Administrator for Space Flight Joseph H. Rothenberg said that the new five-year schedule was "extremely tight and allows for few, if any, launch delays." NASA estimated that the space station would require 43 flights for its assembly, two fewer flights than previously scheduled. Under the revised schedule, Russia planned to launch nine flights, three fewer than originally scheduled, and NASA intended to launch 34 Shuttle flights to transport parts, one more than originally planned. To save money, Russia canceled construction of two life-support modules and a stowage chamber.[495]

---

[492] *Financial Times* (London), "Staring Out Across 14 Billion Light Years: Technology Very Large Telescope: The Earth-Based VLT Brings a Twinkle to a European Astronomer's Eye," 4 June 1998.

[493] Marcia Dunn for Associated Press, "Cosmonauts Replace Mir Computer," 31 May 1998; Marcia Dunn, "Mir Space Station Has a New Problem—Computer Failure Causes Solar Panels To Turn from Sun," *Seattle Times* (WA), 31 May 1998.

[494] Mark Prigg, "Photos Reveal Sun's Power," *Sunday Times* (London), 7 June 1998; Leonard Novarro for Reuters, "Solar Flare May Disrupt Worldwide Communication," 8 June 1998.

[495] Marcia Dunn for Associated Press, "Space Station," 1 June 1998; *New York Times*, "Space Station Partners Approve Assembly Delay," 1 June 1998.

*2 June*

Space Shuttle *Discovery* Mission STS-91 launched on schedule from Kennedy Space Center (KSC) in Cape Canaveral, Florida, despite problems with the automatic steering system on *Mir*. The Shuttle's mission was to retrieve the seventh and last American to live aboard *Mir*. The day's temperature reached 97°F (36°C) during the countdown, making it the "hottest launch day anyone could remember." The mood of the launch was tense, because NASA had never tested the new lightweight fuel tank during flight. According to NASA officials, the new tank worked well. Among the crew members aboard *Discovery* was Valery V. Ryumin, veteran cosmonaut and chief of the Russian operations in the Shuttle–*Mir* program. The Russian Space Agency had selected Ryumin to inspect *Mir*, so that Russian officials could make an informed decision regarding how long cosmonauts could safely continue to inhabit the space station. Other crew members included Commander Charles J. Precourt; Pilot Dominic L. P. Gorie; and Mission Specialists Wendy B. Lawrence, Franklin R. Chang-Dìaz, and Janet L. Kavandi. For the first time, the Shuttle carried the Alpha Magnetic Spectrometer (AMS), designed to search for dark and missing matter (antimatter) in the universe.[496]

*3 June*

At the request of U.S. Senators John McCain (R-AZ) and Bill Frist (R-TN), the U.S. General Accounting Office (GAO) released a report, analyzing the United States' role in the ISS and the cost of the program. GAO estimated that life-cycle costs could reach US$96 billion, US$2 billion more than the 1995 estimate, and that these costs could increase further because of the potential for changes in the program. According to the report, "at the current estimated spending rate, the program would incur additional costs of more than [US]$100 million for every month of schedule slippage." The report cited, as reasons for the cost increases, schedule slippage, delays in production of components, the need for additional flights to and from the space station, and NASA's decision to change its tracking requirement for space debris. GAO's report added as much as US$1 billion to the estimate overall, including crew-return-vehicle testing and resupply flights not part of the original NASA estimate. Although the report cited U.S. Department of Defense as the only agency with appropriate tracking equipment, its equipment would need as much as US$5 billion in upgrades to handle the new NASA requirements. The report also highlighted other issues including insufficient staff at KSC to handle nine Shuttle launches per year.[497]

The Boeing Company announced the creation of a new subsidiary called Boeing Space Operations, based in Houston, Texas, at the Johnson Space Center, which would provide various

---

[496] NASA, "Discovery Launch to Mir on Mission STS-91 Set for June 2," news release 98-85, 20 May 1998; NASA, "Mission Archives: STS-91, Ninth and Final Shuttle–*Mir* Docking," *http://www.nasa.gov/mission_pages/shuttle/shuttlemissions/archives/sts-91.html* (accessed 11 June 2007); *Chicago Tribune*, "Cosmonauts Fix Mir's Steering; Now Shuttle Launch Can Proceed," 2 June 1998; Marcia Dunn for Associated Press, "Discovery Soars on NASA's Last Voyage to Mir," 2 June 1998; Robyn Suriano for Gannett News Service, "Celebrated Cosmonaut Comes Out of Retirement for Shuttle Mission," 1 June 1998.

[497] U.S. General Accounting Office, "Space Station U.S. Life-Cycle Funding Requirements: Statement of Allen Li, Associate Director, Defense Acquisitions Issues, National Security and International Affairs Division," testimony before the House Committee on Science and Technology, report no. GAO/T-NSIAD-98-212, 24 June 1998; Armed Forces Newswire Service, "Space Station Costs Higher than Expected, GAO Says," 17 June 1998; Tamara Lytle, "Space Station's Costs Trouble 2 GOP Senators," *Orlando Sentinel* (FL), 3 June 1998.

services for government and commercial space operations. Rick Stephens, the president of the new subsidiary, stated that the goal of Boeing Space Operations was to make Boeing "the global provider of low-cost, reliable, commercially based space operations services."[498]

Space Shuttle *Discovery* encountered its first malfunction during Mission STS-91, when the KU-band system failed to switch on after *Discovery*'s crew deployed its antenna upon reaching orbit. The KU-band downlink carries broadcasts with a high rate of data. The system failure prevented the Shuttle crew from sending televised images or high rates of information to Mission Control. This problem had never occurred before. NASA spokesperson James Hartsfield stated that it was unclear whether the trouble originated with the antenna, transmitter, or other associated equipment. Hartsfield also mentioned that the failure would "prevent immediate analysis of data from the Alpha Magnetic Spectrometer." The S-band antennas worked normally, and the crew was able to talk with ground controllers as usual.[499]

*4 June*
Space Shuttle *Discovery* docked successfully with *Mir*. Upon the opening of the hatches, astronaut Andrew S. W. Thomas officially ended his 130-day mission aboard *Mir* and became a crew member of *Discovery*. The transfer also marked the completion of a total of 907 days of U.S. astronaut presence aboard the space station. Because of the Ku-band failure, Russian cameras on *Mir* recorded the docking, while Russian ground stations televised live the meeting of the two crews.[500]

During a meeting at Marshall Space Flight Center (MSFC), NASA Administrator Daniel S. Goldin directed MSFC officials to draw up a new plan for meeting the goals of the Bantam Program. The program's original goal had been to develop, by 2001, a rocket capable of launching 150-kilogram or 0.15-tonne (331-pound or 0.2-ton) payloads, at a cost of no more than US$1.5 million. None of the rockets developed with NASA funding would have met that cost target. Bantam Program Manager Sherry Buschmann reiterated NASA's commitment to "achieving a capability to launch small payloads frequently at an affordable cost," saying that the plan would meet a "legitimate need within the science community."[501]

NASA announced the astronaut candidate class of 1998, including 8 pilot candidates and 17 mission specialist candidates, and composed of 21 male and 4 female candidates. Among the candidate class was teacher Barbara R. Morgan, the former understudy to Christa McAuliffe. NASA had designated Morgan as Educator Mission Specialist in January 1998.[502]

*5 June*

---

[498] James Wallace, "Boeing Forms a New Unit To Boost Its Space Business," *Seattle Post-Intelligencer* (WA), 3 June 1998.
[499] NASA, "Mission Archives: STS-91"; Pauline Arrillaga for Associated Press, "Shuttle Soars Toward Final Mir Docking, but Antenna Won't Send TV Signals," 3 June 1998; *Aerospace Daily*, "Lightweight Tank Perfect on First Flight; KU-Band a Problem," 4 June 1998.
[500] NASA, " Mission Archives: STS-91"; Steven Young for Reuters, "Shuttle Collects Last U.S. Astronaut from Mir," 5 June 1998.
[501] Warren Ferster, "NASA Shifts Gears on Low-Cost Launcher: Goldin Renews Commitment to Bantam Goals," *Space News*, 8–14 June 1998.
[502] NASA, "NASA Names Astronaut Class of 1998," news release 98-97, 4 June 1998.

Mark H. Thiemens, professor of chemistry at the University of California at San Diego, and a team of researchers published a study in the journal *Science* disputing NASA's theory that a rock from Mars contained evidence of life. NASA scientists had studied the rock, found in the Allen Hills ice field in Antarctica, concluding that some carbonate mineral structures in the rock were evidence that microbes had lived in the rock before it "blasted away from Mars," during the collision of Mars with an asteroid millions of years ago. By contrast, Thiemens's researchers concluded that the chemical characteristics of oxygen inside the rock originated from the carbon dioxide and ozone of the Martian atmosphere. The team explained that, if a microbe had deposited the oxygen, "the chemical signature of the minerals would have matched that of oxygen in water, not oxygen in the atmosphere." According to the team, this finding did not provide final proof that the rock had never contained life. However, it provided a significant argument against NASA's theory, controversial since its first proposal in August 1996.[503]

Astronauts aboard Space Shuttle *Discovery* attempted, but failed, to fix the spacecraft's high-speed data-transmission system. NASA believed the source of the problem was in an electronics box in the cargo bay, but stated that only a spacewalk would enable the crew to reach the box. NASA Mission Operations representative Robert E. Castle Jr. remarked that, even with a spacewalk, the astronauts would probably not be able to repair the malfunction. Because of the lack of the high-speed data transmission system, scientists could not view the data that the AMS had collected. Although the Shuttle had stored the data on board, NASA was unable to determine whether the data was useful, or whether the instrument's adjustment was correct.[504]

*6 June*
Cosmonauts working aboard *Mir* conducted the first of two tests, using green fluorescent gas made from combining nitrogen, acetone, and biacetyl gases, to search for leaks in the *Spektr* laboratory module damaged in a collision in June 1997. However, they were unable to pinpoint the sources of any seepage. Because *Mir* was nearing its decommission stage, Russia planned no repairs, in the event that the cosmonauts were able to locate the leaks precisely. Instead, the tests were intended as practice for finding leaks on the future ISS.[505]

*8 June*
Space Shuttle *Discovery* undocked from *Mir*, carrying back to Earth Australian-born Andrew S. W. Thomas, the last of seven U.S. astronauts to live aboard the station, and marking the completion of Phase I of the ISS program. Before closing the hatches, *Mir* Commander Talgat A. Musabayev presented Shuttle Commander Charles J. Precourt with a huge wrench that had been used in spacewalks and would be flown to the future ISS.[506]

NASA officials announced that the Space Telescope Science Institute in Baltimore, Maryland, which had operated the science program for the Hubble Space Telescope (HST) since 1983,

---

[503] James Farquhar, Mark H. Thiemens, and Teresa Jackson, "Atmosphere-Surface Interactions on Mars: $\Delta^{17}O$ Measurements of Carbonate from ALH84001," *Science* 280, no. 5369 (June 5, 1998): 1580; *Los Angeles Times*, "New Meteorite Study Casts Doubt on Theory of Martian Life," 5 June 1998.
[504] Marcia Dunn, "Discover Astronauts Unable To Fix Communications Link," *Buffalo News* (NY), 6 June 1998.
[505] Mark Carreau, "Astronauts Unable To Spot Mir Hull Breach," *Houston Chronicle*, 7 June 1998; Marcia Dunn, "Test Fails To Find Leaks in Mir," *Seattle Times* (WA), 7 June 1998.
[506] Pauline Arrillaga for Associated Press, "Discovery Undocks from Mir, Completes Historic Last Linkup," 8 June 1998.

would also manage the science operations for the Next Generation Space Telescope (NGST). As one of the cornerstone missions of the Astronomical Search for Origins and Planetary Systems, a major project within its space science program, NASA expected the NGST "to provide a critical follow-on to the Hubble Space Telescope." With capabilities unavailable in existing ground-based or space telescopes, the NGST's purpose was to observe the first stars and galaxies in the universe, to understand better how the universe formed after the Big Bang. After consulting advisory committees, Wesley T. Huntress Jr., Associate Administrator for NASA's Office of Space Science, had decided that the "most cost-effective and scientifically sound way to proceed" would be "to expand the [Space Telescope Science] Institute's responsibilities to include the management" of the new telescope.[507]

Michael I. Mott, NASA Associate Deputy Administrator (Technical), one of NASA's top three managers, announced his decision to leave NASA to join Boeing Space Transportation in Seal Beach, California, as Vice President of Business Development. Mott had served as NASA Associate Deputy Administrator since January 1994.[508]

*9 June*
The U.S. Senate Appropriations Subcommittee on Veterans Affairs, Housing and Urban Development, and Related Agencies, responsible for allocating space funding, voted to back funding for NASA's ISS obligations but "cautioned the space agency to control costs and ensure Russia's commitment to the expansive project." The subcommittee voted for a US$13.6 billion budget for NASA for the fiscal year, including US$2.3 billion for the space station. Senate Subcommittee Chairperson Christopher Bond (R-MO) said that, although the budget was US$150 million more than President William J. Clinton had requested, it was US$33 million less than NASA's 1998 spending level.[509]

Research scientists working on a US$33 million experiment to search for antimatter in space, spoke at a news conference, announcing that the communication-system breakdown on Space Shuttle *Discovery* had ruined their tests. The astronauts had been unable to repair *Discovery*'s main antenna system, which had broken shortly after launch, rendering the Shuttle unable to transmit high-speed data or video back to Earth. Scientists working on the project needed continuous, high-speed data to calibrate the AMS, which was searching space for rare subatomic particles known as antimatter. However, using a patchwork system, they had only been able to receive a few minutes of data each hour, which meant that, although they could see that the AMS was working, the spectrometer was not providing them with enough data to fine tune the magnet. According to Project Manager James Bates, the scientists had failed to accomplish their objectives because of the malfunction.[510]

*10 June*

---

[507] NASA, "NASA Selects Home for Next Generation Space Telescope," news release 98-96, 8 June 1998.
[508] NASA, "Michael I. Mott To Leave NASA," news release 98-99, 8 June 1998.
[509] Patrick Connole for Reuters, "US Senate Panel Backs Space Station, Cut Superfund," 9 June 1998.
[510] Seth Borenstein, "Shuttle Physics Testing Thwarted; A Glitch Has Undermined an Antimatter Experiment," *Orlando Sentinel* (FL), 10 June 1998.

Aerojet announced it had won a US$485,000 contract to build a rocket injector for NASA's Jet Propulsion Laboratory (JPL) that would help JPL scientists develop propulsion technologies for NASA's Mars Sample Return Vehicle.[511]

*12 June*
Space Shuttle *Discovery* landed safely at KSC in Cape Canaveral, Florida, at 2 p.m. (EDT), returning the U.S. astronaut, Australian-born Andrew S. W. Thomas, to Earth after four and one-half months in space. Thomas, unlike some of his predecessors, took the medical advice typically given to astronauts returning from a lengthy stay in space and allowed medical personnel to carry him off the Shuttle, rather than attempting to walk on his own.[512]

*17 June*
NASA Administrator Daniel S. Goldin submitted a 29-page report to lawmakers of the U.S. House of Representatives, officially responding to "an independent review panel's damning assessment that the [ISS] could cost US$7.3 billion more than NASA estimated." The panel had also reported that NASA would likely complete the ISS more than four years behind schedule. Goldin acknowledged significant budget and schedule problems but maintained that NASA's plan remained manageable, despite its shortcomings.[513]

*19 June*
The U.S. House of Representatives subcommittee responsible for allocating funding for the space program voted on a US$13.3 billion budget for NASA, US$138 million less than President William J. Clinton had requested and US$300 million less than the U.S. Senate subcommittee had approved. Subcommittee Chairperson Jerry Lewis (R-CA) stated that the House had reduced the budget allocation because NASA planned to fly fewer Shuttle missions in the next fiscal years and because of the delays in assembling the ISS. Unlike the Senate, the House did not vote to monitor the budget of the program by creating a separate account for space station funding.[514]

Hughes Global Services Inc. announced the success of its salvage mission to a stranded Asia Satellite Telecommunications Company Ltd. communications satellite. The 3.8-ton (3,500-kilogram or 3.5-tonne) *AsiaSat 3* had remained stranded in a useless orbit since its December 1997 launch. Asia Satellite Telecommunications had received US$200 million from insurers. The insurers had told Hughes Global Services to find a new customer, if the salvage mission was successful. Asia Satellite Telecommunications had originally intended *AsiaSat 3* to provide television and telephone service in Asia, but the Russian-built launcher had malfunctioned, and the satellite's orbit had gone awry, rendering it useless. Hughes Global Services had taken possession of the satellite, implementing a "seemingly improbable" mission to salvage the spacecraft. Hughes had sent the satellite around the Moon twice, using the Moon's gravity to help correct the satellite's orbit around the Earth, and declared the salvage mission a success,

---

[511] *Sacramento Bee* (CA), "Aerojet Lands Contract," 11 June 1998.
[512] Steven Young for Reuters, "U.S. Space Shuttle Returns with Last Mir Astronaut," 12 June 1998.
[513] Larry Wheeler for Gannett News Service, "NASA Chief Admits Space Station Project Has Major Woes," 17 June 1998.
[514] Brett Davis, "House Panel Approves $300 Million Less for NASA Budget than Senate Panel," *Huntsville Times* (AL), 19 June 1998.

even though the satellite was not stationary, which would limit the number of customers who might lease or buy it.[515]

Eighteen members of the U.S. House Committee on Science and Technology submitted a letter to President William J. Clinton, urging him to "abandon plans to reach an agreement in China on expanding space cooperation because of Beijing's weapons transfers and human rights abuses." The letter was the House's response to a 29 April report in the *Washington Times*, disclosing a plan to conclude a space cooperation agreement at the Beijing summit. The article discussed a draft agreement calling for the free and unrestricted exchange of scientific data between NASA and the State Science and Technology Commission of China, a key developer of weapons-related technology. The lawmakers, warning that the pact would increase the likelihood of China's sharing technology with rogue states, referred to a CIA report stating that China had aimed 13 of its 18 long-range strategic missiles at the United States. These comments emphasized the concern of members of Congress that China constituted a direct threat to the United States. Many opponents of the agreement insisted that the United States should not share with China technologies or applications that enemies of the United States might use against the American people or U.S. interests.[516]

James B. Willet, a nuclear astrophysicist who had worked for NASA, died of cancer at his Maryland home. Willet had worked for 20 years on the scientific staff of NASA's JPL in Pasadena, California, and for three years as JPL's liaison to NASA Headquarters before joining the Space Research Association in 1993. At JPL, Willet had acted as a science coordinator and team chief for the Galileo Mission. Galileo was orbiting Jupiter at the time of his death. Willet was also a field operations team leader for the Mars balloon-testing program and had conducted research in gamma-ray astronomy. Willet's duties as liaison officer involved him in the Voyager, Ulysses, and Pioneer programs.[517]

NASA officials at Lewis Research Center (LERC) in Cleveland, Ohio, announced the new five-year collaborative program of LERC and Case Western Reserve University. The program's name, the Glennan Microsystems Initiative, honored the late T. Keith Glennan, the first NASA Administrator, who had served from 1958 to 1961. Glennan was also President Emeritus of Case Institute of Technology, forerunner of Case Western Reserve University. Backed with US$16 million in federal funds and US$4.5 million in state funds, the initiative planned to "develop and apply the technology known as microelectricomechanical systems for industry and for NASA." LERC planned to develop systems for use a decade into the future, including power systems for the next space station.[518]

*20 June*
The Huntsville Alabama L5 Society (HAL5), a chapter of the National Space Society, failed in its second attempt to become the first group of amateurs to launch a rocket into space. The

---

[515] Andrew Pollack for Associated Press, "Rescue Effort Puts a Satellite in Viable Orbit," *New York Times*, 19 June 1998; Associated Press, "US Satellites," 19 June 1998.
[516] Bill Gertz, "House Science Panelists Oppose Space Pact with China," *Washington Times*, 24 June 1998.
[517] *Buffalo News* (NY), "James B. Willet Nuclear Astrophysicist," 20 June 1998.
[518] Jennifer Arend, "Lewis Research Launches Effort To Develop Technology," *Cleveland Plain Dealer* (OH), 20 June 1998.

society attempted to use a hot-air balloon from a NASA barge in the Gulf of Mexico to launch its rocket, a launch concept known as a rockoon. According to the rockoon concept, a small rocket may "obtain very high altitude because there is little air to slow it down during launch." James A. Van Allen had been the first to fly rockoons in the 1950s, as part of a joint U.S. Navy-University of Iowa project. The U.S. Navy had abandoned the technique after large, ground-based, sounding rockets became available. NASA's MSFC signed a cooperative agreement with HAL5 in 1996, continuing the relationship for the High Altitude Lift-Off mission. MSFC had supplied HAL5 with the barge, helium for the balloon, nitrous oxide for the rocket, and some funds for materials. HAL5 sought to "make space more affordable for students, amateurs, experimenters, and researchers." NASA was examining alternate launch sites and methods in an effort to find ways of reducing the cost of rocket launches.[519]

*23 June*
The Lewis Spacecraft Mission Failure Investigation Board, convened by NASA, released its report explaining the reason for the spacecraft's failure, which had cost NASA more than US$70 million. NASA had developed the Earth-orbiting spacecraft to test advanced scientific equipment and technology for measuring land surface changes. The spacecraft, launched on 23 August 1997, had failed to achieve stable orbit. According to the board's findings, faulty attitude-control equipment, together with inadequate monitoring, had caused the craft's destruction upon reentry on 28 September. The investigation board reported that the attitude-control system had not been tested or developed sufficiently. NASA had adapted the system from an earlier design, which had not taken into account the *Lewis* spacecraft's spin axis. The small project crew had been unable to monitor and control the craft adequately during the crucial early flight stage. According to the investigation board's report, this was the primary cause of the mission's failure.[520]

*24 June*
U.S. House Committee on Science and Technology Chairperson F. James Sensenbrenner Jr. (R-WI) and ranking Representative George F. Brown Jr. (D-CA) submitted a letter to President William J. Clinton requesting that he supply a plan within 30 days, "detailing how the [a]dministration expects to resolve the dire problems surrounding the International Space Station." The two members of Congress asked President Clinton to intervene with President Boris N. Yeltsin to "ensure full Russian compliance with its stated commitments to the space station." In the aftermath of the announcement of the independent Cost Assessment and Validation Task Force's findings on the ISS, the members of Congress were attempting to rectify the inadequate budget submission. They asked the President to "direct the Office of Management and Budget to deliver a workable plan to Congress . . . so that the relevant

---

[519] HAL5 Project, "Space Group Hopes To Launch First Amateur Rocket into Space," *Florida Today* (Brevard, FL), 17 June 1998; *Orlando Sentinel* (FL), "Space Buffs in Alabama Can't Get Rocket Launched," 21 June 1998; William R. Corliss, "Chapter 4: Development of the First Sounding Rockets," in *NASA Sounding Rockets, 1958–1968: A Historical Summary* (Washington, DC: NASA Scientific and Technical Information Office, 1971), http://history.nasa.gov/SP-4401/contents.htm (accessed 30 January 2008).
[520] NASA, "Lewis Spacecraft Failure Board Report Released," news release 98-109, 23 June 1998; *Defense Daily*, "NASA Blames Spacecraft Loss on Equipment, Monitoring," 24 June 1998.

[c]ongressional committees can consider OMB numbers when crafting NASA appropriations and authorization bills."[521]

At a hearing of the U.S. House Committee on Science and Technology, NASA Administrator Daniel S. Goldin defended NASA's purpose and role in the ISS project. Representative Nicholas V. Lampson (D-TX) advised Goldin that cost overruns and Russia's failure to meet its agreement had caused some lawmakers to propose cancelling the space station. Goldin responded that canceling the ISS would cancel human spaceflight, relegate the United States to second-class power status, and have international repercussions. Lawmakers at the hearing asked Goldin to furnish NASA's proposal to overcome the problems associated with the ISS, providing estimates of the amount necessary to complete the project. Goldin refused to give specifics until NASA had evaluated several plans and calculated the cost of the alternative plans, including contingency plans in the event Russia could not meet its commitment.[522]

Controllers at NASA's Goddard Space Flight Center (GSFC) sent routine commands to the Solar and Heliospheric Observatory (SOHO) to maneuver the satellite. Instead of responding correctly, SOHO suddenly went into emergency Sun reacquisition mode (ESR), "activated when an anomaly occurs and the spacecraft loses its orientation toward the Sun." Consequently, NASA lost all communication with the satellite, along with the ability to track it. SOHO's mission was to track "major solar activity that may induce power outages and interfere with radio, TV, and other signals from space." The mission was particularly important because of the anticipated solar maximum, expected to occur during the year 2000. During the solar maximum, a phenomenon that peaks every 11 years, the Sun emits intense radiation capable of scrambling signals from communications satellites and causing power failures, as well as brilliant auroras. The European Space Agency (ESA) and NASA had launched SOHO in December 1995, as a joint mission. In April 1998, the spacecraft had successfully completed its initial two-year mission to study the Sun's atmosphere, surface, and interior.[523]

*25 June*
James L. Elliot, an astronomer at the Massachusetts Institute of Technology, and his colleagues, together with Lowell Observatory and Williams College, published findings in the journal *Nature* based on data obtained from NASA's HST. The data indicated that Neptune's largest moon, Triton, had warmed up significantly since the Voyager spacecraft's visit in 1989. Triton's temperature had risen from 37 K (-393°F or -236°C) to 39 K (-389°F or -234°C), a 5 percent increase. The astronomers traced the warming trend to seasonal changes in Triton's polar ice caps attributed to "an extreme southern summer," a cyclical phenomenon occurring every few hundred years. Other possible explanations included a changing frost pattern on Triton's surface, which could have caused the surface to absorb more of the Sun's warmth; or changes in the reflectivity of Triton's ice, which could have caused it to absorb more heat. Because of the warming, Triton's frozen nitrogen surface was turning into gas, making its thin atmosphere

---

[521] Federal Document Clearing House," Committee Asks Clinton for Space Station Plan," government news release, 24 June 1998.
[522] Paul Recer for Associated Press, "NASA Defends Space Station Plans," 24 June 1998; John C. Henry, "NASA Chief Concedes Space Station Delays—Has Contingency Plans If Russia Can't Do Its Part," *San Diego Union-Tribune* (CA), 25 June 1998.
[523] NASA, "SOHO Spacecraft Observations Interrupted," news release 98-112, 26 June 1998; Paul Hoversten, "Sun-Studying Probe Failing To Respond; NASA Fears It's Lost," *USA Today*, 29 June 1998.

denser. The scientists hoped their study of the changes on Triton would provide insight into global warming on Earth, which has a more complicated environment than Triton. HST's "detection of an increase in the moon's atmospheric pressure," measured with one of the space telescope's three fine guidance sensors, had provided the clue to the increased surface temperature.[524]

*30 June*
NASA announced that experts from the ESA and Matra Marconi Space, the prime contractor for the SOHO spacecraft, had met at GSFC in Greenbelt, Maryland, to assess the situation of the noncommunicating spacecraft and to analyze the satellite's status in the event that ground controllers reestablished contact with it. Participants at the meeting named as cochairs of a joint inquiry board to investigate the incident, Professor Massimo Trella, the ESA's Inspector General, and Michael A. Greenfield, Deputy Associate Administrator for the Office of Safety and Mission Assurance at NASA Headquarters.[525]

The European Science Foundation and the U.S. National Research Council published a joint study analyzing 13 projects in astrophysics, planetary science, space physics, earth science, and microgravity research. The report advised the United States and its international partners to "set aside a specific part of their annual space budgets to plan for future critical joint exploration projects," highlighting factors hampering cooperation within the projects analyzed. Hampering factors included cultural differences, poorly defined objectives, and "lukewarm support from the scientific community for projects undertaken more for political than research reasons." The report set forth recommendations for successfully pursuing cooperative space projects within tight budget constraints.[526]

## JULY 1998

*2 July*
PanAmSat Corporation announced the impairment of a critical subsystem of its *Galaxy VII* spacecraft. A spokesperson for the company said that engineers had not yet determined whether the problem was the same as that which caused the failure of the *Galaxy IV* spacecraft on 19 May. Both the primary and backup systems keeping the *Galaxy IV* correctly pointed toward Earth-bound controllers had failed, but although the *Galaxy VII* had lost its primary system, its backup system had continued to function properly. Hughes Electronics Corporation's Space and Communications Company, the manufacturer of the two satellites, was investigating the failures. Both satellites were HS601 models, the world's "best-selling large communications satellite,"

---

[524] NASA, "Hubble Space Telescope Helps Find Evidence that Neptune's Largest Moon Is Warming Up," news release 98-110, 24 June 1998; J. L. Eliot et al., "Global Warming on Triton," *Nature* 393, no. 6687 (25 June 1998): 765–767.

[525] NASA, "Efforts To Recover SOHO Spacecraft Continue as Inquiry Board Co-Chairs Named," news release 98-118, 30 June 1998.

[526] Frank Sietzen Jr. for United Press International, "Study: Space Projects Need Good Science," 30 June 1998; European Space Foundation, "Transatlantic Study Identifies Keys to Success in International Space Collaboration," news release, 30 June 1998, *http://www.esf.org/media-centre/press-releases/ext-single-news/article/transatlantic-study-identifies-keys-to-sucess-in-international-space-collaborations-220.html* (accessed 27 September 2007).

with customers including the U.S. Navy, NASA, the government of Mexico, and communications companies throughout the world.[527]

SPACEHAB Inc., a Virginia-based, commercial space company with a staff of 85, purchased the Johnson Engineering Corporation, a 400-person firm contracted to support NASA's Johnson Space Center (JSC) in Houston, Texas. SPACEHAB's chairperson said the company would "expand the use of Johnson Engineering's larger technical staff to support its plans for expansion into the commercial space arena." Investors had incorporated SPACEHAB in 1984, to "develop a commercial research module that could accommodate small, private, as well as government-financed science experiments and engineering tests during [S]pace [S]huttle mission[s]." However, when the anticipated commercial market had failed to develop, SPACEHAB "changed its strategy," winning NASA contracts. NASA used SPACEHAB modules, stored in the Shuttle's payload bays, as cargo containers for storing "tons of food, clothing, and other gear that was ferried to Russia's *Mir* space station aboard a series of recently concluded U.S. [S]huttle flights."[528]

Russia announced it would retire the *Mir* space station during the summer of 1999, six months earlier than originally planned. The decision resulted from a meeting between Russian Space Agency Director General Yuri N. Koptev and Deputy Prime Minister Boris E. Nemtsov, who agreed that Russia had insufficient funds to continue to fly the station. The agreement allocated approximately US$100 million for *Mir*'s final year, including funds for "a series of rockets to direct *Mir* to an unpopulated area of the Pacific Ocean." Nemtsov also agreed to pay the Russian Space Agency the US$600 million owed for *Mir*'s operations in 1997, using "unspecified 'nonbudgetary sources'."[529]

*4 July*
Due to arrive at Mars in October 1999 after a period of orbiting Earth, Japan's Planet B spacecraft launched from Kagoshima Space Center on the southern Japanese island of Kyushu. The Japanese Education Ministry's Institute of Space and Aerospace Science developed the probe, nicknamed *Nozomi*, or *Hope*. The probe carried 14 scientific instruments developed by Canada, Germany, Japan, Sweden, and the United States. NASA provided two instruments for the probe: the Neutral Mass Spectrometer, designed to measure the gas composition of Mars's upper atmosphere for the first time; and the Ultra Stable Oscillator, deployed to support a radio science experiment, also involving analysis of the Martian atmosphere, and to guide the spacecraft in its orbit around Mars.[530]

*6 July*
U.S. Senate Majority Leader Trent Lott (R-MS) announced that Alliant Techsystems would build composite materials, such as booster nose cones, for the Boeing Delta IV rocket at Iuka, Mississippi, a former site of the Tennessee Valley Authority and NASA. The rocket would

---

[527] Robert Covington for Reuters, "Second PanAmSat Satellite Encounters Problems," 2 July 1998.
[528] Mark Carreau, "Spacehab of Virginia Acquires NASA Contractor," *Houston Chronicle*, 5 July 1998.
[529] Adam Tanner for Reuters, "Russia Agrees to Early End for Mir Space Station," 2 July 1998; Anna Dolgov for Associated Press, "Mir," 3 July 1998.
[530] NASA, "NASA Instruments on Japanese Planet-B Spacecraft Will Aid Studies of Martian Upper Atmosphere," news release 98-119, 1 July 1998; United Press International, "Japan Launches First Mars Mission," 4 July 1998.

"compete for Air Force and commercial contracts for satellite launches," while Senator Lott and U.S. Senate Select Committee on Intelligence Chairperson Richard Shelby (R-AL) would continue to investigate whether launching U.S.-made satellites on Chinese rockets was compromising U.S. security. Rod Bitz, the director of corporate communications for Alliant, said, "Iuka was chosen in part because the large structures can be easily shipped by barge to Decatur." Engineers at the new Boeing plant in Decatur, Alabama, would assemble the rocket.[531]

*7 July*
In a vote of 66 to 33, the U.S. Senate defeated Senator Dale Bumpers's (D-AK) amendment to the appropriations bill for the Veterans Administration, Housing and Urban Development, and Independent Agencies. The amendment would have cut the United States' US$2.3 billion share of the FY 1999 budget for the International Space Station (ISS), cancelling the ISS program. During floor debate preceding the vote, Senator Bumpers, who had been trying to eliminate U.S. funding for the space station for the past eight years, referred to the previous month's U.S. General Accounting Office report. The report had projected that the cost to develop, assemble, and operate the space station over several years would reach US$100 billion, in part because of recurring cost overruns. Senator John H. Glenn Jr. (D-OH) countered Senator Bumpers's arguments against the space station, citing the "incredible scientific possibilities" that could result from research on the ISS, in such areas as neurology, sleep disorders, and balance.[532]

Rocket engineer Max E. Nowak died of natural causes at the age of 89. An original member of Wernher von Braun's German rocket team, which developed the V-1 and V-2 rockets during World War II, Nowak had been among the first of the team members to move to the United States after the war. He had been the assistant to the director of the manufacturing engineering lab during the Apollo program and had directed the assembly of systems for the Saturn I nose cones and Saturn V launch systems.[533]

Japan's National Space Development Agency (NASDA) docked the two satellites—*Orihime* and *Hikoboshi*—in outer space, marking the first time two robotic spacecraft had docked using remote control. U.S. and Russian space programs had used remote control to dock pilotless cargo craft with the piloted *Mir* space station, but had never maneuvered two robotic spacecraft together. NASA spokesperson Brian D. Welch called the accomplishment "an impressive feat." NASDA timed the rendezvous of the two satellites to coincide with Japan's *Tanabata*, or *star festival*, on 7 July, celebrating "a mythical meeting between the princess Orihime and her lover Hikoboshi," associated with the stars Altair and Vega. According to the legend, the pair may meet only once a year on 7 July.[534]

NASA announced that JSC had modified its contract with United Space Alliance, awarding the contractor more Shuttle work. Three other NASA contractors had previously performed the more

---

[531] James W. Brosnan, "Iuka Revving Up for Rockets Again Will Be Part of Delta IV Project," *Commercial Appeal* (Memphis, TN), 7 July 1998.
[532] Frederic J. Frommer for Associated Press, "Space Station," 7 July 1998; Vickie Allen for Reuters, "International Space Station Survives Senate Vote," 7 July 1998; Federal Document Clearing House, "Bumpers Amendment Would Ground Space Station," government news release, 7 July 1998.
[533] Associated Press, "3 Rocket Scientists Die," 12 July 1998.
[534] Associated Press, "Japanese Are Unveiling Stellar Space Program Progress Has Been Made Cheaply and Quietly," *New Orleans Times-Picayune*, 8 July 1998.

than US$900 million of work on the Space Shuttle's solid rocket booster and other elements. USBI Company at Kennedy Space Center had worked on the boosters; Lockheed Martin Corporation in Houston, Texas, had designed and produced the primary Shuttle avionics software; and Boeing Aerospace Operations in Houston had processed flight equipment.[535]

*8 July*
S. George Djorgovski, an astronomer at California Institute of Technology; Dale A. Frail, an astronomer with the Very Large Array (VLA) Observatory in Socorro, New Mexico; and their colleagues issued a notice to fellow astronomers regarding a gamma-ray burst (GRB) detected on 2 July. Frail and Djorgovski had combined the resources of the VLA and the W. M. Keck Observatory in Hawaii to pinpoint the source of the burst, which they calculated to be a galaxy, "alive with new-forming stars," billions of light-years away from Earth. Although scientists did not yet know the cause of GRBs, Frail remarked, "the fact that this burst happened in a hot galaxy where new stars are being rapidly formed is probably the death knell for the theory that collisions between old, dying neutron stars are responsible for the gamma-ray bursts."[536]

*9 July*
Heinrich K. Paetz, an original member of Wernher von Braun's German rocket team, died of natural causes at the age of 88, the second team member to die within the week. Before retiring from NASA in 1970, Paetz had participated in the Redstone, Jupiter, Saturn I, and Saturn V programs as a member of the test branch.[537]

A month after its launch, the National Oceanic and Atmospheric Administration (NOAA) began activating the weather satellite, Geostationary Operational Environmental Satellite-10 (GOES-10), a backup to the failing GOES-9 satellite. The momentum wheels on GOES-9, similar to a gyroscope, which maintain the satellite's stability, were "showing signs of impending failure." GOES-9 was responsible for monitoring weather on the west coast of the United States. GOES-10, placed in storage mode above the center of the country, would need about 30 days, moving approximately 1° of longitude each day, to move to the west coast location. GOES-10's well-timed placement in storage mode would ensure data continuity, saving NOAA a 12- to 15-month wait for a time slot for launch, following the imminent failure of GOES-9.[538]

*10 July*
Rocket scientist Albert E. Schuler died of natural causes at the age of 83, the third original member of Wernher von Braun's rocket team to die in the same week. Schuler had retired from NASA in 1969, after a career overseeing measurement of instrumentation on test stands and rockets during static firings and overseeing flight instrumentation, guidance, and control on rocket flights.[539]

*14 July*

---

[535] NASA, "NASA Awards More Shuttle Work to USA," news release c98-h, 7 July 1998.
[536] John Fleck, "N.M. Scientists Track Galaxy Blast: Explosion Is Billions of Light Years Away," *Albuquerque Journal* (NM), 9 July 1998.
[537] Associated Press, "3 Rocket Scientists Die."
[538] Associated Press, "Failing Weather Satellite Replaced," 10 July 1998.
[539] Associated Press, "3 Rocket Scientists Die."

NASA announced the establishment of a Near-Earth Object Program Office at its Jet Propulsion Laboratory (JPL), responsible for coordinating "NASA-sponsored efforts to detect, track, and characterize potentially hazardous asteroids and comets that could approach Earth," and with facilitating communication between scientists and the public, were NASA to discover any potentially hazardous objects. NASA selected JPL to host the new program, because of "its expertise in tracking precisely the positions and predicted paths of asteroids and comets."[540]

*15 July*
NASA released "unusually detailed" images of Jupiter's moon Ganymede taken by the Galileo probe in June 1996 and June 1997. Although the images did not indicate the existence of any form of life, they showed that life had been possible in Ganymede's oceans billions of years ago. At that time, the moon had water, heat, and organic material, brought by comets from outer space. The images also revealed geological formations, including mountain-like ridges and a valley system of volcanoes. Early in the moon's history, the volcanoes had supplied water for Ganymede's oceans, which are now frozen.[541]

NASA's Inventions and Contributions Board announced the winners of its 1998 Software of the Year Award. The first winner, Tempest, a breakthrough technology "originally developed to support the science experiments on the ISS," had "spawned new markets," and NASA predicted that the technology would continue to do so. NASA's Ames Research Center had designed the second winner, Center TRACON Automation System Software, a set of three software tools for managing air-traffic–control systems at major airports, to optimize flight operations. Analyzing and predicting aircraft paths, the software created visual representations of arriving traffic flow, providing controllers with "up-to-the-second advisories of information" for pilots, thus reducing time between landings to a minimum. The Federal Aviation Administration had chosen the software for immediate implementation at all major airports.[542]

NASA officially opened its new communications terminal on Guam, to provide "global, full-time and real-time communications support for NASA's Space Network customers." The Guam ground station replaced the interim ground terminal in Canberra, Australia, established to provide continuous, full-time, real-time communications support for NASA's Compton Gamma Ray Observatory, after it had "suffered an on-board tape recorder failure in March 1992."[543]

*16 July*
The joint committee of NASA and the European Space Agency investigating the Solar and Heliospheric Observatory (SOHO) incident issued a preliminary report. The committee found that several errors had led to the loss of the satellite's signal. First, Mission Control had "unknowingly activated" two preprogrammed sequences, which lacked commands to reorient the spacecraft properly. The resulting anomalies had caused SOHO to enter into emergency Sun reacquisition mode (ESR). A "rapid decision" to turn off one of the craft's gyroscopes had

---
[540] NASA, "NASA Establishes Near-Earth Object Program Office at Jet Propulsion Laboratory," news release 98-123, 14 July 1998.
[541] Paul Hoversten, "Photos Suggest Life Could Have Existed on Jupiter's Moon," *USA Today*, 16 July 1998; *New York Times*, "Hints of an Ocean on Jupiter's Moon Yield Questions About Life," 16 July 1998.
[542] NASA, "Space Flight, Aviation Programs Receive NASA Software Award," news release 98-124, 15 July 1998.
[543] NASA, "New NASA Facility Will Complete Worldwide Communications Coverage," news release 98-122, 13 July 1998.

ultimately triggered two more ESRs, sending the spacecraft "flailing out of control." Controllers also discovered that "undetected failures," which had occurred sometime earlier in the year, "had taken out three of the four batteries."[544]

*17 July*
U.S. Senator Jeff Sessions (R-AL) offered an amendment to the Subcommittee on Veterans Administration, Housing and Urban Development, and Independent Agencies' FY 1999 appropriations bill, adding US$33 million to the US$150 million budget for NASA already approved by the Senate Committee on Appropriations. The Senate defeated the amendment by a vote of 58 to 37. President William J. Clinton's proposed budget had requested a US$183 million reduction in NASA's budget.[545]

*20 July*
The data transmission system of Galileo malfunctioned during its fifth flyby of Jupiter's moon Europa when it went into safe mode. Although NASA engineers activated a backup system, the transmission system had lost all data collected when the craft passed within 1,141 miles (1,835 kilometers) of the moon. NASA was investigating the malfunction. NASA had launched Galileo in 1989, and the probe had entered Jupiter's orbit in 1995, collecting data from moons Io, Europa, and Callisto during its initial two-year mission, completed in December 1997. NASA had extended the mission for two additional years.[546]

*21 July*
Alan B. Shepard Jr., the first American in space, died in his sleep at the age of 74 after a two-year battle with leukemia. NASA Administrator Daniel S. Goldin remembered him for being "one of the original seven Mercury astronauts, for being the first American to fly in space, and for being one of only 12 Americans to step on the Moon." A former U.S. Navy pilot, Shepard had first traveled to space on 5 May 1961, just 23 days after the Soviet cosmonaut Yuri Gagarin became the first human to orbit Earth. Shepard, the lone astronaut in the small space capsule, *Freedom 7*, rocketed 116 miles (187 kilometers) above Earth, as millions of people watched him live on television. Shepard's spaceflight proved a significant morale-boosting moment for Americans, in the midst of the Cold War with the Soviet Union, which until then had appeared to be winning the space race. After the flight, President John F. Kennedy presented Shepard with the Distinguished Service Medal, challenging the nation to put a man on the Moon by the end of the decade. Shepard made history again in 1971, when he commanded *Apollo 14* on its nine-day mission to the Moon. He delighted Americans watching the expedition on television, when, taking a break from collecting Moon rocks, he hit two golf balls with an improvised club. Shepard, who had retired from NASA and the U.S. Navy in 1974, received the Congressional Space Medal of Honor in 1979. Shepard also engaged in philanthropic pursuits, such as investing

---

[544] NASA, "Efforts To Recover SOHO Spacecraft Continue as Investigation Board Focuses on Most Likely Causes," news release 98-125, 16 July 1998; James Glanz, "Chain of Errors Hurled Probe into Spin," *Science* 281, no. 5376 (24 July 1998): 499; United Press International, "UPI Science News," 28 July 1998.
[545] Federal Document Clearing House, "Senator Sessions Offers Amendment To Increase NASA's Budget by $33 Million," government news release, 17 July 1998.
[546] Agence France-Presse, "Galileo Probe Suffers Malfunction in Jupiter Moon Flyby," 22 July 1998; Associated Press, "Spacecraft Galileo Working Again," 24 July 1998.

in the establishment of the Astronaut Foundation, which provides scholarships. The Astronaut Foundation had also founded the Space Camp program for young people.[547]

*22 July*
NASA named astronaut Terrence W. Wilcutt to replace James D. Halsell Jr. as NASA's Manager of Operational Activities at Star City, Russia. Wilcutt was the 10th astronaut to serve in the rotational position, supporting the training and preparation of NASA astronauts at Gagarin Cosmonaut Training Center and acting as a liaison between NASA and Russian management. Wilcutt had flown on three missions: the first, in 1994, as the pilot on Mission STS-68, to study the Earth's surface; the second, in 1996, as the pilot for the fourth Shuttle–*Mir* docking mission, STS-79; and the third, in 1998, as the commander of Mission STS-89, the eighth docking mission.[548]

*23 July*
NASA announced that the Hubble Space Telescope (HST) had found the youngest cluster of stars ever observed in a galaxy near the Milky Way. The baby stars were "enveloped in a cloud of luminous gases," in the galaxy known as the Small Magellanic Cloud, about 200,000 light-years from Earth. The HST had discovered the 50 new stars, each 300,000 times as bright as the Sun, "concentrated in a region only 10 light-years across in a cloud known as N81." Before the capture of the HST's high-resolution images, astronomers had referred to the area as The Blob.[549]

NASA announced that engineers had repaired the malfunctioning data transmission systems aboard Galileo. Although both subsystems were "working redundantly, as they had been designed to do," it would be up to a week before the probe could resume transmitting scientific data to Earth. Project engineers believed that debris had short-circuited a signal line, causing computer resets.[550]

Two competing teams using NASA's Rossi X-ray Timing Explorer (RXTE) discovered the first known accretion-powered millisecond pulsar. The scientists believed the pulsar to be the missing link between two known types of stars: 1) "old, accreting neutron stars, which are powerful sources of [x]-rays generated from the material they are gobbling up from their companions," and 2) even older "radiowave emitting pulsars that are rotating very rapidly and slowing down gradually." Because the new star, designated SAX J1808.4-3658, was both emitting x-rays and spinning rapidly, scientists believed it was the link between the accreting and the radiowave-

---

[547] NASA, "Statement of NASA Administrator Daniel S. Goldin on the Death of Alan Shepard," news release 98-128, 22 July 1998; NASA, "Alan Shepard, First American Astronaut, Dies at 74," news release 98-131, 22 July 1998; Federal Document Clearing House, "John Glenn Remarks on Alan Shepard," government news release, 22 July 1998; Paul Hoversten, "Shepard's Ride Lifted USA into the Space Age," *USA Today*, 23 July 1998; Jennifer Harper, "Alan Shepard, First American in Space, Dies at 74: 'Pioneer's Pioneer' also Led Apollo 14, Golfed on Moon," *Washington Times*, 23 July 1998.
[548] NASA, "Astronaut Wilcutt Replaces Halsell in Star City, Russia," news release 98-127, 22 July 1998.
[549] NASA, "Nearby Star Cluster Yields Insights into Early Universe," news release 98-132, 23 July 1998; *Washington Times*, "Hubble Telescope Finds Family of Stars," 24 July 1998; *New York Times*, "Genesis in a Cosmic Firestorm," 28 July 1998.
[550] Agence France-Presse, "NASA Repairs Galileo Probe's Glitch," 24 July 1998; Associated Press, "Spacecraft Galileo Working Again," 24 July 1998.

emitting neutron stars. Michiel van der Klis and Rudy Wijnands of the University of Amsterdam found the new pulsar—called the "Holy Grail of X-ray astronomy"—and measured the time between its rapid x-ray pulses, to derive its rotation rate. Van der Klis explained that, based on the fact of accretion, astrophysicists had long theorized that millisecond pulsars existed, but this was the first time that one had been "caught in the act." Accretion, the process of drawing gas from a nearby "companion" star, causes pulsars to heat up, emitting x-rays. After accretion ends, high-velocity beams of subatomic particles continue to "blow material off the companion," eventually causing the companion to vanish. Deepto Chakrabarty and Edward H. Morgan of the Massachusetts Institute of Technology led the team that found the pulsar's 2-hour orbital period, measured the orbit, and inferred the presence of the companion star. Tod E. Strohmayer, a member of the RXTE team at NASA's Goddard Space Flight Center, suggested, "[x]-ray and particle beam ablation may explain why millisecond pulsars are often found alone, despite the fact that they required a companion star to speed up." Pulsars "vaporize" their companions, thus hiding the evidence, which is, as Strohmayer described it, the "stellar version of the perfect crime."[551]

*27 July*

NASA announced that astronomers had located the "wayward" SOHO satellite after losing radio contact on 24 June, when a programming error on the ground caused its solar panels to switch positions. Two U.S. radio telescopes had located the satellite rotating slowly near its original position, approximately 1 million miles (1.6 million kilometers) from Earth. Engineers had calculated its exact location after NASA's Deep Space Network in Goldstone, California, "intercepted the echo of a radio signal transmitted by the giant radio telescope of the National Astronomy and Ionosphere Center in Arecibo, Puerto Rico." The spacecraft's slow rotation on its axis indicated that it had incurred only minor damage and that its solar panels would soon face the Sun again.[552]

*29 July*

Aaron A. Crayford and Cody Grosskopf, the two Cloverdale, California, teenagers accused of hacking into military and government computers in January and February 1998, pled guilty to federal hacking charges. The U.S. Attorney's Office recommended probation and strict restrictions on the boys' computer use and modem access. Crayford and Grosskopf had reportedly hacked into U.S government Web sites, including the Lawrence Livermore National Laboratory, the Naval Undersea Warfare Center, the Naval Postgraduate School, NOAA, NASA, the U.S. Air Force, and Pearl Harbor naval base; Web sites of foreign governments, Taiwan and the United Arab Emirates; and Web sites of a number of universities, including the University of California at Berkeley, the University of California at Santa Barbara, the University of California at Los Angeles, the University of California at Davis, California Polytechnic Institute, Columbia University, and Harvard University.[553]

---

[551] NASA, "Newly Discovered Stellar Cannibal Provides Missing Link," news release 98-129, 22 July 1998; Rudy Wijnands and Michiel van der Klis, "A Millisecond Pulsar in an X-ray Binary System," *Nature* 394, no. 6691 (23 July 1998): 344–346; Deepto Chakrabarty and Edward H. Morgan, "The Two-Hour Orbit of a Binary Millisecond X-ray Pulsar," *Nature* 394, no. 6691 (23 July 1998): 346–348.
[552] Agence France-Presse, "U.S.-European Satellite Located by Engineers," 28 July 1998.
[553] Jody Kleinberg, "Cloverdale Hackers Plead Guilty Probation Recommended for 2 Teens," *Press Democrat* (Santa Rosa, CA), 30 July 1998.

The U.S. House of Representatives voted 323 to 109 against an amendment, sponsored by Representative Timothy J. Roemer (D-IN), which would have eliminated the ISS program before the launch of the first pieces of the laboratory into orbit. Commenting on the hundreds of thousands of pounds of hardware already constructed, Representative David Weldon Jr. (R-FL) remarked that "to pull the plug now seemed inappropriate to a lot of people." Despite bad news throughout the past year, requiring NASA Administrator Daniel S. Goldin to visit Capitol Hill repeatedly to discuss cost overruns and delays, the vote was the "strongest endorsement ever for the station." The station had survived by only one vote in 1993, the year President William J. Clinton had "reorganized the project and invited Russia to join the international consortium" and to build the laboratory.[554]

NASA announced completion of qualification tests of the first wing assembly for its X-34 technology demonstrator, explaining that the prime contractor, Orbital Sciences Corporation, had integrated the wing assembly with the test article's fuselage, marking a major milestone in the program.[555]

*30 July*
NASA announced the selection of veteran cosmonaut Sergei K. Krikalev for the crew of Mission STS-88, the first American ISS assembly mission. In December 1998, Krikalev would join aboard *Endeavour* Commander Robert D. Cabana, Pilot Frederick W. "Rick" Sturckow, and Mission Specialists Nancy J. Currie, Jerry L. Ross, and James H. Newman. NASA had assigned Mission STS-88 the task of attaching the U.S.-built *Unity* module to the Russian-built *Zarya* control module, scheduled to launch in November 1998. Krikalev had been a member of two *Mir* space station crews and had flown aboard the Space Shuttle as a member of the Mission STS-60 crew in February 1994. During Mission STS-60, Krikalev had operated the robotic arm and supported a variety of experiments in materials science.[556]

*31 July*
Kenneth J. Szalai ended his 34-year NASA career to join IBP Aerospace Group Inc. as the company's president and chief operating officer. In his last position at NASA, Szalai was Director of Dryden Flight Research Center (DFRC) in Edwards, California. Szalai had directed DFRC since December 1990. Before becoming the Center's director, Szalai had been head of the research engineering division at DFRC for eight years. NASA had awarded Szalai, an internationally recognized expert on flight control and flight systems, the Exceptional Service Medal in recognition of his work on the first Digital Fly-by-Wire aircraft. He had also received NASA's Outstanding Leadership Medal during his NASA tenure. NASA Administrator Daniel S. Goldin appointed Kevin L. Petersen Acting Director of DFRC, effective 1 August.[557]

## AUGUST 1998

*1 August*

---

[554] Tamara Lytle, "NASA's 'Floating Lemon' Gets Sweet House Boost," *Chicago Tribune*, 30 July 1998.
[555] NASA, "X-34 Completes Critical Milestone," news release 98-136, 29 July 1998.
[556] NASA, "Future Space Station Resident Joins Assembly Crew," news release 98-137, 30 July 1998.
[557] Jim Skeen, "Dryden Director To Leave NASA for Private Sector," *Los Angeles Daily News*, 11 July 1998.

NASA and the Federal Aviation Administration announced the winners of the 1998 National General Aviation Design Competition, in which undergraduate and graduate students, enrolled at U.S. engineering schools "participating in a major national effort to rebuild the U.S. general aviation sector," compete individually or as teams. A 27-member undergraduate team from Virginia Polytechnic Institute won first place for its design, called VicTor, a single-engine, four-seat, high-performance aircraft. VicTor's special design elements included an ergonomic cockpit with adjustable, side-control sticks and dual airbags; a choice between two high-performance engines; and advanced technology instrument displays. The design also included an upgrade option, "to allow autonomous flight if it becomes a reality." A 15-member, Pennsylvania State University team won second place for their high-performance, two-person, single-engine, composite-fuselage, tractor-prop light airplane. Third place went to 13 undergraduate students from the University of Virginia for their computer program predicting drag in the design of new small passenger airplanes.[558]

*3 August*
The Solar and Heliospheric Observatory (SOHO) satellite, out of contact with ground controllers since 25 June, responded to signals sent by NASA's Deep Space Tracking Network antennas in Canberra, Australia. Although the signals contained no data, SOHO transmitted a series of short radio bursts, indicating to space scientists that the craft was still able to receive ground commands and, therefore, that they might still have a chance of salvaging it.[559]

*7 August*
NASA announced that Boeing Company's cost overruns for the International Space Station (ISS) had increased by about 30 percent to US$783 million, and that NASA intended to "further cut the company's bonus fees." In reporting to NASA the latest cost overruns, Boeing had explained that "continued production and testing and the need to set aside money for contingencies" had caused the latest increase, US$183 million more than the US$600 million previously estimated. Under the terms of NASA's cost-plus-fee contract, Boeing would lose an additional US$27.5 million in bonus fees, besides the US$78 million reduction that NASA had imposed.[560]

*8 August*
Following ground control efforts during recent weeks to partially recharge the spacecraft's batteries, SOHO successfully transmitted telemetry data to NASA's Deep Space Tracking Station at Goldstone, California. The 1-minute burst of data contained information about SOHO's temperature and the state of some of its electrical systems, damaged by the extreme cold. The data also indicated that ground control had successfully recharged one of the two batteries. The spacecraft remained in the *L-1 Langrangian point* in space, which is a point 1 million miles (1.6 million kilometers) away from Earth, where gravity forces are stable and "an object will not change its position relative to the Earth."[561]

---

[558] NASA, "NASA and FAA Announce Design Competition Winners," news release 98-140, 1 August 1998.
[559] NASA, "SOHO Spacecraft Contacted," news release 98-145, 4 August 1998; Frank Sietzen Jr. for United Press International, "Contact Re-established with Wayward Satellite," 4 August 1998.
[560] *Seattle Times* (WA), "Boeing Reports More Space Station Overruns," 7 August 1998.
[561] NASA, "New Information from SOHO Increases Chances for Recovery," news release 98-149, 11 August 1998; Frank Sietzen Jr. for United Press International, "SOHO Sun Satellite Comes to Life, Briefly," 12 August 1998; Associated Press, "Fresh Signals from Crippled Satellite Revive Hope for Sun-Studying Mission," 13 August 1998.

*11 August*
Hughes Electronics Corporation concluded its investigation into the 19 May failure of PanAmSat's *Galaxy IV* satellite and the similar problems that later had affected two other satellites. Metallic crystals growing in space, "less than the width of a human hair," had caused tin-plated electrical switches to short, disabling navigational devices that kept the satellites oriented toward Earth-based controllers. However, the investigation had failed to uncover the source of the failure of *Galaxy IV*'s backup system, which led to the widespread blackout on 19 May.[562]

*12 August*
A U.S. Air Force Titan IV rocket, carrying a classified satellite for the National Reconnaissance Office of the Pentagon, exploded 1 minute after liftoff from Cape Canaveral, Florida. More than 20 stories high and weighing nearly 2 million pounds (907,000 kilograms), the Lockheed Martin–built Titan IV was the largest crewless rocket used in the United States. The rocket used solid-fuel boosters similar to those used to launch NASA's Space Shuttle. The explosion destroyed about US$1.3 billion worth of hardware, making it one of the worst American rocket-launch failures to date. The rocket was about 20,000 feet (6,100 meters) above the Atlantic when it exploded, and the ocean carried the debris and toxic substances away from land. This was the second time a Titan IV had exploded during launch; the first explosion occurred in 1993.[563]

*13 August*
A *Soyuz TM-28* spacecraft blasted off from Baikonur Cosmodrome in Kazakhstan, carrying the penultimate crew of the Russian space station *Mir*. The three-member crew was composed of former presidential defense adviser Yuri M. Baturin, Flight Commander Sergei V. Avdeyev, and Engineer Gennady I. Padalka.[564] Baturin was the first Russian bureaucrat to travel into space, and his inclusion in the mission attracted "unusually heavy Russian media coverage." The former Kremlin adviser had trained alongside professional cosmonauts, passing all examinations. Earlier in the year, President Boris N. Yeltsin had fired space physicist Baturin without an explanation, and Baturin had focused since then on training for the mission to *Mir*.[565]

As part of an overall restructuring, the Boeing Company announced plans to reduce its workforce by 12 percent over the next 18 months, through the layoffs of as many as 28,000 workers. Much of the restructuring plan focused on the company's commercial airplane business, affecting those workers who were once part of McDonnell Douglas Corporation, which Boeing had purchased in 1997. In contrast, the company reported it was making strides in its defense and space work.[566]

*18 August*

---

[562] Aaron Pressman for Reuters, "Tiny Crystals Waylaid Satellites, Hughes Says," 11 August 1998.
[563] Marcia Dunn for Associated Press, "Rocket Explosion," 12 August 1998; Kathy Sawyer and William Harwood, "Rocket Explodes, Destroying Spy Satellite," *Washington Post*, 13 August 1998.
[564] Mikhail Metzel for Associated Press, "Next Mission Blasts Off for Mir Space Station," 13 August 1998.
[565] Shavkat Rakhmatullayev for Reuters, "Russia Blasts First Bureaucrat into Space," 13 August 1998; Associated Press, "Mir Mission Under Way Despite Russian Economic Troubles," *Florida Today* (Brevard, FL), 14 August 1998.
[566] Tim Smart, "Boeing To Retool, Lay Off Thousands; Condit Cites Need To Reduce Costs," *Washington Post*, 14 August 1998.

NASA awarded Aerojet, a GenCorp Inc. company, a US$16.4 million contract to provide a deorbit propulsion stage for the X-38 technology demonstrator vehicle. The contract for design, development, manufacture, test, and delivery of one propulsion stage for the first X-38 spaceflight test included a second option for five operational propulsion units for the crew return vehicle (CRV), if NASA selected the X-38 design for the CRV. Engineers at Dryden Flight Research Center were flight-testing the X-38, a prototype spacecraft intended to lead to the development and construction of a CRV for the ISS.[567]

*20 August*
A fire that started in a shed containing an air compressor spread to another building of NASA's Goddard Space Flight Center (GSFC), causing extensive damage. No one was injured, and firefighters were able to stop the blaze before it reached laboratories containing hazardous and flammable materials. Fire investigators had not determined the cause of the fire. NASA did not permit the investigators to survey the damage immediately, because of sensitive material housed in the affected building.[568]

*21 August*
Phillip Z. Tapper, emergency coordinator for NASA's GSFC, announced that exposure to smoke and water during the previous day's fire had damaged parts of two spacecraft. Tapper said that engineers had not yet determined the exact damage to the heat protection panels, which they were planning to install in the Hubble Space Telescope and another spacecraft.[569]

NASA and U.S. Air Force officials announced the hiring of Space Gateway Support of Herndon, Virginia, to run base operations at Kennedy Space Center (KSC), Patrick Air Force Base, and Cape Canaveral Air Force Station in Florida. Johnson Controls and EG&G Florida at KSC had previously held base-operations contracts at the military bases. The new joint base-operations-support contract was worth US$1 billion for five years, with the option to renew for another five years. In consolidating the contracts, the agencies intended to cut costs. According to officials, continuing separate base-operations contracts would have cost 30 percent more. Other primary customers under the contract included government contractors for NASA and U.S. Air Force spaceflight operations, payload ground operations, life sciences, expendable vehicles, and launch operations and support.[570]

Fritz Haber, a German engineer recruited after World War II to work on the American space program, died at the age of 86 in Connecticut. An aeronautical engineer, Haber developed a way to simulate a gravity-free environment by flying a plane in a roller-coaster pattern. NASA first used this simulation to train the astronauts in the Apollo program. Before coming to the United States, Haber had developed a way to transport missiles by piggybacking them on another aircraft. Engineers had never used his piggyback design during World War II, but NASA had

---

[567] *Defense Daily*, "Aerojet Wins $16.4 NASA Contract for X-38 Propulsion," 20 August 1998.
[568] Associated Press, "NASA Fire," 20 August 1998.
[569] Associated Press, "NASA Fire Damages Spacecraft Parts," 21 August 1998.
[570] NASA, "SGS Awarded NASA and Air Force Joint Base Contract," contract announcement C98-m, 21 August 1998; Joni James, "Base Operations Under 1 Contract," *Orlando Sentinel* (FL), 22 August 1998.

implemented his idea, carrying the Space Shuttle on a modified Boeing 747 from its landing site in California back to its launch site in Florida.[571]

*25 August*
Cosmonauts Talgat A. Musabayev and Nikolai M. Budarin returned to Earth from a 207-day sojourn aboard *Mir*, along with former Kremlin aide Yuri M. Baturin, who had accompanied the cosmonauts' replacements to the space station, spending 12 days in space. The two cosmonauts arrived in Russia to find a devalued ruble and the return of Viktor S. Chernomyrdin as Acting Prime Minister. Russian President Boris N. Yeltsin had not yet fired Chernomyrdin's government at the time Musabayev and Budarin left for *Mir*, and the two cosmonauts had missed the entire four-month term of Prime Minister Sergei V. Kiriyenko. Yeltsin had dissolved Kiriyenko's government the week before the cosmonauts' return flight to Earth. Because Yeltsin had devalued the ruble, the two cosmonauts also found that their pay had diminished.[572]

*26 August*
The newly developed Delta III rocket exploded during liftoff, destroying the *Galaxy X* communications satellite it was carrying into orbit. The launch was the maiden flight of Boeing's newest version of the Delta rocket, designed to carry 8,400 pounds (3,800 kilograms), twice the payload of the previous version.[573]

*27 August*
NASA announced the arrival of international astronaut candidates at Johnson Space Center, where they began training as members of the 1998 astronaut class. The international candidates, from Brazil, Canada, France, Germany, and Italy, were Léopold Eyharts, Paolo A. Nespoli, Hans W. Schlegel, and Roberto Vittori of the European Space Agency; Bjarni V. Tryggvason of the Canadian Space Agency; and Marco C. Pontes of the Brazilian Space Agency.[574]

## SEPTEMBER 1998

*2 September*
Pentagon officials announced that, on 12 August, a split-second power outage had caused a Titan IV rocket to explode while launching a secret spy satellite into space. A battery powering the guidance system had failed, interrupting the flow of electricity for a fraction of a second. After the restoration of battery power, the guidance control system had sent a command causing the rocket to pitch down and to the right, while traveling at 675 miles (1,100 kilometers) per hour at 17,000 feet (5,200 meters). The rocket had begun to break up, and the malfunction had triggered the automatic destruct system. In addition, when it had begun to explode, range safety officers had sent the rocket destruct commands, to ensure it would not threaten the coastal population.[575]

*3 September*

---

[571] Ford Burkhart, "Fritz Haber, 86, Dies; Simulated Weightlessness of Space," *New York Times*, 29 August 1998.
[572] Reuters, "Three Russian Cosmonauts Return Safely to Earth," 25 August 1998; Shavkat Rakhmatullayev for Reuters, "Russia's Orbiting Bureaucrat Returns to Earth," 25 August 1998.
[573] Seth Schiesel, "New Type of Rocket To Carry Satellites Explodes in Liftoff," *New York Times*, 27 August 1998.
[574] NASA, "International Candidates Join 1998 Astronaut Class," news release 98-155, 27 August 1998.
[575] Todd Halversen, "Titan 4 Blast Blamed on Brief Battery Power Outage," *Florida Today* (Brevard, FL), 3 September 1998.

The investigation board of engineers and officials from NASA and the European Space Agency released their final report, determining how controllers had lost communication with the Solar and Heliospheric Observatory (SOHO) satellite on 24 June. The board's findings, presented at a joint news conference held in Washington, DC, and Paris, concluded that no anomalies existed on the spacecraft; the loss of SOHO "was a direct result of operational errors, a failure to monitor the status of the spacecraft adequately, and the use of insufficiently tested rescue procedures." However, the board emphasized that no single cause or particular person was responsible for the incident.[576]

NASA announced that Lunar Prospector's gamma-ray spectrometer had returned data about the Moon's elemental composition. Scientists had used the data to develop the first global maps delineating compositional variations of thorium, potassium, and iron ore over the lunar surface. The maps provided insights into how the Moon's crust had developed. Prospector's magnetometer and electron-reflectometer data indicated that the Moon contained magnetized rocks on its upper surface, which had created the two smallest known magnetospheres in the solar system. The craft's Doppler gravity experiment had enabled scientists to develop the "first precise gravity map of the entire lunar surface," showing seven previously unknown lava-filled craters, which cause gravitational anomalies. NASA also announced that Lunar Prospector had supplied additional evidence of water-ice deposits on both poles of the Moon.[577]

*4 September*
NASA awarded contracts to Boeing Information, Space, and Defense Systems; Kelly Space and Technology; Lockheed Martin Astronautics; and Orbital Sciences Corporation Space Access. The five contracting companies would study ways that NASA could meet its requirements for human spaceflight at a lower cost. The studies would examine three scenarios: 1) keeping the Shuttle operational until 2020, 2) replacing the Shuttle when it was cost-effective to do so, and 3) developing an alternative plan in case NASA's funding remained at FY 1999 levels.[578]

The journal *Science* published data gathered from NASA's Lunar Prospector, indicating the presence of hydrogen, often bundled as water molecules, on the north and south poles of the Moon. Alan B. Binder, Chief Scientist for the Lunar Prospector mission, explained that the spacecraft had detected an abundance of hydrogen, and that the science team had interpreted the data to mean that the Moon has 1–10 billion tons (910–9,100 kilograms or 0.9–9.1 billion tonnes) of water. In a previous mission to the Moon to collect data, the spacecraft Clementine had found radar indications of water on the Moon's south pole. Paul D. Spudis, a member of the Clementine science team, remarked that Prospector's discovery was significant because "the presence of lunar water has been confirmed by two different research methods."[579]

---

[576] Warren E. Leary, "Satellite To Study Sun Is Reviving from Dead," *New York Times*, 4 September 1998.
[577] NASA, "Latest Lunar Prospector Findings Indicate Larger Amounts of Polar Water Ice," news release 98-158, 3 September 1998.
[578] NASA, "NASA Contracts for Future Space Transportation Studies," news release c98-n, 4 September 1998; *Defense Daily*, "NASA Awards Future Space Transport Study Contracts," 8 September 1999.
[579] D. J. Lawrence et al., "Global Elemental Maps and the Moon: The Lunar Prospector Gamma-Ray Spectrometer," *Science* 281, no. 5382 (4 September 1998): 1484–1489; Reuters, "Prospector Finds Water on Moon: NASA Spacecraft Locates Ice at the North and South Poles," *Washington Times*, 4 September 1998; Associated Press, "Man in the Moon May Have Plenty To Drink: Abundance of Water Likely; Scientists Foresee Colonization," *Chicago Tribune*, 4 September 1998.

*8 September*
Boeing successfully launched, from a California launch site, five satellites for the Iridium World Communications network, aboard a Delta II rocket. The Delta II launched two weeks after the "disastrous maiden flight" of the new Delta III rocket, which had exploded during liftoff. Officials announced on 5 September that a problem with the rocket's control system had caused the explosion, but that they had not yet determined the exact chain of events. The five satellites launched aboard the Delta II rocket replaced satellites that had malfunctioned and could not be included in the Iridium telecommunications constellation.[580]

*9 September*
A Zenit rocket, launched from Baikonur Cosmodrome in Kazakhstan, failed when a computer error caused a premature engine shutdown during the second-stage burn. The rocket fell, crashing in Siberia. The rocket's failure destroyed 12 of Globalstar Inc.'s communications satellites.[581]

*10 September*
Ghassem R. Asrar, NASA Associate Administrator for Earth Science Enterprise, testified before the U.S. House Science Subcommittee on Space and Aeronautics about budget concerns and delays associated with the Earth Observing System program. House Subcommittee Chairperson Dana Rohrabacher (R-CA) had called the hearing to discuss numerous delays in NASA's Earth Science Enterprise missions. Asrar explained the reasons for the delays, indicating that NASA would work with Raytheon Systems to deliver a scaled-back data system within budget, in time to support two delayed missions. Subcommittee member, Representative David Weldon Jr. (R-FL), "questioned the legitimacy of NASA's ownership of the Earth Science program," suggesting that a different agency, such as the National Academy of Sciences or the National Oceanic and Atmospheric Administration (NOAA), should manage the program. Weldon also questioned whether scientists would be able to make use of the vast amount of data they expected the program to generate. Robert S. Winokur, NOAA's Assistant Administrator for Satellite and Information Services, assured subcommittee members that scientists would be able to use the information. Winokur was responsible for coordinating NOAA's joint efforts with NASA to archive and retrieve environmental data.[582]

NASA announced that Jet Propulsion Laboratory (JPL) had selected Lockheed Martin Missiles and Space of Sunnyvale, California, and TRW Inc. Space and Electronics Group of Redondo Beach, California, as industry team members for the Space Interferometry Mission. The Interferometry Mission would measure the precise location of stars and search for planets orbiting nearby stars. The two contracts, which together exceeded US$200 million, included mission formulation and implementation.[583]

---

[580] Justin Ray, "Delta 2 Lifts Five Satellites into Orbit," *Florida Today* (Brevard, FL), 9 September 1998.
[581] Pavel Polityuk for Reuters, "Ukraine Appeals for Rocket Launch amid Kazakh Row," 13 July 1999; Globalstar Chronology, 13 July 1999, *http://www.astronautix.com/craft/glolstar.htm* (accessed 6 August 2008).
[582] Brian Berger, "Earth Science Faces Hurdles: NASA To Forfeit Data Capabilities To Put Key Program on Track," *Space News*, 14–20 September 1998.
[583] NASA, "Contractors Chosen for Space Interferometry Mission," news release 98-162, 10 September 1998.

*11 September*
NASA researchers released images and temperature readings collected by Mars Global Surveyor. The images revealed that the Martian moon Phobos is covered in a layer of dust at least 3 feet (0.9 meters) deep, the result of the impact of meteoroids, occurring over millions of years. Surveyor's thermal emission spectrometer revealed that, on Phobos, the temperature changes from -170°F to -25°F (-112°C to -32°C) between night and day. Scientists explained that dust accounts for some of the moon's drastic temperature change, because Phobos does not have an atmosphere enabling it to trap heat. Therefore, the surface's small particles have only 7 hours to absorb heat from the Sun, losing heat rapidly after sunset. Images of landslides on steep crater slopes indicated that Phobos's gravity, even at just 1/1,000$^{th}$ of that of Earth, is sufficient to pull objects downward.[584]

NASA Administrator Daniel S. Goldin appointed Arthur G. Stephenson, President of Oceaneering Advanced Technologies, as the new Director of Marshall Space Flight Center (MSFC) in Huntsville, Alabama. MSFC's previous director had retired in January, and at that time, NASA had appointed MSFC's Deputy Director Carolyn S. Griner as Acting Director of the Center. Before his tenure at Oceaneering Advanced Technologies, Stephenson had worked 28 years for TRW Inc. of California. His last position was as Director of TRW Inc.'s Space Transportation and Servicing Advanced Programs. During his 34-year career, Stephenson had worked on various projects related to MSFC, including the Orbital Maneuvering Vehicle, during the 1970s and 1980s; the Gamma Ray Observatory; automated rendezvous and docking; and the space-welding inspection, extravehicular-activity tool. He had also directed the International Space Station (ISS) robotic-system engineering support to the Boeing Company.[585]

Associate Administrator for the Office of Space Science Wesley T. Huntress Jr. announced the selection of two small spacecraft to undertake the first missions of NASA's University-class Explorers program, designed to "provide frequent flight opportunities for highly focused and relatively inexpensive science missions." NASA had selected the Cosmic Hot Interstellar Plasma Spectrometer spacecraft to study the "Local Bubble," a cloud of hot gas surrounding the solar system and extending approximately 300 light-years from the Sun. NASA selected the Inner Magnetosphere Explorer to study how Earth's Van Allen radiation belts respond to variations in the solar wind. NASA had scheduled both missions for 2001.[586]

*14 September*
At a ceremony marking its official opening, Virginia Governor James S. Gilmore III announced that Reston-based DynCorp was the first major investor in the Commercial Spaceport Launch Pad at Virginia Space Flight Center on Wallops Island, the nation's third commercial spaceport. The Virginia Commercial Space Flight Authority, a collaborative effort of the state, the Center for Innovative Technology at Old Dominion University in Virginia, NASA, and private industry, had developed the communications facility, located at one end of NASA's Wallops Flight Facility. The spaceport was one of 18 sites competing to launch the VentureStar™, a privately

---
[584] NASA, "Martian Moon Phobos Hip-Deep in Powder," news release 98-164, 11 September 1998; Jane E. Allen for Associated Press, "Martian Moon," 12 September 1998.
[585] NASA, "Arthur Stephenson Named To Head Marshall Space Flight Center," revised news release 98-163a, 11 September 1998; *Space News*, "Marshall Gets Director After Eight-Month Wait," 14–20 September 1998.
[586] NASA, "NASA Selects First University-Class Explorers," news release 98-165, 11 September 1998.

owned space plane still in development, which could eventually carry construction materials to the ISS.[587]

*16 September*
NASA regained control over SOHO and, after successfully thawing the craft's fuel, determined that its rocket thrusters were useable. NASA spokesperson William A. Steigerwald of Goddard Space Flight Center (GSFC) explained that flight controllers would check each individual system and instrument next, and would then reestablish full automatic gyroscopic control.[588]

Scientists studying images taken by NASA's Galileo spacecraft announced, in a joint teleconference with Cornell University and NASA's JPL, that Jupiter's rings appeared to consist of dust, created in the course of millions of years, as asteroids and comets crashed into the four small moons nearest the planet's surface—Metis, Adrastea, Amalthea, and Thebe. The images, captured during several hours in November 1996, when Galileo was able to observe the rings lit up by the Sun, had provided "unprecedented details of the tiny dust particles." Furthermore, scientists observed that, in the images, "the three rings stopped abruptly at different moons." Images from a second flyby in September 1997 had confirmed that "the depths of the three rings matched the depths of the moons' orbits," and that material in the rings "matched the dark red dust on the moons' surface." The scientists said that more moons than rings exist, because some of the moons supply dust to more than one ring.[589]

*21 September*
NASA announced that California Institute of Technology would continue to manage and operate NASA's JPL under a new five-year contract valued at US$6.25 billion. The contract would support such programs as the Mars Global Surveyor robotic exploration program; the Cassini mission to Saturn; NASA Origins Program missions, including the Space Infrared Telescope Facility; Earth-Observing spacecraft; and the Deep Space Network of communications antennas.[590]

*25 September*
NASA awarded its combined space operations contract to Lockheed Martin, covering operations at Goddard Space Flight Center in Greenbelt, Maryland; Johnson Space Center in Houston, Texas; Kennedy Space Center in Cape Canaveral, Florida; Marshall Space Flight Center in Huntsville, Alabama; and Jet Propulsion Laboratory in Pasadena, California. The contract, worth US$1.9 billion for the first five years, with a five-year renewal worth an additional US$1.5 billion, covered the cost of directing, monitoring, and downloading data from spacecraft. NASA

---

[587] Sonja Barisic, "Firm Gives a Big Boost to Spaceport," *Roanoke Times* (VA), 15 September 1998; Karen Jolly Davis, "$4.5 Million Contribution Gives Wallops Island Spaceport a Boost," *Virginian Pilot and the Ledger-Star* (Norfolk, VA), 15 September 1998.
[588] Frank Sietzen Jr. for United Press International, "Solar Probe Back Under Control," 17 September 1998.
[589] NASA, "Galileo Finds Jupiter's Rings Formed by Dust Blasted Off Small Moons," news release 98-167, 15 September 1998; *New York Times*, "New Images Clear Up a Jupiter Mystery: Its Rings Are Just Dust," 16 September 1998; Dennis Cauchon, "Scientists Solve Mystery of Jupiter's Rings: Pictures Suggest Asteroid Impacts Stirred Moon Dust," *USA Today*, 16 September 1998.
[590] NASA, "NASA Awards New Contract for Operation of Jet Propulsion Laboratory," news release C98-o, 21 September 1998.

estimated that the contract, consolidating work previously done under 17 separate contracts, would save US$1.4 billion over 10 years.[591]

*29 September*
NASA announced that an intense wave of gamma rays, emanating from a magnetar 20,000 light-years away from Earth, had struck Earth's atmosphere on 27 August. Stanford University professor Umran Inan remarked that the occurrence, which had a measurable effect on Earth, was extremely rare, since the event occurred outside the solar system. The radiation wave was "so powerful that it blasted sensitive detectors to maximum or off scale on at least seven scientific spacecraft in Earth-orbit and around the solar system."[592]

## OCTOBER 1998

*1 October*
On NASA's 40th anniversary, President of the Space Frontier Foundation Richard N. Tumlinson and NASA Administrator Daniel S. Goldin spoke before the U.S. House Science Subcommittee on Space and Aeronautics about space exploration and the possibility of privatizing space ventures. Tumlinson's organization advocated limiting NASA's role to supporting technology and research programs. He proposed commercializing the Shuttle fleet and privatizing NASA field centers, to reduce the costs of space exploration and eventual colonization. Goldin advocated reducing costs as well, but did not respond to Tumlinson's proposals. He also outlined new technologies that NASA was developing, to cut the cost of rockets and satellites. Goldin remarked that NASA "would form partnerships with industry for 'cutting-edge research development'," also emphasizing NASA's support of government loans to fuel the growth of the space industry in the United States.[593]

*2 October*
NASA and Russian Space Agency officials announced that Russia had agreed to sell research time aboard the International Space Station (ISS) for US$60 million, with payment dependent on the Russian agency's completion of critical milestones. The ISS collaborators, including representatives of the 16 countries participating in the ISS program, had reached the agreement during a week of meetings in Moscow. Russia had missed three target dates because of insufficient funds. The deal, part of NASA's efforts to bail out its partner and to "prevent costly new delays" in the construction and launch of the ISS, did not release Russia from its obligations to build the service module. As the Moscow meeting concluded, NASA officials also announced that the launches of the first ISS components, the *Zarya* and *Unity* modules, remained on

---

[591] NASA, "Lockheed Martin Space Operations Co. Awarded Contract To Manage NASA's Space Operations," contract announcement c98-p, 25 September 1998; Peter Behr, "Lockheed Team Wins NASA Contract; $3.4 Billion Big Beats Out Boeing," *Washington Post*, 26 September 1998; Anna Wilde Mathews, "Lockheed Group Wins NASA Pact for $3.44 Billion," *Wall Street Journal*, 28 September 1998.
[592] NASA, "Tremendous Gamma-Ray Flare Blasts Earth," news release 98-172, 29 September 1998.
[593] Frank Sietzen Jr. for United Press International, "UPI Science News," 2 October 1998; Frank Sietzen Jr. for United Press International, "NASA Urged To Sell Shuttles," 1 October 1998.

schedule, and that the international partners had agreed to reconvene to refine the ISS assembly sequence, at the launch of the *Unity* module from Kennedy Space Center (KSC).[594]

*5 October*
Pursuant to H. Res. 572, the U.S. House of Representatives agreed to an amended version of H.R. 1702, the Commercial Space Act of 1998.[595] The legislation gave the Federal Aviation Administration the authority to issue licenses to private companies for the launch of reusable space vehicles like NASA's Shuttle. Supporters of the measure stressed that enactment of the law would "better enable American companies to compete in an increasingly competitive international marketplace," warning that, without this law, the United States would risk losing the space industry to overseas markets. The legislation also required NASA to submit a report and an independently conducted market study to Congress, identifying commercial opportunities and evaluating commercial interest in the development of the ISS.[596]

George W. Lewis Jr., NASA aeronautical engineer and son of George W. Lewis, former research director of NASA's predecessor agency, the National Advisory Committee for Aeronautics, died at his home at the age of 81. He had worked for 30 years in Cleveland, Ohio, at Lewis Research Center, which NASA had named for his father.[597]

*6 October*
The Russian government agreed to pay a US$44.5 million debt to the Russian Space Agency, to ensure that the agency would be able to complete a long-delayed component of the new space station. Director General of the Russian Space Agency Yuri N. Koptev said that the government had agreed to pay the debt in three installments, in October, November, and December.[598]

*7 October*
U.S. Representative F. James Sensenbrenner Jr. (R-WI) convened a hearing before the House Committee on Science and Technology to discuss NASA's request for an additional US$60 million to purchase hardware from Russia for the ISS, over and above the US$1.2 billion in space-related funds already paid to Russia. NASA Administrator Daniel S. Goldin indicated that the additional funding was essential and that, without it, the program would risk termination. Sensenbrenner refused to support approval of the funding without a detailed explanation from U.S. Department of State and White House officials of why the program needed additional funds. Sensenbrenner had asked Jacob Lew of the Office of Management and Budget and Deputy Secretary of State Strobe Talbott to speak at the hearing, but both failed to appear. Sensenbrenner said he could not "go along with NASA's request to start bailing out the Russian

---

[594] NASA, "Statement Following Conclusion of Moscow Meetings," news release 98-176, 2 October 1998; Vladimir Isachenkov for Associated Press, "Russian Space Agency Makes Sale," 5 October 1998; William J. Broad, "To Buy Some Time, Russia To Sell U.S. Some Time in Space," *New York Times*, 5 October 1998.
[595] On 28 October 1998, Congress enacted H.R. 1702 into law with Pub. L. No. 105-303, the Commercial Space Act of 1998.
[596] Jim Abrams for Associated Press, "House Bills Help Federal Workers," 5 October 1998; Associated Press, "House Sends Senate Bill To Allow Private Space Launches," 5 October 1998.
[597] *Cleveland Plain Dealer* (OH), "George W. Lewis, 81, Was An Engineer at NASA," 9 October 1998.
[598] Dow Jones Newswire, "Russian Government To Pay $44.5 Mln Debt to Space Agency," 6 October 1998.

space program," stating that he had seen no convincing evidence that either NASA, or the White House, or Russia, "would make good use of the money."[599]

*8 October*
NASA announced that a long-exposure, infrared image, taken with a camera on the Hubble Space Telescope (HST), had uncovered the most distant galaxies ever seen. HST's Near Infrared Camera and Multi-Object Spectrometer (NICMOS) had detected galaxies that might be over 12 billion light-years away, depending on which cosmological models scientists used. NASA's Acting Associate Administrator for Space Science, Edward J. Weiler, deemed the discovery a major step toward fulfilling one of HST's key objectives: to search for the "faintest and farthest objects in the universe."[600]

*13 October*
NASA announced that two of its research pilots had been the first Americans to fly a modified Russian Tu-144 supersonic jetliner, participating in three evaluation flights over a two-week period during mid-to-late September, at Zhukovsky Air Development Center near Moscow. Pilots Robert Rivers of Langley Research Center and C. Gordon Fullerton of Dryden Flight Research Center had flown the jetliner as part of a jointly funded activity under the auspices of NASA's High-Speed Research Program and the Boeing Commercial Airplane Group. The purpose of the test flights was to gain operational experience and experimental flight data to develop an "environmentally friendly, second-generation, supersonic transport" in the United States.[601]

*14 October*
NASA announced that it had restructured and renamed its Office of Aeronautics and Space Transportation Technology, known as Code R, appointing retired U.S. Air Force Lieutenant General Spence M. Armstrong to head the newly reorganized Office of Aero-Space Technology. NASA's spokesperson said that NASA's management had created three new divisions within the Office—Goals, Programs, and Institutions—to help NASA achieve its goals. Major goals included accessing space more cheaply, with reusable launch vehicle programs; creating greater synergy between aeronautics and space transportation technology; decreasing aircraft accident rates, while tripling the nation's aviation-system capacity; diminishing aircraft noise and emissions; and reducing the cost of air travel and the cost of placing payloads in orbit. The role of the Goals Division would be to implement the challenging development of new technologies; that of the Programs Division would be to manage the various programs within the Office, and the Institutions Division would oversee "institutional problems at the field centers."[602]

*20 October*
NASA officials announced that U.S. Senator John H. Glenn Jr. would not participate in the melatonin portion of a sleep study planned for the upcoming Shuttle mission. The study's

---

[599] Paul Recer, "Angry Congressman Threatens To Withhold Space Station Money," *San Diego Union-Tribune* (CA), 8 October 1998.
[600] NASA, "Hubble Goes to the Limit in Search of Farthest Galaxies," news release 98-179, 8 October 1998.
[601] NASA, "NASA Pilots Fly Russian Tu-144LL Flying Laboratory," news release 98-186, 13 October 1998.
[602] NASA, "Aeronautics Enterprise Reorganizes and Changes Name," news release 98-187, 14 October; *Aerospace Daily*, "NASA Renames, Reorganizes Code R under Armstrong," 15 October 1998.

principal investigator, Charles A. Czeisler of Brigham and Women's Hospital in Boston said that Glenn had failed to meet one of the medical criteria for the study. Czeisler did not specify the reason, and Glenn declined to elaborate. Researchers had rated the melatonin experiment as one of two top priorities among the human studies to take place during Glenn's Shuttle flight. Four members of the Shuttle flight for the Neurolab mission, which occurred earlier in 1998, had participated in the melatonin study, and Chiaki Mukai, a member of Glenn's crew, would be a fifth participant. However, Glenn's disqualification meant that the scientists would be unable to test two of their four hypotheses.[603]

*21 October*
After experiencing two disastrous launch failures in 1996 and 1997, the European Space Agency (ESA) successfully launched its French-made Ariane 5 rocket and recoverable space capsule from Korou, French Guiana. The 1.6 million-pound (800-ton, 726,000-kilogram, or 726-tonne) rocket carried a 2.6-tonne (2.9-ton, 5,700-pound, or 2,600-kilogram) mock-up of a telecommunications satellite and a 2.8-tonne (3.1-ton, 6,200-pound, or 2,800-kilogram) recoverable capsule. With the successful recovery of the capsule from the Pacific Ocean, France became the fourth country to develop such technology, following the United States, Russia, and China. French scientists had designed the Ariane 5 rocket to carry payloads of 5.9–6.8 tonnes (6.5–7.5 tons, 13,000–15,000 pounds, or 5,900–6,800 kilograms) into geostationary orbit, using twice the power of the Ariane 4; to provide launches for low-altitude orbits and interplanetary space probes; and to lift the cargo vehicle that would be the ESA's contribution to the ISS.[604]

*22 October*
Scientists studying Galileo data published findings in the journal *Nature* describing the likely presence of a subsurface, liquid, salty ocean on Callisto, Jupiter's second largest moon. Margaret G. Kivelson, space physics professor at the University of California at Los Angeles and NASA's Principal Investigator for Galileo's magnetometer instrument, explained that data showed variability in Callisto's magnetic field similar to that of another Jupiter moon, Europa. However, neither Callisto's atmosphere, which lacks charged particles, nor Callisto's icy crust, is a good conductor for generating a magnetic field. Kivelson explained that a layer of salty melted ice could carry "sufficient electrical currents to produce the magnetic field." In addition, she noted that Galileo data indicated electrical currents flowing in opposite directions at different times, a "key signature consistent with the idea of a salty ocean." However, scientists had concluded that, unlike Europa, Callisto was unlikely ever to support life, even if it had an ocean, because of its insufficient energy sources. Callisto's only apparent source of heat is radioactive elements, whereas Europa's closer position to Jupiter means it possesses tidal energy.[605]

*24 October*

---

[603] Lawrence K. Altman, "Glenn Unable To Perform Experiment Planned for Space Flight," *New York Times*, 21 October 1998.
[604] John-Thor Dahlburg, "Third Try's the Charm for European Rocket," *Los Angeles Times*, 22 October 1998; Frank Sietzen Jr. for United Press International, "Ariane Capsule Test Called a Success," 21 October 1998.
[605] NASA, "Jupiter's Moon Callisto May Hide Salty Ocean," news release 98-192, 21 October 1998; K. K. Khurana et al., "Induced Magnetic Fields as Evidence for Subsurface Oceans in Europa and Callisto," *Nature* 395, no. 6704 (22 October 1998): 777–780.

NASA successfully launched Deep Space 1, a spacecraft that could "think for itself." Driven by ion propulsion, Deep Space 1 was the first craft to rely on this technology for its primary propulsion. The craft's destination was an asteroid 120 million miles (193 million kilometers) away from Earth. Although tracking stations encountered initial difficulties receiving signals from the spacecraft, Deep Space 1 communicated 2 hours after liftoff that it was on "the right course" and "in good shape." NASA scientists expected the spacecraft to travel 450 million miles (724 million kilometers) to catch up with the moving asteroid. Without the ion-propulsion technology, it would have required 10 times more fuel to reach its target. The craft was also flying with 10 other "futuristic technologies," such as lens-covered solar arrays to generate additional power, and a radio beacon to communicate with ground controllers without their prompting the probe from Earth.[606]

*27 October*
The third and final Progress cargo spacecraft of the year, Progress M-40, arrived at *Mir*, docking smoothly and delivering food, water, fuel, a giant space mirror, and other scientific equipment. NASA had delayed the cargo craft's launch several times since its originally scheduled launch date in August 1998.[607]

Officials from the National Oceanic and Atmospheric Administration reported that they had placed GOES-8, the weather satellite responsible for monitoring the East Coast, in a "safe hold shutdown," after it had developed problems with its attitude sensors. GOES-10 was then required to watch the entire country, although it could provide images only every 30 minutes, instead of every 15 minutes, as it did when each of the two satellites was able to monitor half of the United States. NASA officials were especially concerned about monitoring East Coast weather, because of worry that winds from Hurricane Mitch, located over the coast of Central America, could interfere with the upcoming launch of the Space Shuttle carrying veteran astronaut, U.S. Senator John H. Glenn Jr.[608]

NASA's Ames Research Center awarded a five-year, US$99.3 million contract to Sverdrup Technology Inc. of Tullahoma, Tennessee, to provide testing and facility operation, including wind tunnels and arc-jet testing.[609]

*28 October*
NASA awarded KSC Small Expendable Launch Vehicle Services (SELVS-KSC) contracts to Coleman Research Corporation of Orlando, Florida, and Orbital Sciences Corporation of Dulles, Virginia, to provide launch services for NASA and for NASA-sponsored small-class payloads. These indefinite delivery–indefinite quantity contracts, ranging in value from US$100,000 to US$400,000, were part of NASA's effort to fulfill the goals and objectives of its Human

---

[606] *New York Times*, "U.S. Launches a Spacecraft That Thinks for Itself," 25 October 1998.
[607] Associated Press, "Mir," 27 October 1998; *Spacewarn Bulletin*, no. 540, 1 November 1998, *http://nssdc.gsfc.nasa.gov/spacewarn/spx540.html* (accessed 28 October 2008).
[608] Randolph E. Schmid for Associated Press, "East Weather Satellite Shuts Down," 28 October 1998.
[609] NASA, "NASA Awards $99 Million Contract to Tennessee Firm," contract announcement C98-q, 27 October 1998.

Exploration and Development of Space Enterprise. NASA expected to award up to 16 missions over the contract's five-year period.[610]

*29 October*
Space Shuttle *Discovery* Mission STS-95 launched from KSC in Cape Canaveral, Florida, carrying Commander Curtis L. Brown Jr.; Pilot Steven W. Lindsey; Mission Specialists Scott E. Parazynski, Stephen K. Robinson, and Pedro Duque; and Payload Specialists Chiaki Mukai and John H. Glenn Jr. The primary objectives of STS-95 included conveying a variety of science experiments inside the SPACEHAB module; deploying and retrieving the *Spartan* free-flyer payload; performing operations with the HST Orbiting Systems Test; and transporting the International Extreme Ultraviolet Hitchhiker payloads. The historic mission also returned veteran astronaut John H. Glenn Jr. to space. Thirty-six years after Glenn had first orbited Earth, the seventy-seven-year-old astronaut became the oldest person to fly in orbit. As a veteran of NASA's early Mercury program, Glenn had been the first American to orbit Earth. On 20 February 1962, in a flight lasting 5 hours, Glenn had orbited Earth three times in a tiny capsule. After he had passed a battery of medical tests and participated in astronaut training, NASA had selected U.S. Senator John H. Glenn Jr. as a crew member of STS-95, so that scientists could study "how space travel affects an aging astronaut's body."[611]

## NOVEMBER 1998

*2 November*
NASA's Ames Research Center awarded a six-year, US$90 million contract to Logicon Syscon Inc. of Falls Church, Virginia, to provide operations, development, maintenance, and modification of the Center's aeronautical simulation facilities. The contract included the operation and maintenance of the world's largest motion-base simulator, the Vertical Motion Simulator, designed to aid in the study of helicopter and vertical short takeoff and landing issues, as well as the Crew-Vehicle Systems Research Facility, designed for the study of human factors in aviation safety.[612]

*3 November*
Astronauts aboard Space Shuttle *Discovery* successfully retrieved the *Spartan* spacecraft, thus completing a priority project of the nine-day mission. The crew had deployed the satellite on 1 November, to record observations of the Sun's outer atmosphere and the solar wind. *Spartan* had lagged 40 miles (64 kilometers) behind the Shuttle, in free flight, as it recorded the data. The flight was the last in a series of four flights begun in 1993 to "gather observations at different points in the [S]un's 11-year cycle of magnetic activity." NASA had originally scheduled the mission for November 1997, but astronauts had "failed to send a crucial command" before

---

[610] NASA, "NASA Awards Launch Services Contracts for Small Payloads," contract announcement c98-r, 28 October 1998.
[611] Bill Sammon, "Senator Upholds Status as Hero with His Return to Orbit," *Washington Times*, 30 October 1998; John Noble Wilford, "Pioneer Returns as Crewman in the Shuttle," *New York Times*, 29 October 1998; NASA, "Mission Archives: STS-95, John Glenn's Flight; SPACEHAB," http://www.nasa.gov/mission_pages/shuttle/shuttlemissions/archives/sts-95.html (accessed 20 October 2008).
[612] NASA, "NASA Awards $90 Million Contract to Virginia Firm," news release c98-s, 2 November 1998.

releasing the satellite, so *Spartan* had gathered no data. The crew had also tested an experimental laser range-finding device mounted in the Shuttle, aiming at targets on *Spartan*.[613]

*7 November*
Astronaut John H. Glenn Jr. and his crewmates returned to Earth aboard Space Shuttle *Discovery*. Despite concern that the landing might be rough, because of a missing door on *Discovery*'s drag chute, the craft landed safely at Kennedy Space Center in Cape Canaveral, Florida. The door had fallen off during the 29 October launch. Mission Control had warned Pilot Steven W. Lindsey that a braking chute might deploy unexpectedly, but it remained safely in place throughout the landing process.[614]

*10 November*
The ion-propulsion engine on NASA's Deep Space 1 probe shut down suddenly after running for 4½ minutes, the first time the probe had turned on its engine since the spacecraft's launch three weeks before. Ground controllers did not know the cause of the shutdown and made several unsuccessful attempts to restart the engine.[615]

*11 November*
Ground controllers in Madrid, Spain, lost contact with the Voyager 2 spacecraft as it approached the edge of the solar system but were able to reestablish communication. The probe was "powering down to conserve its plutonium energy source" when controllers lost contact. NASA had launched Voyager 2 in 1977. After the craft had toured Jupiter, Saturn, Uranus, and Neptune, NASA located it traveling toward the edge of the solar system, at about 5.2 billion miles (8.4 billion kilometers) from Earth. Scientists did not know why controllers had temporarily lost communication with the craft.[616]

*15 November*
Dr. William K. Douglas, the physician who had overseen the care of the seven Mercury astronauts, died of a viral infection at the age of 76. As part of his assignment to NASA's Manned Spacecraft Center at Langley Air Force Base in Virginia, Dr. Douglas had been "personally responsible for the medical care and flight preparation of the seven Mercury astronauts." He had cared for Alan B. Shepard Jr., Virgil I. "Gus" Grissom, John H. Glenn Jr., and M. Scott Carpenter when they undertook the nation's first four human spaceflights. After retiring from the U.S. Air Force in 1977, Dr. Douglas had joined the McDonnell Douglas Corporation, where he had helped design an orbiting space station before retiring in 1988. In

---

[613] *Washington Times*, "Glenn Reports No Space Sickness as Discovery Releases Sun Gazer," 2 November 1998; *Washington Times*, "Astronauts Snag Sun Observer To Complete a Primary Mission: Glenn, 77, Undergoes More Tests of His Response to Space," 4 November 1998; *New York Times*, "Shuttle Easily Regains Satellite, and Photos," 4 November 1998.
[614] Kathy Sawyer, "Glenn Floats Back to Earth on the Wings of Discovery," *Washington Post*, 8 November 1998; NASA, "NASA Mission Archives: STS-95," http://www.nasa.gov/mission_pages/shuttle/shuttlemissions/archives/sts-95.html (accessed 11 June 2007).
[615] *Washington Times*, "Engine Failure Strikes Space Probe," 12 November 1998; *New York Times*, "Innovative Engine on U.S. Spacecraft Mysteriously Shuts Down," 13 November 1998.
[616] Associated Press, "Voyager Glitch," 18 November 1998.

1992 Douglas was inducted into the International Space Hall of Fame in Alamogordo, New Mexico.[617]

*16 November*
NASA Administrator Daniel S. Goldin appointed Edward J. Weiler Acting Associate Administrator for NASA's Office of Space Science, effective immediately. Weiler had acted in the position since Wesley T. Huntress Jr.'s departure on 28 September. Since March 1996, Weiler had served as Science Director of the Astronomical Search for Origins and Planetary Systems within the Office of Space Science. He had been a program scientist for the Hubble Space Telescope (HST) since 1979.[618]

*17 November*
Satellites survived the annual Leonid meteor storm with no major damage. Astronomers had predicted the storm would be the most intense in three decades, because Comet Tempel-Tuttle had swung closer to the Sun in February, as it does every 33 years, creating a greater-than-average amount of comet dust. Scientists had feared that particles traveling at 43 miles (69 kilometers) per second could shred satellites' solar arrays, damage sensitive instruments, or cause circuits to short.[619]

*19 November*
NASA's ER-2 aircraft, a "close relative" of the U-2 spy plane, officially broke the altitude record for medium-weight planes when it flew to 68,700 feet (20,900 meters), nearly twice the cruising altitude of commercial airliners, during a mission to measure components of the atmosphere. A Canadian P-42 aircraft had set the previous record of 62,500 feet (19,050 meters) in 1988. NASA officials said the craft had flown at such altitudes in the past, but this flight was the first time that scientists had documented its altitude and made it public.[620]

*20 November*
After a delay of more than a year, a Russian Proton booster rocket launched from Baikonur Cosmodrome in Kazakhstan, carrying into orbit the first module of the International Space Station (ISS), the 41.2-foot (12.6-meter) *Zarya*. The role of *Zarya*, translated as *Sunrise*, would be to act as a "space tugboat" in the early stages of construction, providing propulsion, power, and communications.[621]

*22 November*
Russia's first privately owned satellite, a *Bonum-1* communications satellite, launched successfully from Cape Canaveral, Florida, aboard a Delta rocket. NASA had canceled the first three attempts to launch the satellite because of numerous technical problems. The launch was also the first time a U.S. rocket had placed a Russian commercial satellite in orbit and the first

---

[617] Wolfgang Saxon, "William Douglas, 76, the Doctor for the Seven Early Astronauts," *New York Times*, 24 November 1998.
[618] NASA, "Weiler Named Associate Administrator for Space Science," news release 98-204, 16 November 1998.
[619] Paul Hoversten, "Space Storm Fizzles; Satellites Survive 'Drizzle'," *USA Today*, 18 November 1998.
[620] Agence France-Presse, "NASA Airplane Flies High into Sky: ER-2 Able To Break Record for Altitude," *Washington Times*, 23 November 1998.
[621] Vladimir Isachenkov for Associated Press, "Russia Puts First Segment of International Space Station in Orbit," 20 November 1998.

time an American company had built a satellite for a Russian company. Hughes Space and Communications had built the *Bonum-1* for Moscow-based Media Most, owner of Russia's largest television network. The chairperson of the board of Media Most, Vladimir A. Goussinsky, described the event as a "revolution," because the satellite would provide up to 50 television channels for 200 million viewers in Russia, western Siberia, and Eastern Europe.[622]

*23 November*
NASA released results from the Hubble Deep Field South (HDF-S) to the worldwide astronomy community for research, as well as to the public. HDF-S was a 10-day observation that had taken place in October 1998. With all of its new cameras and other instruments trained simultaneously down the same 12 billion–light-year–long "corridor," the team of astronomers at the Space Telescope Science Institute and NASA's Goddard Space Flight Center had captured a "core sample" of the universe, doubling the number of far-flung galaxies available for astronomers to study. The astronomers had studied a region in the constellation Tucana, near the south celestial pole, complementing a previous deep-field study carried out in late 1995, when scientists aimed the HST at a "small patch of space near the Big Dipper."[623]

*25 November*
NASA's Jet Propulsion Laboratory announced that the ion engine responsible for the Deep Space 1 probe's primary propulsion was back in operation after its unexpected shutdown two weeks earlier. Scientists believed that metallic grit or another type of contaminant located between two high-voltage grids had caused the engine to shut down after running for 4½ minutes.[624]

Russian Space Agency officials reported, for the first time since *Zarya*'s launch, that the module had experienced malfunctions. At a press conference, Director General of the Russian Space Agency Yuri N. Koptev explained that none of the module's three minor malfunctions had any impact on the module's overall operations. Koptev reported that one of the eight batteries storing energy collected from the module's solar panels had not performed properly, but he emphasized that *Zarya* needed only four of the eight batteries to operate. Mission Control had detected an "unexpected abundance of humidity" inside the module, but engineers did not yet know its cause. The third problem involved a malfunctioning antenna in the craft's automatic docking system, a device that the first ISS crew would not need until after their arrival in January 2000.[625]

**DECEMBER 1998**

*1 December*
NASA announced its selection of Litton/PRC of Mclean, Virginia, to perform work under NASA's sounding rocket operations contract, consolidating several previous contracts. The four-year contract, an indefinite delivery–indefinite quantity contract worth between US$11.9 million

---

[622] Robyn Suriano, "Delta Lifts Russian Satellite," *Florida Today* (Brevard, FL), 23 November 1998; *Orlando Sentinel* (FL), "U.S. Broadcast Satellite Was Launched for Russia," 23 November 1998.
[623] NASA, "The Universe 'Down Under' Is the Target of Hubble's Latest Deep-View," news release 98-206, 23 November 1998.
[624] NASA, "Deep Space 1 Ion Propulsion System Starts Up," news release 98-215, 25 November 1998; Associated Press, "Spacecraft's Troubled Ion Engine Restarts," *Washington Post*, 26 November 1998.
[625] Adam Tanner for Reuters, "Minor Problems Plague New Space Station," 25 November 1998.

and US$211.8 million, included two three-year options worth up to an additional US$572.5 million. Goddard Space Flight Center's (GSFC's) Wallops Flight Facility, at Wallops Island, Virginia, was to provide services, including design, fabrication, integration, testing, and performance of mission operations for sounding rocket missions. NASA considered the contract an integral part of the Wallops 2000 strategic plan, initiated in 1997 to ensure the facility's future "stability and vitality."[626]

NASA awarded US$33 million in grants to 48 researchers, to conduct microgravity biotechnology research through NASA's Office of Life and Microgravity Science and Applications. To those investigators awarded the 40 grants for ground-based research, NASA provided access to its microgravity research facilities, including aircraft flying parabolic trajectories and sounding rockets. NASA awarded eight grants to flight-definition investigators, to prepare for experiments on the International Space Station (ISS).[627]

*2 December*
NASA modified a contract with Boeing Information, Space and Defense Systems, increasing by US$163.4 million the company's contract for work on the ISS. Under the modified contract, Boeing would supply additional engineering support and prelaunch testing for the ISS. The original eight-year prime contract, effective since 1995, had a value of US$7.1 billion and covered the construction and integration of the ISS. The modification covered sustained engineering for station elements, engineering support following a component's delivery to NASA, and support of multielement integrated testing.[628]

NASA announced that Raytheon STX Corporation of Lanham, Maryland, would provide routine data operations, research, and development support for the Space Science Data Operations Office and the National Space Science Data Center at GSFC, under a two-year contract with three one-year options. Valued at more than US$33 million, the contract covered the processing of space-science data and the acquisition, modeling, analysis, archiving, and dissemination of NASA's archival data to the scientific community, educators, and the public.[629]

*4 December*
After a one-day delay, Space Shuttle *Endeavour* launched from Kennedy Space Center (KSC) in Cape Canaveral, Florida, carrying Commander Robert D. Cabana, Pilot Frederick W. "Rick" Sturckow, and Mission Specialists Nancy J. Currie, Jerry L. Ross, James H. Newman, and Sergei K. Krikalev. The purpose of STS-88 was to connect the U.S.-built *Unity* module of the ISS to the first ISS component, *Zarya*. The Russian Space Agency had placed *Zarya* in orbit in November. Launch controllers had scrubbed the liftoff of *Endeavour* planned for 3 December, after the master alarm sounded in the cockpit with only 4½ minutes remaining in the 10-minute countdown window. Engineers had determined that the alarm indicated momentary low pressure

---
[626] NASA, "NASA Selects Litton/PRC for NASA Sounding Rocket Program," contract announcement C98-t, 1 December 1998.
[627] NASA, "NASA Announces Research Grants in Microgravity Biotechnology," news release 98-217, 1 December 1998.
[628] NASA, "Engineering and Testing Support To Be Supplied by International Space Station Contract Modification," contract announcement c98-u, 2 December 1998.
[629] NASA, "NASA Selects Raytheon STX To Provide Goddard Support," contract announcement c98-v, 2 December 1998.

in the hydraulics system. The crew experienced no problems with the second launch, until another alarm sounded shortly after they had reached orbit, indicating that one of three hydraulic power units had overheated. The crew then switched to a backup controller to allow the overheated unit to cool before shutting down each unit as planned.[630]

*5 December*
A modified L-1011 jetliner flying at 40,000 feet (12,000 meters) dropped a Pegasus XL rocket carrying the Submillimeter Wave Astronomy Satellite (SWAS), designed to study the formation of stars and planets. The launch marked the beginning of a two-year, US$64 million mission to learn more about the "composition of interstellar clouds and monitor how they cool as they collapse to form stars and planets." The 397-pound (179-kilogram) SWAS would study water, molecular oxygen, atomic carbon, and isotopic carbon monoxide, by means of submillimeter wave radiation, which scientists cannot observe from the ground. NASA's GSFC managed the mission.[631]

*6 December*
Using precision flying, Space Shuttle *Endeavour*'s robotic arm, cameras, and a computerized vision system, the U.S.-Russian crew of *Endeavour* successfully joined the U.S. module *Unity* to the Russian module *Zarya*. With a weight of 21 tons (19,100 kilograms or 19.1 tonnes), *Zarya* was the "most massive object" that the robotic arm had ever lifted. Because *Unity* blocked the astronauts' view from the Shuttle's window, they relied on a system of cameras, as well as on the new Canadian Space Vision System, to provide visual cues concerning the two units' exact positions.[632]

At an American Geophysical Union meeting, a team of research scientists led by David E. Smith of GSFC discussed new research findings based on data collected in the spring and summer by the Mars Orbiter Laser Altimeter (MOLA). The data indicated that the ice cap on Mars's north pole, although definitely water ice, is much smaller than previously thought. The scientists concluded that the amount of water in the ice cap is not voluminous enough to have created the "deep gullies that scar the surface of the planet," nor could it have filled an ancient ocean. MOLA images showed the ice cap resting in a deep basin, possibly created by an asteroid impact. Scientists described this formation as unique in the solar system. The team of researchers remarked that their findings raised more questions about Mars's formation and conditions than they answered.[633]

---

[630] Marcia Dunn for Associated Press, "Space Station," 4 December 1998; Brad Liston for Reuters, "NASA Hopes for Another Shuttle Try on Friday," 3 December 1998; NASA, "Mission Archives: STS-88, First International Space Station Flight," http://www.nasa.gov/mission_pages/shuttle/shuttlemissions/archives/sts-88.html (accessed 28 October 2008).
[631] Associated Press, "NASA Launches Probe To Study Stars," 5 December 1998; *CNN.com*, "NASA Launches Satellite To Study Star Formation," http://www.cnn.com/TECH/space/9812/05/swas.launch/ (accessed 27 October 2008).
[632] Kathy Sawyer, "U.S., Russian Modules Are Linked in Orbit: Foundation for New Space Station Formed," *Washington Post*, 7 December 1998.
[633] NASA, "Laser Provides First 3-D View of Mars' North Pole," news release 98-219, 7 December 1998; Joseph B. Verrenga for Associated Press, "Icecap Seen on Mars' North Pole," 6 December 1998; Maria T. Zuber et al., "Observations of the North Polar Region of Mars from the Mars Orbiter Laser Altimeter," *Science* 282, no. 5396 (11 December 1998): 2053.

*7 December*
Astronauts Jerry L. Ross and James H. Newman undertook a 4½-hour-long spacewalk, to install 40 electrical and data connectors linking the U.S.-built connecting module *Unity* with the Russian-built power module *Zarya*. The pair also bolted handrails to the side of *Unity* and photographed two of *Zarya*'s backup docking antennas, which had failed to deploy. Russian ground controllers sent commands to power up *Zarya*, jump-starting *Unity*'s electrical systems. Ross and Newman then removed thermal covers from *Unity*'s outside computers to prevent overheating.[634]

*8 December*
Intel Corporation agreed to grant NASA, the U.S. Department of Energy, Sandia National Laboratory, and the U.S. Armed Forces a royalty-free license to adapt the Pentium chip to withstand extreme radiation, so that other agencies could use it in spacecraft and military equipment.[635]

*9 December*
Astronauts Jerry L. Ross and James H. Newman undertook their second spacewalk, lasting 7 hours, to install two 100-pound (45.4-kilogram) communications antennas on each side of the *Unity* module. They hooked one of the antennas to a video cable running from the *Zarya* module. Then the two astronauts erected a folding sunshade over a computer outside of *Unity*, bundling loose cables that ran along the exterior of the module. Ross and Newman also removed the restraints that had stabilized the four *Unity* hatchways during the launch. Finally, Newman successfully tapped free one of *Zarya*'s stuck antennas, using a 10-foot-long (3-meter-long), extendable grappling hook.[636]

*11 December*
*Endeavour*'s Commander Robert D. Cabana and Russian cosmonaut Sergei K. Krikalev, leading the four other crew members, entered the *Unity* module for the first time, via an airlock attached to the Shuttle. The crew checked air pressure and quality, turned on lights and systems, detached components that were stored in the modules, and set up a communications system, connecting internal components with the two antennas outside *Unity*. Astronauts Jerry L. Ross and James H. Newman had installed the two antennas on 9 December. The Shuttle crew spent 28 hours inside the modules before turning off the lights and closing the station.[637]

After a one-day delay to correct a software problem, Mars Climate Orbiter launched from Cape Canaveral, Florida, atop a Delta II rocket. NASA had scheduled the 1,400-pound (635-kilogram) robotic craft, paired with Mars Polar Lander, to launch on 3 January 1999. Mars Climate Orbiter and Mars Polar Lander, the second pair headed to study Mars following Mars Pathfinder and Mars Global Surveyor, were part of NASA's "long-term strategy for Mars exploration," calling for small robotic craft "to leave Earth two at a time," about every two years. According to

---
[634] Paul Hoversten, "Spacewalkers Complete a Tricky Wiring Job," *USA Today*, 8 December 1998.
[635] Dean Takahashi, "Intel Corp. To Send Its Pentium Chip into Outer Space: Firm To Grant Royalty-Free License to U.S. To Adapt Product for Spaceships," *Wall Street Journal*, 9 December 1998.
[636] Warren E. Leary, "2 Astronauts Make 2d Space Walk To Install Antennas on Station," *New York Times*, 10 December 1998.
[637] Warren E. Leary, "Astronauts Enter International Space Station," *New York Times*, 11 December 1998; Paul Hoversten, "Shuttle Finishes Linkup with Space Station," *USA Today*, 14 December 1998.

NASA's plan, Climate Orbiter's first task would be to provide a two-way communications relay between ground controllers and Polar Lander after Lander's arrival and during its planned 60–90 day mission. Thereafter, Orbiter would generate weather maps and profile the Martian atmosphere for a full Martian year, or 687 days.[638]

*12 December*
Jerry L. Ross and James H. Newman made their final spacewalk to free the second Russian docking antenna on *Zarya*, which had failed to unfurl. The pair also finished outfitting the station. The astronauts had required a total of 21 hours over three spacewalks to complete the work.[639]

*15 December*
Ending one of NASA's "most successful and historic missions" with a rare night landing, Space Shuttle *Endeavour* landed at KSC in Cape Canaveral, Florida, after its weeklong linkup with the new ISS. Before departing the ISS on 13 December, Pilot Frederick W. "Rick" Sturckow flew around the station slowly, while the crew photographed the ISS, permitting ground controllers to see the station for the first time. NASA's Deputy Station Chief Frank L. Culbertson Jr. remarked that the mission was as historically significant as that of *Apollo 8*, the first piloted mission to the Moon, and that the hardware of ISS was the foundation of "what will put humanity in space forever."[640]

*17 December*
NASA's tailless, single-engine, 18-foot-long (5.5-meter-long) X-36 airplane resumed flight-testing a system to help a damaged airplane continue its flight. Over a 20-minute period, a pilot used remote control from the ground to fly the plane, sending signals to simulate damage. The first in a series of five to 10 flights, the test was part of Reconfigurable Control for Tailless Fighter Aircraft (RESTORE). RESTORE was testing neural-network software that permitted a plane to reconfigure flight-control surfaces automatically. The software would have both civil and military applications.[641]

*18 December*
KSC officials broke ground on a complex for testing a NASA experimental rocket, the X-34. Engineers had designed the vehicle—a small, winged rocket—to launch into orbit from an airplane and to return to Earth, landing like an airplane. Planning flight tests to begin in 2000, KSC hoped the new facility would attract commercial space enterprises, such as Lockheed's VentureStar™, still under development. Facilities anywhere in the country would be able to

---

[638] Kathy Sawyer, "Mars Probe To Examine Climate; NASA Launch Is First of Two Designed To Shed Light on Potential for Life," *Washington Post*, 12 December 1998.
[639] Paul Hoversten, "Shuttle Finishes Linkup."
[640] Paul Hoversten, "Shuttle Finishes Linkup"; United Press International, "Shuttle Lands in Florida," 15 December 1998; NASA, "NASA Mission Archives: STS-88," *http://www.nasa.gov/mission_pages/shuttle/shuttlemissions/archives/sts-88.html*.
[641] Jim Skeen, "A Bird Without a Tail; Craft Will Mimic Damaged Airplane," *Los Angeles Daily News*, 18 December 1998.

launch the VentureStar™, once completed, because it would not use boosters or fuel tanks and, therefore, would not require a coastal launch site.[642]

*21 December*
NASA announced it had renamed the Advanced X-ray Astrophysics Facility the Chandra X-ray Observatory, in honor of the late Nobel Laureate Subrahmanyan Chandrasekhar. NASA chose the name Chandra because it was the late scientist's preferred name among friends and colleagues, and because it means *moon* or *luminous* in Sanskrit. NASA Administrator Daniel S. Goldin commented that Chandrasekhar had made "fundamental contributions to the theory of black holes and other phenomena," which the Chandra X-ray Observatory would study. NASA also announced that it had set a new launch date for the observatory—no earlier than 8 April 1999.[643]

*29 December*
NASA announced that Colorado State University would lead a US$145 million study of clouds, to better understand global warming and to "dramatically improve" weather forecasts. The five-year project would use a satellite named *CloudSat* to capture three-dimensional global images of clouds.[644]

## JANUARY 1999

*3 January*
NASA launched Mars Polar Lander from Cape Canaveral, Florida, atop a Boeing Delta II rocket, three weeks after the launch of its companion craft, Mars Climate Orbiter. The craft's main purpose was to search Mars for signs of water. Lander carried machinery and two microprobes, as well as a microphone to record the sounds of Martian wind. Mars Polar Lander and Mars Climate Orbiter formed a pair, the second launched in a 12-year series of dual missions to explore Mars, with the ultimate goal of collecting a sample of Martian soil to bring to Earth by 2008.[645]

*6 January*
An independent study group commissioned by the U.S. Air Force announced the findings of its investigation into the loss of business at Florida's Cape Canaveral Air Force Station. Retired U.S. Air Force General Richard C. Henry had headed the group, comprising 11 government agencies and 19 aerospace firms. After surveying commercial launch companies and satellite manufacturers, the group had measured the cost and ease of doing business at Cape Canaveral against three prime competitors: the European Space Agency's Kourou Space Center in French Guiana, South America; Kazakhstan's Baikonur Cosmodrome, primarily used by Russia; and China's Xichang Space Center. General Richard Henry summarized the crux of the group's findings: "companies regard Canaveral as superb in location and facilities, but rank it second

---

[642] Robyn Suriano, "NASA Breaks Ground for $8 Million Complex To Serve New Rocket, Attract VentureStar," *Florida Today* (Brevard, FL), 19 December 1998.
[643] NASA, "NASA Renames Telescope and Sets New Launch Date," news release 98-225, 21 December 1998.
[644] Bill Scanlon, "CSU Will Lead NASA Cloud Study $145 Million Project Is Designed To Improve Weather Forecasts, Study Global Warming," *Rocky Mountain News* (Denver), 30 December 1998.
[645] Reuters, "U.S. Mars Probe Launched from Florida," 3 January 1999; Beth Dickey, "Craft Joins Mars Quest for Water," *New York Times*, 4 January 1999.

primarily because of the perception that the commercial customer is second to the government customer." Companies had sought launch services elsewhere to avoid the bureaucratic red tape associated with Cape Canaveral Air Force Station, operated by the U.S. Air Force. With increasing demand, spaceports in South America, Russia, and China had been doing more business during the previous two decades, at Florida's expense. The group issued three recommendations to correct the situation and to attract business to Florida: 1) boost the number of launches staged from Cape Canaveral, 2) shift to customer-friendly operations, and 3) modernize launch-support systems at the spaceport, many of which the United States had first implemented in the late 1950s and early 1960s.[646]

*7 January*
NASA Inspector General Roberta L. Gross released a report estimating that NASA had "spent [US]$3 million in 'excessive costs'" on its Houston-to-Moscow air-charter program. The investigators criticized the program, saying it was more expensive than commercial travel. An investigation had revealed that the flights, which typically used Boeing 727s seating 143 travelers, sometimes carried as few as nine people. The charter flights originated in Houston and usually stopped in Huntsville, Alabama, and Washington, DC, before continuing on to Moscow. A few weeks before the release of the report, NASA's Office of Space Flight had rejected a recommendation from the Inspector General to terminate the service, arguing that the program "provided 'tangible and intangible' benefits to workers on the International Space Station [ISS]." NASA had developed the program as a cost-saving measure and as an alternative to contracting with the U.S. Department of Defense to transport employees involved with the multibillion-dollar space station program. The Inspector General found that the chartered flights averaged around 50 people; NASA had estimated a 90-passenger threshold for the program to realize savings. In response to the demand of U.S. Senate Committee on Governmental Affairs Chairperson Fred Thompson (R-TN) that NASA "halt this waste immediately," Administrator Daniel S. Goldin announced that NASA would terminate the program "as soon as practical."[647]

*11 January*
Mario Dario Grossi, radio physicist and engineer, died at the age of 74 in Boston, Massachusetts. Grossi developed the concept of tethered satellites and the hardware to make them possible. Together with a colleague, Grossi had conducted most of his work at the Harvard-Smithsonian Center for Astrophysics. Grossi had first approached NASA with his concept in 1972, and in 1982 NASA had issued a contract to build a system. In July 1992, NASA had tested the tether in space when astronauts aboard Space Shuttle *Atlantis* Mission STS-46 "reeled out a half-ton [450-kilogram or 0.45-tonne] satellite of the Italian Space Agency dangling on a spaghetti-thin cord 12 miles [19 kilometers] long." The successful test had provided data that would assist engineers in applying the tether satellite technology to build space stations or in "operating instrument-laden satellites far from the [S]huttle."[648]

*12 January*
The National Research Council Committee on Space Shuttle Upgrades released a report cautioning NASA "not to wait too long to choose between keeping the Shuttle until 2020 or

---

[646] Todd Halvorson, "Study: Spaceport Must Change or Lose," *Florida Today* (Brevard, FL), 7 January 1999.
[647] Michelle Mittelstadt for the Associated Press, "NASA Ends Moscow Charter Flights," 7 January 1999.
[648] Wolfgang Saxon, "M. Grossi, 74; Created Tether for Satellites," *New York Times*, 18 January 1999.

scrapping it after 2012 in favor of a new vehicle like Lockheed Martin's proposed VentureStar." The committee, led by former astronaut and former head of the Space Shuttle Program Bryan D. O'Connor, advised NASA to make a timely decision to either phase out the Shuttle upgrade program or make the major investments necessary for the Shuttle to carry out reliably its long-term mission. The panel specifically addressed the Liquid Fly Back Booster (LFBB), the most expensive Shuttle upgrade under consideration, with development costs estimated at between US$2 billion and US$7 billion. The panel's report revealed that NASA had not investigated enough alternatives to the LFBB to ensure that it was "the most appropriate way to meet the needs of the Shuttle and other programs."[649]

*13 January*
NASA's Jet Propulsion Laboratory (JPL) announced that the plutonium-powered Cassini spacecraft, en route to Saturn, had entered safe mode on 11 January. The craft, NASA's largest and most complex interplanetary probe, had detected a possible error in its orientation, which prompted it to halt all "non-critical" activity, while maintaining its communications link with Earth.[650]

JPL appointed Glenn E. Cunningham, Manager of NASA's Mars Surveyor Operations Project, as the new Deputy Director of JPL's Mars Exploration Directorate. In the newly created position, JPL tasked Cunningham with overseeing the implementation and operations of all Mars missions through 2013. As Cunningham's successor, JPL appointed Richard A. Cook, former Flight Operations Manager for the 1997 Mars Pathfinder Lander and Rover Mission, to manage the Mars Surveyor Operations Project.[651]

*15 January*
At a Boeing Company awards ceremony, NASA Administrator Daniel S. Goldin and U. S. Representative Dana Rohrabacher (R-CA) recognized "members of the Delta launch team who were instrumental in sending NASA spacecraft to Mars," awarding Jay L. Witzling, Vice President of Delta Programs, with NASA's Public Service Medal. Other critical leaders in the program also received awards for their dedication and participation. In his remarks, Goldin commented on the characteristics that made Boeing's Delta rocket "more and more the launch vehicle of choice." Goldin explained that the Delta rocket was the least expensive launch vehicle, had the fastest turnaround, and had demonstrated the highest degree of reliability. Goldin also commented that the four launches to Mars aboard Delta rockets had "changed the face of the space program."[652]

NASA announced that NASA scientists and other weather researchers had "gained intriguing new information about upper-level winds that drive hurricanes and about the storms' devastating impact as they collide with mountains." The research findings were the result of a seven-week study conducted by NASA, the National Oceanic and Atmospheric Administration, and several

---

[649] *Aerospace Daily*, "NRC Panel Worried NASA Hasn't Studied Alternatives to LFBB," 13 January 1999.
[650] Associated Press, "Saturn-Bound Spacecraft Develops Problem, JPL Says," *Los Angeles Times*, 14 January 1999.
[651] NASA Jet Propulsion Laboratory, "Cunningham, Cook Named to Mars Posts," news release 99-03, 13 January 1999, *http://www.jpl.nasa.gov/releases/99/marsappts.html* (accessed 23 January 2008).
[652] PR Newswire, "Mars Missions Earn Honors for Boeing Delta Employees, NASA's Goldin and Congressman Rohrabacher Present Awards," 27 January 1999.

universities during the previous summer, as part of the Third Convection and Moisture Experiment (CAMEX-3). The study's purpose was to gauge the strength of Atlantic hurricane winds and rainfall. CAMEX-3 researchers had used data collected with special laser instruments aboard NASA's DC-8 aircraft, which flew through hurricanes Bonnie, Danielle, Earl, and Georges; data collected from NASA's ER-2 high-altitude aircraft, which flew above the hurricanes to collect "first-of its-kind data"; and data from the Tropical Rainfall Measuring Mission satellite. Edward J. Zipser of Texas A&M University remarked, "the multi-aircraft datasets obtained by NASA aircraft in these hurricanes are unprecedented in their comprehensiveness."[653]

*22 January*
After having agreed in June 1998 to abandon the space station within a year, Russian Prime Minister Yevgeny M. Primakov issued a decree to keep *Mir* in space for three more years, using private funds, so that Russia could dedicate its insufficient space budget to its ISS commitments. The decree reportedly mentioned that Russian and foreign contractors and investors would develop a plan allowing *Mir* to remain in orbit without using ISS-allocated funds.[654]

*23 January*
Astronomers captured for the first time optical images of a gamma-ray burst (GRB) as it was occurring. Detectors of the Burst and Transient Source Experiment (BATSE) on NASA's orbiting Compton Gamma Ray Observatory detected the GRB as it began and radioed its position to the Gamma Ray Burst Coordinates Network (GCN) at NASA's Goddard Space Flight Center (GSFC). GSFC immediately forwarded the position to astronomers around the world. In response to this signal, the Robotic Optical Transient Search Experiment (ROTSE) team in Los Alamos, New Mexico, photographed the patch of sky where the GRB was occurring. Using precise information from instruments aboard the Italian-Dutch BeppoSAX satellite, the ROTSE team was able to locate the GRB within the images they had captured. Astronomer Scott D. Barthelmy of GSFC remarked that capturing a burst as it unfolded was the "holy grail" of GCN, and that, previously, optical telescopes had only seen the afterglow of a burst, never the burst itself.[655]

*25 January*
NASA Administrator Daniel S. Goldin named Lori B. Garver as Associate Administrator for NASA's Office of Policy and Plans, effective immediately. Garver had served as Acting Associate Administrator since September 1998. She had begun working for NASA in 1996 as Special Assistant to the Administrator. Before joining NASA, she had been Executive Director of the National Space Society, since 1987, and a president of Women in Aerospace. Goldin remarked that Garver's experience inside and outside NASA had "proven invaluable," and that she had demonstrated the management skills and policy leadership that her new position required.[656]

---

[653] NASA, "NASA Hurricane Study Reveals Intriguing Results," news release 99-4, 15 January 1999.
[654] Peter N. Spotts, "Saving Mir: At What Cost to New Space Station? Russians Decree That They Will Keep Aging Orbital Outpost Raises Questions in U.S. About Their Ability To Fund International Project," *Christian Science Monitor*, 25 January 1999.
[655] NASA, "Gamma Ray Burst Imaged for the First Time," news release 99-8, 27 January 1999.
[656] NASA, "Garver Named Associate Administrator for Policy and Plans," news release 99-6, 25 January 1999.

*26 January*
NASA announced that it had selected, from among 35 proposals, five candidates for participation in its Medium-Class Explorer (MIDEX) program. Each of the candidates received US$350,000 to conduct a four-month implementation feasibility study. NASA planned to study these proposals rigorously over the following five months, before choosing the two MIDEX program participants. The five proposals selected for further study were 1) the Swift Gamma Ray Burst Explorer, a three-telescope space observatory for studying the position, brightness, and physical properties of GRBs; 2) the NGSS or Next Generation Sky Survey, a four-channel, supercooled, infrared telescope designed to survey the entire sky with 1,000 times more sensitivity than previous missions; 3) FAME, the Full-Sky Astrometric Mapping Explorer, a space telescope designed to obtain highly precise position and brightness measurements of 40 million stars; 4) the AMM or Auroral Multiscale MIDEX Mission, a formation of four identically instrumented small satellites in a near-polar, highly elliptical orbit; and 5) ASCE, Advanced Solar Coronal Explorer, a powerful solar telescope, which would reveal the physical processes in the Sun that lead to the solar wind and explosive coronal mass ejections. NASA had also selected instruments from two proposed MIDEX missions for technology-development funding. NASA awarded US$700,000 each to Richard E. Rothschild of the University of California at San Diego, to develop an x-ray detector for studying black holes of all sizes, and to Gary R. Swenson of the University of Illinois at Urbana-Champaign, to develop detectors for studying waves in Earth's upper atmosphere.[657]

## FEBRUARY 1999

*2 February*
McDermott International Inc., a worldwide energy services company, announced the election to its board of directors of Kathryn D. Sullivan, President and CEO of the Ohio Center for Science and Industry since 1996. Before serving in that position, Sullivan had been a chief scientist at the National Oceanic and Atmospheric Administration, and before that, she had been an astronaut and mission specialist at NASA. In 1978 NASA had selected Sullivan as a mission specialist for the first class of Space Shuttle astronauts. In 1984, while serving on a mission aboard Space Shuttle *Challenger*, she had become the first American woman to walk in space. Sullivan had also flown aboard Space Shuttle *Discovery* on the Hubble Space Telescope (HST) deployment mission in 1990 and aboard Space Shuttle *Atlantis* with the ATLAS-1 Spacelab flight in 1992.[658]

*4 February*
The Aerospace Safety Advisory Panel, established by Congress after the *Apollo 1* spacecraft fire on 27 January 1967, submitted its annual report to NASA. The panel found Shuttle safety satisfactory, but cautioned that cutbacks, such as hiring freezes and budgetary restrictions, had created a "looming crisis" of future shortfalls in the workforce. The panel warned that NASA was at risk of "losing the core competencies needed to conduct the nation's spaceflight and aerospace programs in a safe and effective manner." Budgetary constraints had also caused NASA to postpone planned improvements to the Shuttle fleet and the International Space Station (ISS). The panel's chairperson remarked that both programs were operating at an "acceptable

---

[657] NASA, "Five Explorer Mission Proposals Picked for Feasibility Studies," news release 99-7, 26 January 1999.
[658] Business Wire, "Kathryn D. Sullivan Elected to McDermott International Board of Directors," 2 February 1999.

level of risk," but that updating Shuttle computers and refurbishing the supply of spare parts would make the programs safer.[659]

NASA's Mars Global Surveyor successfully completed the aerobraking phase of its mission, raising its orbit completely out of the Martian atmosphere and preparing the craft to begin its primary mapping mission. Glenn E. Cunningham, Deputy Director of the Mars Exploration Program at NASA's Jet Propulsion Laboratory (JPL), described the Mars spacecraft's "pioneering operation" of aerobraking as a long and arduous task, which had become a valuable learning experience for all involved—engineers and scientists alike. The technique required the spacecraft to use frictional drag "as it skim[med] through the planet's thin atmosphere," altering the shape of the craft's orbit and providing an innovative method to change the orbit using less on-board fuel. NASA had first tested aerobraking in 1994, during the final days of the Magellan mission to Venus. A structural problem with Mars Global Surveyor's solar panel had delayed the final burn of aerobraking by approximately one year, so that the flight team could ensure that the process would not overstress the weakened panel. The extended aerobraking phase had enabled Global Surveyor to acquire bonus science data, yielding "some spectacular new findings about Mars," such as a "profile of the planet's northern polar cap and information about the unique nature of its remnant magnetic fields."[660]

*5 February*
An international team of scientists published findings based on research using data from NASA's Solar and Heliospheric Observatory (SOHO) spacecraft. The team had sought to locate the origin of solar wind, which interacts with Earth's magnetic field, creating auroral displays like the Northern Lights, as well as disrupting satellites and communications equipment. The SOHO team believed that the solar wind originated from honeycomb-shaped areas called convection cells, located beneath the coronal holes. Using data from the spectrometer, the scientists had measured for the first time the structure and motion of gases inside the coronal holes. The team reported that, in some places above coronal holes, magnetic field lines do not loop back to the Sun's surface, but rather, the magnetic lines shoot upward without returning to the Sun, thereby originating the solar wind.[661]

The Russian Space Agency's "space mirror" failed to unfurl properly at the *Mir* space station, and Russia officially terminated the experiment. Crew attached the mirror to a cargo spacecraft filled with trash, sending it to burn up in Earth's atmosphere. Engineers had intended the 83-foot (25-meter) mirror to "work like an artificial [M]oon," acting as a prototype for larger mirrors that "could illuminate [S]un-starved northern cities through the long Arctic nights and spotlight disaster areas." The experiment's failure was a "big disappointment" for the Russian Space Agency, which had hoped to display "its ability to conduct pioneering, ambitious projects despite a lack of funds."[662]

---

[659] NASA, "Aerospace Safety Advisory Panel To Present Report to NASA," news release N99-8, 4 February 1999; Marcia Dunn for the Associated Press, "Space Safety," 5 February 1999.
[660] NASA Jet Propulsion Laboratory, "Mars Global Surveyor Successfully Completes Aerobraking," news release 99-06, 4 February 1999, http://www.jpl.nasa.gov/releases/99/mgsbrake.html (accessed 23 January 2008).
[661] Donald Hassler et al., "Solar Wind Outflow and the Chromospheric Magnetic Network," *Science* 283, no. 5403 (5 February 1999): 810; United Press International, "Source of Solar Wind Believed Found," 4 February.
[662] Vladimir Isachenkov for the Associated Press, "Russia-Space Mirror," 5 February 1999.

*7 February*
After a one-day delay, a Delta II rocket launched from Cape Canaveral, Florida, carrying the US$128 million Stardust spacecraft. Stardust launched on a mission described as NASA's first attempt to "return to Earth solid extraterrestrial material from beyond our [M]oon's orbit." Astronomers expected that, after traveling for about five years, the craft would rendezvous with a medium-sized comet. A device aboard the Stardust would collect tiny particles from the comet before returning to Earth on 15 January 2006.[663]

*8 February*
NASA Administrator Daniel S. Goldin named Kevin L. Petersen, who had been serving in an acting capacity since August 1998, as Director of NASA's Dryden Flight Research Center (DFRC) in Edwards, California. Petersen had joined Dryden as an aerospace engineer in 1974 and had worked on flight-research projects, including the F-8 Digital Fly-by-Wire, the HiMAT (the Highly Maneuverable Aircraft Technology), and the X-29 forward-swept wing projects. He had served as Chief of the Vehicle Technology Branch, Chief of the National Aerospace Plane Projects Office, and, since January 1996, as DFRC's Deputy Director.[664]

NASA released findings from its Near Earth Asteroid Rendezvous (NEAR) spacecraft's flyby of asteroid 433, known as Eros. The flyby had occurred on 23 December 1998. The NEAR spacecraft's instruments had collected data revealing that the asteroid was slightly smaller than previously thought, with at least two medium-sized craters and a density comparable to that of the Earth's crust.[665]

*9 February*
An Israeli court indicted Israeli teenager Ehud Tenenbaum and four accomplices—Guy Fleisher, Ariel Rosenfeld, Barak Abutbul, and Rafael Ohana—for illegal entry of computers in the United States and Israel. The group allegedly hacked into the computer systems of the Pentagon and NASA, as well as those of academic institutions in Israel and Israel's Knesset.[666]

NASA released images taken with cameras aboard the HST, revealing the infancy of planets. NASA described the images as a "glimpse of what our solar system looked like 4.5 billion years ago when the Earth and other planets began to condense." The images did not show actual planets, but rather the discs of dust that surround young stars. The HST images of six young stars, located 450 light-years away in the constellation Taurus, showed important details never before revealed in infrared or radio observations, such as the size, shape, thickness, and orientation of the discs.[667]

NASA announced the selection of Italian astronaut Umberto Guidoni to fly aboard Shuttle Mission STS-102. Guidoni was the first astronaut named to Mission STS-102 and the first European astronaut selected for an ISS mission. The primary purpose of STS-102 would be to

---

[663] Michael Cabbage, "Stardust Is Launched To Sweep Comet Debris," *New Orleans Times-Picayune*, 8 February 1999.
[664] NASA, "Petersen Named Dryden Director," news release 99-14, 8 February 1999.
[665] NASA, "NEAR Spacecraft Reveals Major Features of Eros," news release 99-13, 8 February 1999.
[666] *Chicago Tribune*, "5 Are Indicted for Hacking in the U.S.," 10 February 1999.
[667] Associated Press, "Hubble May Have Early Planet Images," 10 February 1999.

deliver a 21-by-15-foot (6.4-by-4.6-meter), multipurpose logistics module, named Leonardo, to the ISS. The Italian Space Agency and NASA had constructed Leonardo under a bilateral agreement that included a flight opportunity for an Italian astronaut. Before his assignment on STS-102, Guidoni had flown in 1996 as a payload specialist aboard Mission STS-75. That 16-day mission had included working with the U. S. Microgravity Payload and testing the NASA-Italian Space Agency Tethered Satellite System.[668]

*11 February*
NASA announced its selection of a team of industry partners to develop a "highway in the sky system," allowing the average person to travel in small, easy-to-fly, personal aircraft at four times the speed of a car. NASA selected Avidyne Corporation of Lexington, Massachusetts; AvroTec Inc. of Portland, Oregon; Lancair of Redmond, Oregon; Raytheon Aircraft of Wichita, Kansas; Rockwell Collins of Cedar Rapids, Iowa; Seagull Technologies of Los Gatos, California; and Allied Signal of Olathe, Kansas, as the industry partners. NASA tasked the team with completing hardware and software development of an entirely new concept for a cockpit display system, intended to guide the pilot to a preprogrammed destination.[669]

*12 February*
NASA officials announced that the Galileo spacecraft was operating normally again after entering safe mode on 31 January. Galileo's computer had become confused while flying past Jupiter's moon Europa. David A. Senske, a member of the Galileo imaging team at NASA's JPL, explained that the unusual angle Galileo required for studying Europa had caused the computer to malfunction.[670]

*18 February*
Scientists studying images obtained by NASA's Mars Global Surveyor published two reports in the journal *Nature* about Martian volcanic activity. One study reported that the images contained evidence that "ten times more lava" than previously thought had erupted onto Mars's surface. The other study suggested that at least one of Mars's volcanoes was still active.[671]

*20 February*
*Soyuz TM-29* launched on a mission to *Mir* with Russian cosmonauts Viktor M. Afanasyev and Ivan Bella. Jean-Pierre Haigneré of the European Space Agency was also on board *Soyuz*.[672]

*23 February*

---

[668] NASA, "Guidoni To Accompany First Italian Space Station Element to Orbit," news release 99-18, 11 February 1999; Reuters, "Italian Astronaut To Take Part in Discovery Mission," 9 February 1999.
[669] NASA, "General Aviation To Get a NASA-Industry 'Lift'," news release 99-17, 11 February 1999.
[670] Associated Press, "Galileo Craft Resumes Its Jupiter Mission," 13 February 1999.
[671] William Hartmann et al., "Evidence for Recent Volcanism on Mars from Crater Counts," *Nature* 397, no. 6720 (18 February 1999): 586; Alfred S. McEwen et al., "Voluminous Volcanism on Early Mars Revealed in Valles Marineris," *Nature* 397, no. 6720 (18 February 1999): 584; Rick Callahan, "Mars Was Truly Red Planet—as in Red-Hot Molten Lava—Scientists Say One Volcano Would Have Coated U.S. 4 Miles Deep," *Seattle Times* (WA), 17 February 1999.
[672] NASA, "U.S. and Russian Human Space Flight 1961—September 30, 2000," in *Aeronautics and Space Report of the President: Fiscal Year 2000 Activities* (Washington, DC, 2001).

Boeing launched a Delta II rocket from California's Vandenberg Air Force Base, carrying the first science satellites from Denmark and South Africa and the U.S. Air Force's Advanced Research and Global Observation Satellite (ARGOS). The Air Force intended the Boeing-built ARGOS to spend three years in space, "collecting data on the Earth's global environment and performing technology demonstrations for top-priority military space programs," as well as testing advanced space technologies for use on the ISS. The launches of Denmark's *Orsted* and South Africa's SUNSAT (Stellenbosch University, South Africa, satellite) were critical milestones for both countries' space programs. SUNSAT carried remote-sensing experiments to produce images from space at lower cost than larger systems, and Denmark deployed *Orsted* to study Earth's magnetic fields and electrical properties.[673]

*25 February*

The Russian Space Agency signed a cooperative agreement with Russian arms exporter Rosvooruzheniye, "intended to raise Russia's earnings on the world market and help it develop space hardware." Russia had earned only US$800 million, compared to the United States' US$50 billion, in commercial satellite launches during 1998. Rosvooruzheniye hoped the deal would help Russia break into the international space markets.[674]

*26 February*

A European Ariane 4 rocket launched, carrying two communication satellites from French Guiana. The rocket placed into orbit an *Arabsat 3A* satellite, providing television broadcasting and telephone communications to the Middle East, North Africa, and Southern Europe, as well as the *Skynet 4E* satellite, a British armed forces strategic and tactical communications satellite. Both craft would move into geostationary orbits.[675]

## MARCH 1999

*1 March*

NASA Administrator Daniel S. Goldin officially announced the change of the name of Lewis Research Center in Cleveland, Ohio, to "John H. Glenn Research Center at Lewis Field." George W. Lewis had been the research director for the National Advisory Committee for Aeronautics (NACA), NASA's predecessor agency, which built the research facility in 1941. Shortly after Lewis's death in 1948, NACA had named the facility "Lewis Flight Propulsion Laboratory" to commemorate him. NASA had modified its name to "Lewis Research Center" in 1958, when the facility became part of the newly formed National Aeronautics and Space Administration. U.S. Senator R. Michael DeWine (R-OH) had proposed the most recent name change for the facility in the FY 1999 appropriations bill for the Departments of Veterans Affairs, Housing and Urban Development, and Independent Agencies. Congress had enacted the bill into law on 21 October 1998 (Pub. L. No. 105-276). Goldin remarked that naming the Center for both George W. Lewis and John H. Glenn Jr. was an appropriate tribute to "two of Ohio's famous names—one an aeronautic researcher and the other an astronaut legend and lawmaker."[676]

---

[673] *CNN.com*, "Rocket Carries Trio of Science Satellites into Orbit," 23 February 1999, http://www.cnn.com/TECH/space/9902/23/argos.launch/index.html (accessed 10 April 2008).
[674] *CNN.com*, "Russia's Space Agency, Arms Exporter Sign Cooperation Deal," 25 February 1999.
[675] *BBC News*, "Ariane Blasts Off Successfully," 27 February 1999.
[676] NASA, "NASA Announces Field Center Name Change," news release 99-29, 1 March 1999.

*3 March*
Medialink Worldwide Inc., a New York–based company specializing in audio-video production and satellite-distribution services, announced it had signed a contract with the European Space Agency (ESA) to help the agency broadcast its news worldwide, via television. The two-year contract required Medialink to "establish and improve the space organization's relations with broadcast media across Europe," as well as to produce video news releases and other products "intended to raise the profile of the multinational agency."[677]

*4 March*
The Federal Laboratory Consortium appointed Donald J. Campbell, Director of Glenn Research Center (GRC), as the 1998 Laboratory Director of the Year for Technology Transfer. The Consortium honored Campbell, NASA's first African American center director, for his "exemplary contributions to the overall enhancement of technology transfer for economic development." Industries had created at least 20 new products from GRC-developed technologies in the previous five years. Campbell had also led the Center in establishing the Lewis Incubator for Technology, to help entrepreneurs and start-up companies commercialize NASA-developed technologies, as well as the Garrett Morgan Commercialization Initiative, to help small businesses in Ohio and the Great Lakes region use NASA technologies to become more competitive.[678]

NASA launched its Wide-Field Infrared Explorer (WIRE) aboard the three-stage Pegasus-XL rocket. Orbital Sciences Corporation's L-1011 jet carried the rocket to an altitude of 40,000 feet (12,200 meters) before releasing it. Shortly after WIRE's release into orbit, it encountered difficulties maintaining its stability, and NASA immediately began to investigate the malfunction. The satellite's scientific instrument was a "cryogenically cooled telescope designed to shed light on the history of star formation in the universe," as part of NASA's Origins program. The 12.5-inch-aperture (31.8-centimeter-aperture) telescope aboard WIRE had no moving parts and was enclosed in a cooling system to keep its mirrors below -436°F (-260°C), to prevent its own heat emission from overwhelming the light it would detect from space.[679]

*5 March*
The NASA WIRE team declared a spacecraft emergency, as ground controllers worked to recover the spacecraft. The satellite continued to spin, instead of maintaining a stable position in orbit, and had a higher temperature than expected. NASA formed a spacecraft-recovery team headed by David F. Everett of NASA's Goddard Space Flight Center (GSFC), as well as an anomaly-investigation board.[680]

*8 March*

---

[677] Medialink Worldwide Inc., "European Space Agency Names Medialink for TV Contract," news release, 3 March 1999.
[678] NASA, "Campbell Named Laboratory Director of the Year," news release 99-32, 4 March 1999.
[679] Associated Press, "NASA's Star-Gazing Wire Satellite Successfully Launched," 4 March 1999; Associated Press, "Satellite Launch," 5 March 1999.
[680] NASA, "NASA Working To Correct Spin Rate of Wire Spacecraft," news release 99-34, 5 March 1999.

NASA announced the end of the scientific mission of its WIRE satellite, after the craft released into space its entire supply of frozen hydrogen, a critical element of the mission. Preliminary reports indicated that the satellite had released the primary telescope cover three days earlier than planned, exposing the telescope's cryostat to sunlight. The sunlight had warmed the frozen hydrogen in the cryostat, which was supposed to maintain the telescope's temperature, causing the hydrogen to vent into space at a much higher rate than planned. Ground controllers believed the fast venting of the hydrogen had caused the satellite to spin, but they did not yet know what had caused the release of the telescope cover.[681]

*9 March*
At a briefing at NASA Headquarters, scientists explained how sigmoids—fiery S-shapes on the Sun's surface—would help them "predict when potentially dangerous and disruptive blasts of electrically charged gas will come hurtling toward Earth." These solar explosions, or coronal mass ejections (CMEs), have the power to disable power grids, damage satellites, and scramble communications networks. Previously, scientists could detect solar explosions once they had occurred, providing two or three days for Earth to prepare for potential problems, but they were unable to predict the CMEs. With the Japanese *Yohkoh* spacecraft's discovery of sigmoids, scientists would be able to issue a warning five or six days in advance of a solar explosion. Sarah E. Gibson of Cambridge University in England described the sigmoids as slinky-like structures, related to the Sun's underlying magnetic field. Alphonse Sterling of the Institute of Space and Astronautical Science of Japan likened the structures to "loaded guns," with a high probability of discharging. Scientists explained further that "not every CME is presaged by a sigmoid, but most sigmoids signal very large CMEs." The *Yohkoh* images showed that the sigmoid's S-shape is the precursor to a CME, confirmed by the appearance of an arch shape following the blast.[682]

Joseph G. Beerer of NASA's Jet Propulsion Laboratory stated that all of the instruments aboard Mars Global Surveyor were operating, and that the craft had begun mapping Mars, approximately one year later than originally planned. NASA had launched the craft in November 1996, and Surveyor had entered orbit around Mars in September 1997. NASA had delayed the mapping portion of the mission when scientists detected a structural problem with one of the craft's solar panels. To avoid placing excessive stress on the panel, the flight team had opted to proceed cautiously in aerobraking. Aerobraking, a pioneering operation, was necessary to position the spacecraft for its Mars mapping mission.[683]

*12 March*
NASA selected winners of its Government Inventor of the Year and Commercial Invention of the Year awards. Charles E. Clagett, Associate Head of the Component and Hardware Systems Branch at GSFC, won the Government Inventor of the Year Award for his Apparatus for Providing Torque and for Storing Momentum Energy. Clagett had developed the apparatus, known more commonly as the SMEX Reaction/Momentum Wheel, for NASA's Small Explorer program. NASA had used the apparatus successfully in the Transition Region and Coronal Explorer and the Submillimeter Wave Astronomy Satellite. Paul M. Hergenrother, Joseph G.

---

[681] NASA, "Wire Spacecraft Instrument Runs Out of Hydrogen," news release 99-37, 8 March 1999.
[682] Deborah Zabarenko for Reuters, "Sun's Fiery S-Shapes Foretell Massive Solar Blasts," 9 March 1999.
[683] NASA Jet Propulsion Laboratory, "Mars Global Surveyor Successfully Completes Aerobraking," news release 99-006, 4 February 1999, *http://www.jpl.nasa.gov/releases/99/mgsbrake.html* (accessed 27 December 2007).

Smith Jr., and Brian J. Jensen of Langley Research Center won the Commercial Inventor of the Year Award for their Phenylethynyl Terminated Imide Oligomers, fifth composition, known as PETI-5, which is a type of glue that holds fibers together. The Langley team had developed PETI-5 as an adhesive for use in various aerospace and commercial applications. Used in high-speed, high-temperature aircraft, PETI-5 had netted US$10 million in sales since becoming commercially available.[684]

*15 March*
After assessing its safety aspects, officials from the Federal Aviation Administration's (FAA's) commercial space office approved the first space launch from a platform in the sea. The FAA issued the launch license to an international launch consortium known as Sea Launch. The Boeing Company, based in the United States, was a 40 percent partner in the consortium. The Sea Launch Company planned a demonstration launch for 27 March 1999, from a converted, self-propelled, oil-drilling platform in the Pacific Ocean. The consortium's initiative, to launch communications satellites from a mid-ocean location at the equator, offered several advantages, such as the capability to lift heavier-than-normal payloads.[685]

*26 March*
Lockheed Martin and its Russian business partners announced the appointment of Wilbur C. Trafton as President of International Launch Services (ILS); Trafton had been Acting President since the previous December. Lockheed Martin, Russia's Khrunichev State Research and Production Space Center, and S. P. Korolev Rocket and Space Corporation Energia had established ILS in 1995, as a joint venture to market commercial launch services on Russian Proton rockets and Lockheed Martin Atlas rockets. Before joining ILS, Trafton had been Associate Administrator for NASA's Office of Space Flight. In that position, he had headed NASA's Human Exploration and Development of Space Enterprise. Ray Colloday, President of Lockheed Martin Astronautics, remarked that Trafton had brought to ILS "years of experience working on joint space programs with Russia." Colladay said that Trafton's background, together with his knowledge of the Atlas product line, made him "particularly well suited to lead ILS."[686]

Researchers reported in the journal *Science* that NASA's Galileo spacecraft had detected hydrogen peroxide on Jupiter's moon Europa. Hydrogen peroxide does not occur naturally on Earth. A scientific instrument aboard the craft—the Near Infrared Mapping Spectrometer (NIMS)—had aided scientists in their study of Europa's surface. Working like a prism, the instrument had broken up infrared light, detecting dark areas of hydrogen peroxide. The scientists reported that the chemical was forming constantly on Europa, through the process of radiolysis; in this case, the process involved "intense particle radiation" emanating from Jupiter. Principal Investigator for NIMS Robert W. Carlson explained that hydrogen peroxide begins breaking down almost as soon as it forms and, therefore, its life on Europa spans no more than a

---

[684] NASA, "NASA Selects Top Inventions of the Year," news release 99-42, 12 March 1999.
[685] Federal Aviation Administration, "FAA Issues Launch License to First International Sea Launch Consortium," news release APA 34-99, 15 March 1999, *http://www.faa.gov/news/press_releases/news_story.cfm?newsId= 4952* (accessed 27 December 2007); Associated Press, "FAA Gives Its OK to 1st Spacecraft Launched from Sea," *Salt Lake Tribune* (UT), 16 March 1999.
[686] Lockheed Martin, "Wilbur Trafton Named President, International Launch Services," news release, 26 March 1999," *http://www.lockheedmartin.com/news/press_releases/1999/WilburTraftonNamedPresidentInternat.html* (accessed 21 December 2007).

few months. The short lifespan of the chemical limited the scientists' ability to study the long-term chemical history of Europa. However, scientists would be able to apply their observations of short-term chemical changes on Jupiter to the study of how the moons of Jupiter interact with the planet, as well as to the study of similar processes elsewhere in the solar system.[687]

*29 March*
A dummy satellite reached its orbit, 23,000 miles (37,000 kilometers) above Earth, signaling an important success for the Sea Launch Company. The international consortium, which had invested US$500 million to develop the first commercial, marine-based launch system, launched the dummy satellite from the Odyssey, a converted oil rig located 1,400 miles (2,300 kilometers) south of Hawaii. A Zenit-3SL rocket, built jointly by Ukraine and Russia, carried the satellite aloft. One hour after launch, the satellite separated from the third stage of the rocket at an altitude of 1,200 miles (1,900 kilometers), to continue to its ultimate position.[688]

*31 March*
The ESA signed a 60 million Euro contract with Matra Marconi Space, to design and build the Mars Express spacecraft, with a planned launch date of June 2003. The agency chose that target date because the planets would be in optimal alignment, allowing for minimum travel time to Mars with maximum payload capacity. Major space missions can require upwards of 11 years from concept to launch, and only six years remained before the 2003 date, so the ESA selected a contractor with the means to develop missions quickly and inexpensively. Matra Marconi demonstrated that it would be able to streamline the development of Mars Express, building and launching the mission for about half the budget required for similar efforts.[689]

**APRIL 1999**

*5 April*
Advanced Communication Systems Inc. announced that the Naval Air Systems Command had awarded its Aerospace Division a five-year contract with option years. The Naval Air Systems Command intended the US$46 million Aircraft Structural Life Surveillance contract to assist the U.S. Navy in developing, verifying, executing, and enhancing methods and processes for monitoring aircraft structural fatigue. Advanced Communication's Aerospace Division Vice President Thomas M. Brennan remarked that the contract covered air vehicle technology that would "have application across all military services, as well as NASA and the FAA [Federal Aviation Administration]."[690]

*6 April*
NASA appointed Space Shuttle Program Manager Thomas W. Holloway as Manager of the International Space Station (ISS) Program, effective 19 April. Holloway replaced Randolph H.

---

[687] NASA Jet Propulsion Laboratory, "NASA's Galileo Finds 'Bottle Blonde' Chemical on Europa," news release 99-022, 25 March 1999, *http://www.jpl.nasa.gov/releases/99/europaperoxide.html* (accessed 21 December 2007); R. W. Carlson et al., "Hydrogen Peroxide on the Surface of Europa," *Science* 283, no. 5410 (26 March 1999): 2062–2064.
[688] Associated Press, "Satellite Reaches Orbit After First Commercial Launch from Ocean," 30 March 1999.
[689] European Space Agency, "Mission to Mars Set To Revolutionize ESA's Working Methods," *Florida Today* (Brevard, FL), 31 March 1999.
[690] PR Newswire, "ACS Wins $46 Million Aircraft Structures Contract," 5 April 1999.

Brinkley, who departed NASA for the private sector. NASA named Ronald D. Dittemore to replace Holloway as Manager of the Space Shuttle Program. Holloway had begun his NASA career in 1963, working in the Mission Control Center, where he planned activities for Gemini and Apollo flights. He had served as a flight director in Mission Control early in the Space Shuttle Program, becoming Chief of the Flight Director Office in 1985. In the 1990s, he had served as Deputy Manager for Program Integration with the Space Shuttle Program, followed by an appointment as Director of the Phase I Program of Shuttle–*Mir* dockings. Holloway had begun managing the Space Shuttle Program in 1995.[691]

*8 April*

NASA announced the departure of Gretchen W. McClain, a senior space station official at NASA Headquarters, naming as her replacement in an acting capacity, W. Michael Hawes, NASA's Chief Engineer for the ISS. In 1997 NASA had appointed McClain as Deputy Associate Administrator for Space Development (Space Station). In that position, she was responsible for directing the space station budget, establishing and implementing station policy, coordinating external communications, and undertaking liaison activities with the executive branch, Congress, industry, and international partners. McClain left NASA to return to private industry.[692]

*10 April*

U.S. Air Force officials announced that, despite initial media reports of a successful launch, the previous day's launch of the USA 142 aboard a Titan IVB rocket had failed. The USA 142, also called the DSP 19, an American geosynchronous military satellite, had launched from Cape Canaveral, Florida. The failed launch had placed the Defense Support Program satellite in a useless orbit after an attached rocket engine failed. This was the first launch attempt since the explosion of a Titan IVA in August 1998. Engineers had designed the Titan IVB, an improved version of the Titan IV, to survive the electrical malfunction that had led to its predecessor's demise. Because of a policy of secrecy concerning military spy satellites, the U.S. Air Force had delayed informing its public affairs office about the launch failure. Although senior Air Force officials knew of the loss of the satellite, they were unable to make the event public until they were certain that the failure had not entailed the release of classified information. The Air Force had designed the satellite, a highly sensitive, 2-ton (1,800-kilogram or 1.8-tonne), infrared telescope, to detect missile launches and nuclear detonations.[693]

*12 April*

Celebrating the anniversary of the flight of Yuri Gagarin on 12 April 1961, Russian President Boris N. Yeltsin presented state medals to several dozen cosmonauts and space officials during a Cosmonaut's Day ceremony. Yeltsin briefly praised *Mir*'s achievements in his remarks at the Kremlin, but when Director General of the Russian Space Agency Yuri N. Koptev presented him with a model of the new ISS, Yeltsin asked Koptev whether *Mir* would stay aloft. Koptev

---

[691] NASA Johnson Space Center, "Holloway Named Space Station Manager, Dittemore To Head Shuttle; Brinkley Leaves NASA for Private Sector," news release J99-9, 6 April 1999, http://www.nasa.gov/centers/johnson/news/releases/1999_2001/j99-9.html (accessed 15 January 2008).

[692] NASA, "McClain To Leave NASA; Hawes Named Acting Chief of Space Station," news release 99-50, 8 April 1999.

[693] Robyn Suriano, "Air Force Says Security Questions Delayed Announcement," *Florida Today* (Brevard, FL), 13 April 1999; *Spacewarn Bulletin*, no. 546, 1 May 1999, http://nssdc.gsfc.nasa.gov/spacewarn/spx546.html (accessed 28 October 2008).

responded in the affirmative, but explained during a press conference afterward that private investors had not yet produced funds for *Mir*'s continued operation. Vitaly I. Sevastyanov, a lawmaker and former cosmonaut, called on Russians to contribute money to a charity formed to keep *Mir* in orbit.[694]

## 13 April

NASA announced the discovery of a "mysterious class of 'middleweight' black holes." Two teams of astronomers studying x-ray light at NASA and Carnegie Mellon University had independently found evidence of this new class of black hole. Astronomers did not know what process had formed the newly discovered black holes, 100 to 10,000 times as massive as the Sun and located in spiral-shaped galaxies throughout the universe. Before this discovery, astronomers had known of only two types of black holes: 1) stellar black holes, formed from the "remains of dead stars several times heavier than the Sun" and compressed to a "diameter of a few miles or less"; and 2) supermassive black holes, which have masses equal to 1 million to 1 billion Suns and likely formed in the "early universe from giant gas clouds or from the collapse of clusters of immense numbers of stars." Edward H. Colbert and Richard F. Mushotzky of NASA's Goddard Space Flight Center (GSFC) had observed hints of the black holes while studying x-rays from 39 nearby galaxies. Andrew Ptak and Richard Griffiths of Carnegie Mellon University had studied x-ray light in galaxy M82, which was not one of the set of galaxies that the Goddard team had studied. Both teams had identified unique x-ray light indicative of a class of black holes that was neither stellar nor supermassive.[695]

## 15 April

*Landsat-7* launched successfully from Vandenberg Air Force Base. Although Lockheed Martin, Missiles and Space, had built *Landsat-7*, the newest land-surface observation satellite, with a design life of five years, the company's spokesperson remarked that 15 years after its 1984 launch, *Landsat-5* continued in operation. The upper stage of the rocket that launched the commercially built *Landsat-6* in October 1993 had failed to fire the satellite into orbit, destroying the craft and prompting NASA and the U.S. Geological Survey to resume control of the Landsat program in 1994. The latest model in the satellite series carried "a new generation of remote-sensing devices far more sophisticated than the equipment on satellites mapping Earth over the past decade." Raytheon Remote Sensing of Santa Barbara, California, had built the craft's enhanced thematic mapper to measure solar radiation reflected off the Earth's surface. The device was capable of capturing images of Earth's surface in 114-mile-wide (183-kilometer-wide) swaths and could resolve images as small as 50 feet (15 meters) across.[696]

## 16 April

*Mir* Commander Viktor M. Afanasyev and French cosmonaut Jean-Pierre Haigneré replaced scientific equipment on *Mir*'s exterior during a planned 5-hour spacewalk that went overtime. The pair overcame "numerous technical blips," to complete the French portion of the tasks and most of the Russian portion. Malfunctions in their spacesuits' ventilation systems delayed both men in leaving the station. Other problems included equipment failure during a simulated repair to an imaginary hole on *Mir*'s superstructure; an unsuccessful attempt to retrieve data from a

---

[694] Vladimir Isachenkov for Associated Press, "Mir," 12 April 1999.
[695] NASA, "Astronomers Discover 'Middleweight' Black Holes," news release 99-51, 13 April 1999.
[696] Frank Sweeney, "New Earth Camera Launched into Orbit," *San Jose Mercury News* (CA), 20 April 1999.

pollution study in process around the station; and the failure to install a new Russian-sponsored experiment to measure cosmic rays. Cosmonauts succeeded in other tasks during the spacewalk, including the recuperation of a French-sponsored test and the installation on the station's hull of comet technology designed to gather space dust for analysis. Haigneré was on his second mission to *Mir*, but this was his first spacewalk.[697]

*22 April*
At the 14th Annual NASA Continual Improvement and Reinvention Conference on Quality Management, Administrator Daniel S. Goldin presented the George M. Low Award, NASA's highest honor for quality and technical performance, to four U.S. companies: 1) Barrios Technology Inc. of Houston, Texas, for its small-business product; 2) Kay and Associates of Edwards, California, for small-business service; 3) Raytheon Support Services Company of Annapolis Junction, Maryland, for the large-business service category; and, 4) Thiokol Space Operations of Brigham City, Utah, for the large-business product category. NASA had evaluated the recipients for performance, delivery according to agreed cost and schedule, innovation, management leadership, alignment of organizational goals with NASA's strategic plans, customer orientation, and adherence to the Total Quality Management philosophy. Thiokol Corporation, which had won the Low Award in 1991 and Marshall Space Flight Center's Contractor Excellence Award in March 1999, had manufactured the reusable solid rocket motor for the Space Shuttle, as well as providing launch support and refurbishment services for the motor. Over the past seven years, Thiokol had never delayed its delivery of hardware. Its cost reduction efforts had saved NASA US$152 million.[698]

*23 April*
Scientists Peter Smith of the University of Arizona and Justin Maki of NASA's Jet Propulsion Laboratory published a report in the *Journal of Geophysical Research* summarizing the results of the Mars Pathfinder Mission. After an exhaustive review of more than 17,000 images from the Pathfinder's 1997 mission, the report had concluded that Mars's air and dirt are yellowish-brown in color, not red. Viking probes of the 1970s had likewise indicated that Mars is not red, but rather yellowish-brown. However, Mars appears red to the naked eye from Earth, and the Hubble Space Telescope (HST) data had indicated that Mars is red. The scientists reported that the detailed analysis using Pathfinder data had revealed that the HST, not Pathfinder, needed to be recalibrated.[699]

*27 April*
Lockheed Martin launched from Vandenberg Air Force Base an Athena II rocket, carrying the *Ikonos I* satellite, a civilian satellite capable of capturing the detailed images that only spy satellites could produce in the past. Launch officials had planned to cease communications with the craft 8 minutes after liftoff and to reestablish contact later in the flight, but had not been able to reestablish contact once initial communications ended. Officials were unable to determine

---

[697] Agence France-Presse, "Cosmonauts Struggle Through Grueling Space Walk Program," 19 April 1999.
[698] NASA, "Four Aerospace Firms Win NASA's Highest Honor for Quality," news release 99-53, 22 April 1999, *http://www.nasa.gov/home/hqnews/1999/99-053.txt* (accessed 7 February 2008); Cordant Technologies Inc., "Thiokol Propulsion Wins NASA's Highest Honor for Quality," news release, 3 May 1999.
[699] *Salt Lake Tribune* (UT), "The Red Planet? Mars Is Really Yellowing Brown: Pathfinder Discovers What Hubble Missed," 24 April 1999.

whether the craft had remained in orbit. Lockheed Martin Commercial Space Systems had built the satellite; Raytheon had built the communications, image processing, and other elements of the system; and Eastman Kodak had built the digital camera system. The camera on board *Ikonos I* was capable of resolving objects as small as 1 square meter (11 square feet), enabling the satellite to distinguish between a car and a truck. Before the manufacture of *Ikonos*, only military satellites had possessed the ability to photograph Earth in such detail.[700]

*28 April*
NASA announced it had completed negotiations on a US$625.6 million contract with Lockheed Martin Michoud Space Systems to purchase the last set of materials required to build 60 new Space Shuttle external fuel tanks. Although it was the sixth and last in a series of purchases of fuel tanks, this was the first purchase NASA had made consisting entirely of super lightweight tanks. The tank's design was the same as that of the lightweight tank design NASA had previously used, but the new tanks were 7,500 pounds (3,400 kilograms) lighter, and used a new aluminum lithium alloy. NASA and industry partners had developed the lighter alloy, 30 percent stronger than the material used to manufacture the lightweight tank, to enable the Shuttle to deliver ISS components into the proper orbit.[701]

*29 April*
NASA announced that the magnetometer on board Mars Global Surveyor had revealed "banded patterns of magnetic fields on the Martian surface," bearing a "striking similarity to patterns seen in the crust of the Earth's sea floors." Scientists interpreted the patterns as evidence that the Martian crust had shifted in the past. The discovery of banded patterns on the floor of Earth's oceans had provided a record of Earth's magnetic history, validating the theory of plate tectonics. Jack Connerney of NASA's GSFC, a member of Global Surveyor's magnetometer team, explained, "if the bands on Mars are an imprint of crustal spreading, they are a relic of an early era of plate tectonics on Mars." He added that such tectonic activity on Mars is likely extinct, unlike on Earth. The spacecraft's magnetometer had been able to capture the magnetic field observations because of the extended aerobraking phase, devised to protect against further damage to one of the spacecraft's solar panels. The period of aerobraking had permitted the magnetometer to "obtain better-than-planned regional measurements of Mars." Each time it reached the lowest point of its elliptically shaped orbit, below the planet's ionosphere, the probe encountered minimal magnetic interference and enabling it to obtain additional data.[702]

*30 April*
In its third consecutive failure, the second loss within one month, the U.S. Air Force launch of a Titan IV rocket from Cape Canaveral, Florida, placed a geosynchronous military communications satellite, the USA 143, also known as *Milstar 2*, in a useless low-Earth orbit.[703]

---

[700] John Antczak for Associated Press, "Contact Fails During Flight: Civilian Satellite Lost After Launch," 29 April 1999.
[701] NASA Marshall Space Flight Center, "NASA Completes Purchase of Material for 60 Shuttle External Tanks," news releases 99-069, 28 April 1999, http://www.msfc.nasa.gov/news/news/releases/1999/99-069.html (accessed 5 February 2008).
[702] NASA, "Magnetic Stripes Preserve Record of Ancient Mars," news release 99-56, 29 April 1999.

## MAY 1999

*1 May*

After a search of nearly two weeks, a team of salvagers funded by the Discovery Channel located the *Liberty Bell 7*, the Mercury space capsule flown by astronaut Virgil I. "Gus" Grissom, which had been lost at sea on 21 July 1961. Grissom had survived the splashdown and had been rescued from the Atlantic Ocean. He had maintained until his death in the 1967 *Apollo 1* launchpad fire that he had done nothing that could have caused the hatch to blow out following the splashdown. The cause of the accident remained a mystery and "forever marred" Grissom's otherwise successful 15-minute suborbital flight, the nation's second piloted spaceflight. The salvagers' remotely operated submersible had located and recorded video of the capsule about 300 miles (480 kilometers) southeast of Cape Canaveral, Florida, 3 miles (4.8 kilometers) beneath the surface of the ocean. Expedition leader Curt Newport said the video showed that the capsule was in "amazingly good condition." It was "still shiny in spots," the window and parachute liner were intact, its periscope was extended, and the words "United States" and "Liberty Bell" were "plainly visible." Newport explained the difficulty of locating the capsule, comparing the attempt to find it to searching for the Titanic—the *Liberty Bell*'s capsule, smaller than one of the Titanic's boilers, was hidden in water 0.5 miles (0.8 kilometers) deeper than the doomed ocean liner. Although the submersible located two cameras and a tape recorder with the capsule, which might possibly help to explain why the hatch blew open prematurely, it was doubtful that the film from those items would be salvageable after 38 years underwater. Although the discovery of the hatch itself would more likely provide an answer, Newport estimated that the hatch could be a mile from the capsule. The capsule remained in the ocean, for recovery at a later unspecified date.[704]

*4 May*

A Boeing Delta III rocket, carrying an *Orion* satellite for Loral Space and Communications, malfunctioned shortly after launch, leaving the satellite in the wrong orbit. In the inaugural launch in August 1998 of a Boeing Delta III, a rocket designed to carry twice the payload of the Delta II, the rocket had exploded in midair. In this second attempt at flight, the rocket had survived the initial launch, but malfunctioned when the second stage failed to ignite properly, leaving *Orion* in an orbit lower than planned. Engineers were uncertain whether it would be possible to boost the satellite using on-board fuel and thrusters.[705]

Lockheed Martin Corporation named former Martin Marietta President and Chief Operating Officer A. Thomas Young to lead an independent review of the management, engineering, manufacturing, and quality control processes of its space and missile business, specifically Astronautics, Missiles and Space, and Michoud Space Systems. The company had experienced a "series of expensive and well-publicized failures of its rockets and satellites," including two in the previous week, when a Titan IV placed one military satellite in the wrong orbit and when

---

[703] Associated Press, "Air Force Declares Military Satellite Dead," 12 May 1999; *Spacewarn Bulletin*, no. 546, 1 May 1999, http://nssdc.gsfc.nasa.gov/spacewarn/spx546.html (accessed 28 October 2008).
[704] *Washington Times*, "Salvagers Able To Locate Grissom's Space Capsule," 3 May 1999; Marcia Dunn for Associated Press, "Sunken Spacecraft," 3 May 1999.
[705] Marcia Dunn for Associated Press, "New Boeing Rocket Malfunctions, Satellite Enters Wrong Orbit," 5 May 1999.

another satellite, the remote-sensing *Ikonos*, lost contact with the ground. The company instructed the panel to submit a report to senior Lockheed management by 1 September.[706]

The U.S. Air Force officially declared the US$1.2 billion launch of a communications satellite on 30 April a "mission failure." A malfunction in the upper stage booster of the Titan IVB rocket had stranded the *Milstar* satellite in the wrong orbit. The Air Force established an investigation into the failure, which was its "third costly space failure in nine months."[707]

*11 May*
Following a series of launch failures, Lockheed Martin Corporation announced management changes in three of its business sectors. Raymond S. Colloday, President of Lockheed Martin Astronautics, retired, effective immediately, and G. Thomas Marsh, Executive Vice President of Missiles and Space, succeeded him. Lockheed named the President of Missiles and Space, K. Michael Henshaw, as President and Chief Operating Officer of the Energy and Environment Sector, effective 1 June. In that position, Henshaw succeeded Robert J. Stevens, who had served in dual positions but was directed to "devote his full attention" to his role as Corporate Vice President of Strategic Development. Albert E. Smith, President of Aerospace Electronics Systems and of Sanders, a Lockheed Martin company in Nashua, New Hampshire, succeeded Henshaw as President of Lockheed Martin Missiles and Space of Sunnyvale, California. Lockheed named Walter P. Havenstein President of Sanders and Acting President of Aerospace Electronics System, a promotion from his former position as Executive Vice President. In addition to the recent launch failures, Lockheed had been "bedeviled" by problems in its space business, including a 10 percent reduction in sales in its Space and Strategic Missiles Sector and a 17 percent decline in operating profit, after inadequate preparation and cancellations had caused the delay of commercial satellite launches.[708]

*12 May*
The U.S. Air Force "officially declared dead" its US$800 million *Milstar* communications satellite. Launched atop a Titan IV rocket on 30 April and placed in the wrong orbit, the *Milstar* would "never reach its intended 22,300-mile-high [35,900-kilometer-high] orbit." U.S. Air Force representative Aaron Renenger stated that the useless satellite "could remain in orbit for hundreds of years," but posed no threat to other spacecraft.[709]

*17 May*
The U.S. Air Force announced that it had postponed indefinitely Brigadier General F. Randall Starbuck's departure from his command of the 45th Space Wing, with responsibility for all

---

[706] Lockheed Martin, "Lockheed Martin Announces Comprehensive Space Sector Review, Former Martin Marietta President Will Chair Independent Panel," news release, 4 May 1999, *http://www.lockheedmartin.com/news/press_releases/1999/LockheedMartinAnnouncesComprehensiv.html* (accessed 7 February 2008); Tim Smart, "Panel To Review Lockheed Failures: Former Company President To Head Probe of Aerospace Foul-Ups," *Washington Post*, 5 May 1999.
[707] Reuters, "U.S. Air Force Investigates Costly Satellite Loss," 4 May 1999.
[708] Lockheed Martin, "Lockheed Martin Announces Management Changes," news release, 11 May 1999, *http://www.lockheedmartin.com/news/press_releases/1999/LockheedMartinAnnouncesManagementCh.html* (accessed 6 February 2008); Tim Smart, "Lockheed Revamping 2 Units: New Leaders Named After Series of Launch Failures," *Washington Post*, 12 May 1999.
[709] Associated Press, "Air Force Declares Military Satellite Dead," 12 May 1999.

military and commercial launches from Cape Canaveral Air Force Station in Florida. The Air Force spokesperson did not state the reasons for the change. Starbuck was to report on 3 June as Director of the Expeditionary Forces Management Team at Langley Air Force Base in Virginia. His replacement, Brigadier General Kevin P. Chilton, a former Space Shuttle Commander, was to report on 3 June to Patrick Air Force Base, to replace Starbuck as Commander of the 45th Space Wing.[710]

*18 May*
NASA announced the appointment of Baruch S. Blumberg, winner of the 1976 Nobel Prize for Physiology or Medicine, as Director of NASA's Astrobiology Institute, effective immediately. NASA had created the Astrobiology Institute in July 1998, a "virtual organization" comprising NASA's centers, universities, and other scientific entities for the study of "the origin, evolution, distribution, and destiny of life in the universe." The Institute, with a multidisciplinary focus, sought to bring together astronomers, biologists, chemists, exobiologists, geologists, and physicists, to search for the origins of life. NASA Administrator Daniel S. Goldin made the announcement of Blumberg's appointment at Ames Research Center in Mountain View, California, the site of the Institute's headquarters. Goldin stated that NASA had tasked the Astrobiology Institute with "providing the 'intellectual underpinnings' for building new types of instruments and space probes for finding life in the solar system and beyond."[711]

*20 May*
President of the Aerospace Industries Association John W. Douglass testified before the U.S. Senate Commerce Subcommittee on Science, Technology, and Space that the nation needed an "enlightened" national space program, combining "public and private investment in synergistic ways." Douglass stated that the government needed to "remove barriers to the growth of commercial space." He said that the most critical action needed for the industry's growth was renewal of the indemnification provisions of the Commercial Space Launch Act, because without such provisions, U.S. companies are unable to compete with the government-backed programs of foreign competitors. Douglass also testified that U.S. launch ranges lacked modern facilities, causing launch delays. Such delays "wreak havoc" in the commercial world, because launch sites require more time to reconfigure a launch than convenient for commercial enterprises. He cited the export licensing process as a serious problem, because it was causing the U.S. commercial space industry to lose market share in both satellite and launch services. Additionally, Douglass called the "dramatic drop" in funding for aerospace research and development over the last two decades a "serious structural problem," which the United States should correct, to remain competitive in the world market.[712]

U.S. Senate Committee on Governmental Affairs Chairperson Fred Thompson (R-TN) announced the findings of a U.S. General Accounting Office (GAO) investigation into NASA's computer security. Thompson had requested the study, which found that government specialists were able to penetrate several mission-critical systems and could have stolen, modified, or destroyed system software and data. GAO had found that 135 of the 155 mission-critical systems

---

[710] Robyn Suriano, "Starbuck Staying Put at Helm of 45th Space Wing," *Florida Today* (Brevard, FL), 18 May 1999.
[711] NASA, "Nobel Prize Winner To Lead NASA Astrobiology Institute," news release 99-61, 18 May 1999; Warren E. Leary, "Search for Life Beyond Earth Gets a Leader," *New York Times*, 19 May 1999.
[712] "AIA Urges Congress To Remove Barriers to Commercial Space Industry," *Aerospace News*, 20 May 1999.

reviewed did not meet all of NASA's requirements for risk assessment. Specifically, NASA had not conducted an Agency-wide review of information-technology security since 1991; 60 percent of reviewed systems had not been independently audited for security or weaknesses; NASA had no security training program; and NASA field centers did not report security incidents to a central location. GAO had also found that NASA had no policy for determining what type of information its staff could post on public Web sites, rendering it vulnerable to Internet attacks. NASA also had no policy for protecting mission-critical systems from well-known Internet threats.[713]

The U.S. Air Force announced that Brigadier General Kevin P. Chilton would not replace Brigadier General F. Randall Starbuck as Commander of the 45th Space Wing at Patrick Air Force Base, but would lead the 9th Reconnaissance Wing at Beale Air Force Base near Sacramento, California. Starbuck was to remain in command of the 45th Space Wing indefinitely until the Air Force had chosen another successor. The change followed "the U.S. launch industry's worst string of accidents in more than a decade," including "three consecutive failures of Air Force Titan rockets" launched from Cape Canaveral, Florida. The Air Force also reported that, earlier in May, a military navigation spacecraft had sustained US$51 million in damage during a thunderstorm. The Air Force announced that it had opened a formal investigation into that incident, in addition to the investigation President William J. Clinton had ordered into the string of launch failures. The President had appointed the CIA, the U.S. Department of Defense, and NASA to carry out that investigation.[714]

*25 May*
An international team of astronomers, led by Wendy L. Freedman of the Carnegie Observatories, held a news conference to announce their findings regarding the age of the universe. Freedman's team had concluded that the universe was between 12 billion and 13.4 billion years old, at least 1 billion years younger than predicted. Freedman's international team had studied eight years of calculations, made observations using NASA's Hubble Space Telescope (HST), and used a range of celestial measurements to recalculate the age of the universe. Among those measurements was the "Hubble constant," used in mathematical equations to "gauge the speed at which galaxies are accelerating away from each other." Named for Edwin Hubble, who had discovered 70 years ago that the universe is expanding, the Hubble constant is an essential piece of the equation determining the age and size of the universe. The team's findings drew criticism from other astronomers. Although the Australian physicist Charles H. Lineweaver had conducted separate research with the same result as that of Freedman's team, some astronomers associated with the Carnegie Observatories and Harvard University did not agree with the Freedman team's conclusions. Allan Sandage of the Carnegie Observatories stated that he believed that the team's methodology contained systematic errors and that the final number was not correct. Harvard University astronomer Robert P. Kirshner had led a third team, using the Hubble constant to study the age of the universe, but although he calculated a different age than the other teams,

---

[713] Federal Document Clearing House, "Thompson Says NASA Must Tighten Computer Security," government news release, 20 May 1999; U.S. General Accounting Office, "Information Security: Many NASA Mission-Critical Systems Face Serious Risks" (report no. GAO/AIMD-99-47, Washington, DC, May 1999), *http://www.gao.gov /archive/1999/ai99047.pdf* (accessed 7 February 2008).
[714] Robyn Suriano, "Chilton Not Coming to Cape; Starbuck To Remain Indefinitely," *Florida Today* (Brevard, FL), 21 May 1999.

Kirshner downplayed the differences in their results, remarking that "what was once a very big disagreement is now narrowing down."[715]

*26 May*
India's government-run space agency began commercial operations with the launch, from an island in the Bay of Bengal, of its Polar Satellite Launch Vehicle C-2 (PSLV), carrying Indian, German, and South Korean satellites. PSLV carried into orbit *Oceansat-1*, India's remote-sensing satellite, intended to carry out oceanographic research; KITSAT-3 of South Korea's Satellite Technology Research Center; and the German Aerospace Center's *Tubsat*. Chairperson of Indian Space Research Organization Krishnaswamy Kasturirangan stated that India could "offer to launch satellites at a much cheaper rate than other countries with launch facilities." Without naming how much India was charging for its newly inaugurated launch services, Kasturirangan said that in the international market it would cost US$1 million to piggyback a satellite.[716]

*27 May*
Space Shuttle *Discovery* launched successfully from Kennedy Space Center in Cape Canaveral, Florida, after a one-week delay and a six-month "launch drought." NASA had delayed the launch of Mission STS-96, scheduled for 20 May, after a violent thunderstorm earlier in the month damaged the Shuttle's external fuel tank, requiring repairs. The previous six months had been the longest period in NASA's history without any launches since the *Challenger* disaster, which grounded the Shuttle fleet for two and one-half years. However, problems with the Shuttle fleet had not caused the delay during recent months. Delays in the construction of a critical Russian-built component for the International Space Station (ISS), as well as ongoing problems with a grounded NASA x-ray telescope, had disrupted the launch schedule. *Discovery*'s crew included Commander Kent V. Rominger, Pilot Rick D. Husband, and Mission Specialists Ellen Ochoa, Tamara E. Jernigan, Daniel T. Barry, Julie Payette, and Valery I. Tokarev. During the 10-day ISS Assembly Mission 2A.1, the crew planned to deliver supplies to the ISS; to repair U.S. and Russian equipment; to install mufflers that would reduce the noise of the Russian fans; and to conduct a spacewalk, attaching one crane and part of another to the outside of the station. STS-96 (ISS 2A.1) was the first mission to dock with the new ISS.[717]

NASA released its first three-dimensional global map of Mars. The Mars Orbiter Laser Altimeter (MOLA), an instrument aboard NASA's Mars Global Surveyor, had generated the data for the high-resolution map, created from measurements gathered in 1998 and 1999. David E. Smith of NASA's Goddard Space Flight Center, Principal Investigator for MOLA and the lead author of a study published in the journal *Science*, said that, according to the new topographic map, Mars is slightly lopsided—the northern hemisphere is low and smooth, but the southern hemisphere had

---

[715] NASA Johnson Space Center, "Hubble Completes Eight-Year Effort To Measure Expanding Universe," news release H99-65, 25 May 1999, *http://spaceflight.nasa.gov/spacenews/releases/1999/H99-65.html* (accessed 2 January 2008); Robert Lee Hotz, "Scientists Calculate Most Precise Age Yet for Universe; Astronomy: New Data, Hubble Telescope Help Two Separate Teams Put Range at 12 Billion to 13.4 Billion Years," *Los Angeles Times*, 26 May 1999.
[716] Dow Jones Newswire, "Indian Rocket with Korean and German Satellites Takes Off," 26 May 1999.
[717] NASA, "Mission Archives: STS-96, Second International Space Station Flight," *http://www.nasa.gov/mission_pages/shuttle/shuttlemissions/archives/sts-96.html* (accessed 6 February 2008); Marcia Dunn for Associated Press, "Space Shuttle," 27 May 1999.

many craters and an elevation of about 5 kilometers (3.1 miles) higher than that in the north. Additionally, Smith noted that the depression in Mars's northern hemisphere is "distinctly not circular," indicating that it formed from internal geologic processes rather than an external impact. Carl B. Pilcher, NASA's Science Director for Solar System Exploration, commented that the data MOLA had collected enabled scientists to know "the topography of Mars better than many continental regions on Earth." Pilcher said that the data would "serve as a basic reference book for Mars scientists for many years," inspiring "new insights about the planet's geologic history."[718]

## JUNE 1999

*3 June*
NASA announced that spacecraft engineers who had worked on NASA's new Quick Scatterometer (QuikSCAT) had won an American Electronics Association Technical Achievement Award for the "development of a spacecraft at a record-setting pace of one year." Ball Aerospace and Technologies Corporation of Boulder, Colorado, had built the satellite for NASA's Jet Propulsion Laboratory (JPL), assembling the QuikSCAT faster than any construction of a spacecraft since the 1958 *Explorer I* satellite. To accomplish this fast pace, NASA's Goddard Space Flight Center had developed a new procurement system called the Rapid Spacecraft Acquisition process, to accelerate NASA's purchase procedure. The new acquisition process had enabled NASA to take advantage of "low-cost commercial technology from the burgeoning spacecraft industry," to develop satellite systems such as the QuikSCAT. NASA had been able to reduce the time needed to select a contractor and to initiate spacecraft development, from one year to 30 days. Chip Barnes, a QuikSCAT spacecraft system engineer at Ball Aerospace commented that the production had involved a "remarkable effort to get the spacecraft built, integrated, and tested in an 11-month time frame."[719]

*6 June*
Space Shuttle *Discovery* landed safely at Kennedy Space Center in Cape Canaveral, Florida. Shortly before returning to Earth, astronauts aboard *Discovery* had completed their last assignment, releasing *Starshine*, an educational satellite that 25,000 students worldwide would track over the following seven to eight months, to calculate the atmosphere's density. The US$1 million *Starshine*, a 19-inch (48.3-centimeter) sphere, "covered with 878 small, circular mirrors" that schoolchildren had polished, had become visible from the ground once it had "popped" from its canister in the Shuttle's cargo bay. In addition, the astronauts had spent six of their 10 days in orbit at the new International Space Station (ISS), where the crew had performed maintenance tasks. *Discovery*'s crew had delivered 2 tons (1,800 kilograms or 1.8 tonnes) of tools, water, and other supplies for the first permanent crew of the ISS, scheduled to arrive in the spring of 2000.[720]

---

[718] NASA, "First Global 3-D View of Mars Reveals Deep Basin and Pathways for Water Flow," news release 99-66, 27 May 1999, http://www.nasa.gov/home/hqnews/1999/99-066.txt (accessed 7 February 2008); David E. Smith et al., "The Global Topography of Mars and Implications for Surface Evolution," *Science* 284, no. 5419 (28 May 1999): 1495–1503.
[719] NASA Jet Propulsion Laboratory, "Quikscat Team Wins American Electronics Achievement Award," news release 99-045, 3 June 1999, http://www.jpl.nasa.gov/releases/99/aeaaward.html (accessed 12 February 2008).
[720] Associated Press, "Shuttle Astronauts Release Satellite, Head for Home," *Washington Post*, 6 June 1999.

*7 June*
AlliedSignal Inc. agreed to acquire Honeywell Inc., in a stock deal worth US$13.8 billion. Aerospace business, with revenues totaling about 40 percent of the enterprise, was the centerpiece of the agreement, in which AlliedSignal added Honeywell's business in cockpit electronics (avionics) and global-positioning equipment to its own market in aircraft engines and collision-avoidance systems. An analyst with Frost and Sullivan, a California-based, market-research firm, remarked that the merger put the combined company ahead of Rockwell Collins as "the no. 1 player in avionics." The new company would retain the name of Honeywell and maintain its headquarters at AlliedSignal's base in New Jersey.[721]

*8 June*
Following a series of recent launch failures, Boeing Company announced the convening of an independent panel of experts to examine its rocket programs, specifically Boeing's Delta, Sea Launch, and Inertial Upper Stage programs. To head the panel, Boeing had selected Sheila E. Widnall, former Secretary of the U.S. Air Force and a professor at the Massachusetts Institute of Technology. The panel was tasked with examining organizational roles and responsibilities within Boeing; recent launch failure investigation findings; processes used for government, civil, and commercial launches; acceptance processes for major subsystems and complex assemblies from suppliers; and manufacturing, assembly, transportation, and storage activities of Boeing launch programs.[722]

*10 June*
A Boeing Delta II rocket launched from Cape Canaveral Air Force Station in Florida, carrying four Globalstar cellular telephone satellites. Counting this launch, Globalstar Inc. had placed in orbit 24 of the 32 satellites the company needed to begin regional cellular phone service. The company had planned to begin offering global cellular service in the latter part of 1998, but in September 1998, a failed launch destroyed a dozen satellites aboard a Ukrainian Zenit II rocket. Globalstar had launched its first eight satellites on Delta II rockets, during February and April 1998, and 12 more satellites, early in 1999, on three Russian Soyuz rockets. Globalstar was competing with its industry rival Iridium World Communications Inc., which had placed a complete constellation of 66 satellites in orbit and had begun offering global cellular service before the end of 1998. To meet Globalstar's launch demands over the summer of 1999, Boeing had increased the pace of manufacturing at its Pueblo, Colorado, rocket factory. In addition, Boeing brought 50 to 70 workers from its Delta launch team at Vandenberg Air Force Base in California to join its Cape Canaveral workforce in Florida.[723]

*11 June*
Taiyuan Satellite Launch Center in China's Shanxi Province launched two Iridium World Communications satellites into orbit aboard a Chinese Long March 2C rocket, replenishing the 66-satellite Iridium constellation, which provides telephone, paging, and fax communication services worldwide. The satellite was the 57th launched from Taiyuan Satellite Launch Center

---

[721] Ameet Sachdev, "Aircraft Supply Titan Is Created," *St. Petersburg Times* (FL), 8 June 1999.
[722] Associated Press, "Boeing Forms Independent Panel To Review Space Failures," 9 June 1999.
[723] Todd Halvorson, "Boeing and Globalstar Begin Flurry of Launches," *Florida Today* (Brevard, FL), 11 June 1999.

and the 15th consecutive successful launch since October 1996 for the Long March rocket series.[724]

*15 June*
The Boeing Company named former cosmonaut Vladimir G. Titov as the company's director for Space and Communications, Russia and the Commonwealth of Independent States, located in Boeing's Moscow office. With the appointment, Titov became responsible for initiating new business opportunities, working closely with Boeing's ISS program members and with Russian and Ukrainian partners on the Sea Launch program. A cosmonaut since 1976, Titov had commanded several *Soyuz* missions, served as a mission specialist on two Space Shuttle missions, and lived aboard Russia's *Mir* space station, logging 387 days in space, including nearly 19 hours of extravehicular activity.[725]

Keith R. Hall, Assistant Secretary of the U.S. Air Force for Space and Director of the National Reconnaissance Office, and Edward C. "Pete" Aldridge Jr., Chief Executive Officer of the Aerospace Corporation, which provides technical analysis for the United States and international space programs, testified before the U.S. House Permanent Select Committee on Intelligence's Subcommittee on Technical and Tactical Intelligence, regarding three rocket-launch failures. All of the launch failures—an explosion after launch in August 1998, a missile-warning satellite launched into the wrong orbit on 9 April 1999, and a military communications satellite launched into the wrong orbit on 30 April 1999—had involved Lockheed Martin Titan IV rockets. The three mission failures had cost taxpayers "at least [US]$3 billion." Hall explained that the U.S. Air Force had traced the three launch failures to human error. The investigators had traced the most recent failure specifically to a misplaced decimal point. Aldridge, whose company had been responsible for checking U.S. Air Force rockets for defects prior to launch, explained that the mistake "got through," despite the system in place to prevent such occurrences. Aldridge remarked that his company's workforce of engineers, scientists, and support personnel had decreased by 30 percent since 1993, and that the recent launch failures had caused further delays, resulting in increased costs for the Titan program. Both Aldridge and Hall requested federal funding. House Subcommittee Chairperson Michael N. Castle (R-DE) asked whether the military rocket program, which receives less supervision than NASA, needed greater congressional oversight. Castle announced that the subcommittee would hold further public hearings on the failures.[726]

The National Oceanic and Atmospheric Administration (NOAA) detached the Geostationary Operational Environmental Satellite-L (GOES-L) from an Atlas IIA rocket, returning the satellite to the integration facility Astrotech, where engineers planned to recondition its batteries and purge it of gaseous nitrogen, to prevent degradation. NOAA had planned to store the satellite in orbit, ready to replace one of the weather satellites, GOES-8 or GOES-10, but GOES-L had been on the launchpad since 6 May. On 15 May, NASA and NOAA had decided to delay the launch

---

[724] *Florida Today* (Brevard, FL), "Iridium Launches Two More Satellites," 11 June 1999.
[725] The Boeing Company, "Director Space & Communications for Russia and CIS Named," news release, 15 June 1999, http://www.boeing.com/news/releases/1999/news_release_990615r.html (accessed 8 February 2008).
[726] Stephen Sobek for Gannett News Service, "Human Error Called Culprit in 3 Rocket Launch Failures," *Florida Today* (Brevard, FL), 16 June 1999.

of GOES-L until they had received the results of the investigations into the recent Titan and Delta launch failures.[727]

*16 June*
NASA Associate Administrator for Space Flight Joseph H. Rothenberg announced several immediate management changes in the Office of Space Flight. NASA appointed William F. Readdy, former Director for Space Shuttle Requirements, as Deputy Associate Administrator, with primary responsibility for daily management of personnel and program activities. Having participated in Shuttle missions in 1992, 1993, and 1996, Readdy was an astronaut on flight status, eligible for future Shuttle missions. NASA appointed W. Michael Hawes, a former chief engineer for the ISS, as Deputy Associate Administrator for Space Development (Space Station), responsible for directing the space station budget, establishing and implementing station policy, coordinating external communications, and serving as a liaison to the executive branch, Congress, industry, and NASA's international partners. Norman B. Starkey replaced Readdy as Director for Space Shuttle Requirements, assuming responsibilities that included directing the Space Shuttle budget, establishing and implementing Shuttle policy, coordinating external communications, and serving as a liaison to the executive branch, Congress, industry, and NASA's international partners.[728]

*17 June*
NASA disclosed that a sequence of computer commands issued by flight controllers on the ground had failed to fire the ISS's engines, placing the uninhabited space station in potential danger. The cause of the failure was human error. Flight controllers had sent commands directing the new space station to move out of the path of a piece of space junk, which posed a danger to the craft. A U.S. military organization tracking such objects had predicted that the "fairly large" piece of space junk of Russian origin would pass within 0.6 miles (1 kilometer) of the space station on 13 June. However, the debris came no closer to the new ISS than 4.5 miles (7.2 kilometers). A collision could have destroyed the uninhabited space station. NASA's Deputy Program Manager for Space Station Operations, Frank L. Culbertson Jr., explained that, because the space station's computers had rejected the flight controllers' faulty commands, the station had no motion control for an entire orbit.[729]

*19 June*
A Lockheed Martin Titan II rocket launched successfully, carrying JPL's QuikSCAT satellite. Ball Aerospace and Technologies Corporation had built QuikSCAT in 11 months, setting an industry record. JPL stated that the satellite had "opened its solar arrays as planned an hour after launch, and a tracking station in Norway acquired the first signal from the spacecraft 18 minutes later." NASA had originally scheduled the satellite's launch for November 1998, but delayed it when a Titan IVA exploded after launch, carrying a payload for the National Reconnaissance Office. NASA had postponed the QuikSCAT launch until engineers could determine whether Titan II used any of the hardware that had caused the explosion on the Titan IVA. Following the

---

[727] NOAA, "Weather Satellite De-mated from Launch Vehicle, NOAA and NASA Announce," news release, 15 June 1999, *http://www.publicaffairs.noaa.gov/releases99/june99/noaa99GOESDemate.html* (accessed 12 February 2008).
[728] NASA, "NASA Selects Key Space Flight Managers," news release 99-71, 16 June 1999.
[729] Marcia Dunn for Associated Press, "Space Station," 18 June 1999.

launch of a Titan IVB rocket that placed a *Milstar* satellite in a faulty orbit, NASA had again delayed the QuikSCAT's launch, pending investigation of the failure. The successfully launched QuikSCAT satellite was on a two-year mission with the option of a third year, to "gather about 400,000 detailed measurements of the speed and direction of winds over the ocean's surface." NASA intended the US$98 million mission to "improve weather forecasting and detect the onset of conditions like El Niño." Engineers had designed the QuikSCAT to radiate microwave pulses over wide areas and to listen for the pulses' echo. Those return signals would "allow scientists to determine wind speeds and directions at the surface of the oceans. Understanding the interaction [of wind and ocean circulation] is important for weather prediction." Under the terms of NASA's first contract using the Rapid Spacecraft Acquisition process, Ball Aerospace had provided the QuikSCAT spacecraft bus, launch interface, system integration, test, launch support, and two years of mission operations.[730]

*21 June*
NASA announced the appointment of N. Jan Davis, Director of the Human Exploration and Development of Space Independent Assurance Office, as Deputy Director of the new Flight Projects Directorate at NASA's Marshall Space Flight Center (MSFC) in Huntsville, Alabama, effective July 1999. NASA had tasked the Directorate with overseeing "development of crucial parts of the ISS," such as connecting Nodes 2 and 3, the Multi-purpose Logistics Modules, commercial EXPRESS (EXpedite the PRocessing of Experiments to the Space Station) racks, the environmental and life support systems, and the Payload Operations Integration Center. Davis had begun her NASA career at MSFC in 1979, leading a team "responsible for structural analysis of the Hubble Space Telescope, the telescope-servicing mission, and the Chandra X-ray Observatory." She had also served as "lead engineer for the redesign of the Shuttle Solid Rocket Booster External Tank attach ring" before becoming an astronaut in 1987. Davis had flown on three Shuttle missions, logging more than 670 hours in space.[731]

*22 June*
NASA's JPL announced that *Discover Magazine* had awarded NASA's Solar Electric Propulsion Technology Application Readiness (NSTAR) program team its Award for Technological Innovation in the exploration category. The annual awards "honor teams whose innovations improve the quality of everyday life." The NSTAR program team won the award for its work on the futuristic ion-propulsion system used on NASA's Deep Space 1 spacecraft.[732]

*24 June*

---

[730] Dow Jones Newswire, "Lockheed Martin's Titan Launches with QuikScat Satellite," 20 June 1999; "Satellite Launched on Weather Mission," *New York Times*, 20 June 1999; *Aerospace Daily*, "QuikSCAT Makes It to Space After Six-Month Delay," 22 June 1999; Ball Aerospace and Technologies Corp., "BATC Launches Its First Commercial Spacecraft," news release, 19 June 1999, http://www.ballaerospace.com/page.jsp?page=30&id=30 (accessed 25 February 2008).

[731] NASA Marshall Space Flight Center, "Astronaut Jan Davis Returns to Marshall Space Flight Center," news release 99-102, 21 June 1999, http://www.msfc.nasa.gov/news/news/releases/1999/99-102.html (accessed 11 February 2008); James McWilliams, "Ex-Astronaut Named Marshall Executive," *Huntsville Times* (AL), 22 June 1999.

[732] NASA Jet Propulsion Laboratory, "Ion Propulsion System Wins Discover Magazine Award," news release 99-054, 22 June 1999, http://www.jpl.nasa.gov/releases/99/discoverawards.html (accessed 11 February 2008).

NASA successfully launched its Far Ultraviolet Spectroscopic Explorer (FUSE) telescope aboard a Delta II rocket. FUSE's three-year mission was to measure the abundance of deuterium, or heavy hydrogen, which stars consume every day and convert to helium. Astronomers hoped that the study of deuterium would help them understand better "what the universe was like moments after creation." The FUSE scientists sought to determine how much deuterium still exists and how much of it stars consume regularly, thereby calculating "the original makeup of the universe." NASA had nearly cancelled the FUSE project in the early 1990s, because of cost overruns and missed deadlines, but in 1994 NASA had asked Johns Hopkins University to manage the project. Project managers at Johns Hopkins had scaled down the scientific instrument and used existing equipment instead of developing new technologies. The Canadian Space Agency had provided the camera, the French Space Agency had provided the spectrograph, and other U.S. universities had provided various parts for the telescope. Orbital Sciences Corporation had built the spacecraft using an existing design to reduce costs further. NASA considered the spacecraft and its launch a success; NASA's Science Director Edward J. Weiler remarked, "We have never had a satellite as sensitive as FUSE."[733]

*25 June*
NASA announced that the WIRE Mishap Investigation Board had concluded that NASA's Wide-Field Infrared Explorer (WIRE) satellite had failed after its 4 March 1999 launch, "because of an incorrectly designed electronics box." The premature firing of explosive devices—the pyrotechnics—had caused the instrument's telescope cover to eject too early in the mission, thus exposing the instrument's frozen hydrogen to the Sun. The frozen hydrogen was necessary to cool the telescope's infrared detectors. The satellite had lost the frozen hydrogen within 48 hours of its launch, rendering the instrument incapable of carrying out its scientific mission. Darrell R. Branscome, Deputy Associate Administrator (Enterprise Development) for NASA's Office of Space Flight and chairperson of the eight-member investigation team, emphasized that there had been no component failure, but "simply a case of a design error that allowed power to get to the explosive charges before it should have." The board's report also concluded that engineers had failed to identify the design errors in the circuitry controlling the pyrotechnical functions. Unlike the other systems in the satellite, the electronics box design had received no peer review.[734]

*28 June*
NASA announced its decision to end a US$240 million project to land a small spacecraft on a comet, so that scientists could learn how to destroy a comet if it were on course to collide with Earth. NASA had planned to launch the probe in 2003 and to land it on the surface of Comet Tempel 1 in 2005. After landing, the probe was to drill into the nucleus of the comet, helping scientists understand the comet's composition. Manager of the Space Technology 4 project Brian Muirhead remarked, "we know very, very little about how comets are formed and what their constituency is," adding that anyone serious about planetary protection must learn more about comets. NASA's JPL was in the early stages of developing the Space Technology 4/Champollion project and had only spent US$10 million. NASA had also considered canceling Mars Surveyor 2001, but had already spent US$100 million on that project. Other projects were experiencing

---

[733] *Washington Times*, "Special Telescope To Look for First Signs of Creation," 25 June 1999; Steven Young for Reuters, "NASA Launches Telescope To Seek Universe's Origin," 24 June 1999.
[734] NASA, "Investigation Finds Design Errors Caused WIRE Spacecraft Failure," news release 99-74, 25 June 1999, *http://www.nasa.gov/home/hqnews/1999/99-074.txt* (accessed 11 February 2008).

cost overruns also, such as the US$1.5 billion Chandra X-ray Observatory, which needed an extra US$60 million to cover last-minute development problems. NASA intended to redirect some funds from Space Technology 4 to other missions experiencing funding difficulties, such as Chandra.[735]

## JULY 1999

*5 July*
The Russian Defense Ministry launched *Raduga-1*, an early-warning-system satellite for detecting missile launches, aboard a Proton-K booster rocket from Baikonur Cosmodrome in Kazakhstan. Ground controllers lost communication with the satellite minutes after its launch, and just as it reached its preliminary orbit, both the satellite and the second stage of the rocket crashed in a remote region of Kazakhstan. Russian space officials initially stated that the satellite had crashed in an uninhabited area, but a Russian television network reported that a large piece of the Proton rocket had fallen onto farmland in the Karaganda region of Kazakhstan, nearly hitting a house. Nobody was injured in the crash.[736]

*6 July*
Pending the outcome of an investigation into the crash of a Russian Proton rocket the previous day, the Kazakhstan government banned Russian launches from Baikonur Cosmodrome, causing a potentially significant disruption to Russia's commercial satellite launch program. Kazakhstan's foreign ministry informed the Russian Ministry of Foreign Affairs that launches from the facility would cease until investigators had identified the causes of the crash and evaluated the damage sustained. The ministry did not indicate how long the ban would remain in effect. Kazakhstan investigators collected soil and water samples to test for possible hazardous material dispersed in the crash.[737]

*8 July*
Astronaut Charles P. "Pete" Conrad Jr. died at age 69 of injuries sustained in a motorcycle accident. George W. S. Abbey, Director of Johnson Space Center, described Conrad, the third man to walk on the Moon, as a person who "combined skill and ability with wit and humor to become one of the courageous pioneers who took humankind beyond the bounds of our planet." As an example of Conrad's wit, Abbey quoted his parody of the famous words Neil A. Armstrong had uttered when taking his first steps on the Moon. Upon leaving the lunar module *Intrepid* on 19 November 1969 for his own moonwalk, Conrad had quipped, "Whoopee! Man, that may have been a small one for Neil, but that's a long one for me." Conrad had been a member of NASA's second candidate class of astronauts in 1962 and had flown on the Gemini 5 mission with L. Gordon Cooper Jr. The pair had spent a record eight days in orbit in August 1965, perfecting techniques for later lunar missions. In September 1966, Conrad had commanded the Gemini 11 mission; during that mission, he and Richard F. Gordon Jr. had linked their *Gemini* spacecraft with an Agena target vehicle, establishing a record for the fastest space rendezvous and docking in history. Conrad had also served as Commander of the Apollo 12 mission in November 1969, with Alan Bean and Richard Gordon as crewmates. During his final

---
[735] Matthew Fordahl for Associated Press, "Comet Lander," 29 June 1999.
[736] United Press International, "Russian Satellite Crashes into Kazakhstan," 6 July 1999.
[737] Sujata Rao for Reuters, "Kazakhstan Halts Russian Space Launches," 6 July 1999.

spaceflight in 1973, Conrad had been the first Commander of Skylab, the first American space station. The Skylab crew had spent 28 days in space, establishing another endurance record. Conrad had won the Congressional Space Medal of Honor, two NASA Distinguished Service Medals, two NASA Exceptional Service Medals, two Navy Distinguished Service Medals, and two Distinguished Flying Crosses.[738]

Ukraine's Prime Minister Valery Pustovoitenko appealed to Kazakhstan's Premier Nurlan Balgimbayev to permit the launch of the *Okean-O* research satellite aboard a Ukrainian Zenit rocket, despite the launch ban following the 5 July crash of a Russian rocket. Ukrainian officials expressed concern that a delay beyond 15 July, with the rocket's lengthy stay on the launchpad, would endanger the prospect of a successful launch. The Ukrainian space program, seeking to compete in the lucrative commercial space market, was under pressure to make a successful launch after the crash in September 1998 of a Zenit rocket, destroying 12 of Globalstar Inc.'s communications satellites.[739]

*10 July*
After a two-day delay because of high-altitude winds, the Boeing Company successfully launched a Delta II rocket from Cape Canaveral Air Force Station in Florida, carrying four Globalstar communications satellites. An hour after launch, the rocket deployed the 988-pound (448-kilogram) satellites in pairs, bringing the total number of satellites in Globalstar Inc.'s constellation to 28, 4 less than the 32 needed to begin network service.[740]

*13 July*
Donald D. Engen, Director of the Smithsonian Institution's National Air and Space Museum, and William Ivans, an internationally known, award-winning pilot from La Jolla, California, died when the motorized glider that Ivans was piloting broke apart and crashed. Investigators were attempting to determine the cause of the accident. Larry Sanderson, President of the Soaring Society of America, remarked that "both victims were top pilots in 'an extremely well-built aircraft. So it had to be a very unusual set of circumstances that stressed the craft'." Both Engen and Ivans were officers of the Soaring Society, and Ivans, a pioneer in the field, had won many awards for high-altitude soaring. Engen had retired from the U.S. Navy in 1978, with the rank of vice admiral and had been a test pilot for many years. After retiring from the Navy, he had served for two years on the National Transportation Safety Board and as administrator of the Federal Aviation Administration between 1984 and 1987. He became Director of the National Air and Space Museum in 1996, following the resignation of Martin O. Harwit.[741]

President Nursultan Nazarbayev set Kazakhstan's conditions for permitting Russia to launch a Progress cargo craft carrying supplies to the *Mir* space station, scheduled to lift off from Baikonur Cosmodrome on 14 July. Kazakhstan had suspended all Russian launches from the facility, following the crash of a Russian Proton-K rocket on 5 July. Nazarbayev's terms for

---

[738] NASA Johnson Space Center, "Third Man To Walk on Moon Dies in Motorcycle Accident," news release J99-24, 9 July 1999, http://www.nasa.gov/centers/johnson/news/releases/1999_2001/j99-24.html (accessed 2 January 2008).
[739] Pavel Polityuk for Reuters, "Ukraine Appeals for Rocket Launch amid Kazakh Row."
[740] Justin Ray, "Third Try Was the Charm for Delta 2 Rocket," *Florida Today* (Brevard, FL), 11 July 1999.
[741] *Washington Times*, "Air and Space Director Killed in Glider Crash," 14 July 1999.

permitting the launch of the Progress craft included Russia's payment of its US$300 million debt for the lease of Baikonur Cosmodrome and a visit of high-level Russian officials to the Proton-K rocket's crash site.[742]

Kenneth R. Timmerman, President of Middle East Data Project Inc., testified before the U.S. House Science Subcommittee on Space and Aeronautics that Iran was designing a new missile named Kosar, capable of reaching the continental United States. Timmerman stated that NASA had given Russian aerospace entities millions of dollars for the Russian space program, but that Russia had diverted those funds to support Iran's missile program. House Committee on Science and Technology Chairperson F. James Sensenbrenner Jr. (R-WI) remarked that each new report of Russian proliferation activities raised the possibility that NASA was inadvertently subsidizing Russian industries, which the United States believed were helping Iran to threaten the United States' friends and allies in the Middle East and in Europe. Henry D. Sokolski, Executive Director of the Nonproliferation Policy Education Center, testified that legislation sponsored by Representative Benjamin A. Gilman (R-NY) was critical to ensuring Russia's cooperation with the nonproliferation efforts of the United States. Gilman's legislation would require the President of the United States to determine whether Russia was "assisting Iran's programs to develop weapons of mass destruction and ballistic missiles." If the President determined that Russia was furnishing such aid to Iran, the federal government would prohibit NASA from "transferring U.S. tax dollars to the [Russian Space Agency] or any enterprise under the [Russian Space Agency] jurisdiction." An unnamed NASA official testified that to replace the operational capabilities that Russia provided to the International Space Program would cost as much as US$5 billion.[743]

*14 July*

The Boeing Company announced that it had completed negotiations on a cooperative agreement with NASA to develop an experimental space plane. The US$173 million contract stipulated that Boeing and the federal government share costs equally. Boeing and NASA hoped that the X-37 space plane, previously called the Future-X Pathfinder, would "serve as a test bed for new reusable launch technologies," helping to achieve the "goal of reducing the cost of placing space vehicles and cargo into orbit." Boeing envisioned a vehicle that would be "unpiloted, autonomously operated, and capable of speeds up to Mach 25, while demonstrating aircraft-like operations." NASA's Marshall Space Flight Center led the X-37 government team, comprising the U.S. Air Force Flight Test Center at Edwards Air Force Base and NASA facilities, including Ames Research Center, Kennedy Space Center, Goddard Space Flight Center, Langley Research Center, and Dryden Flight Research Center.[744]

Kazakhstan partially lifted its ban on launches from Baikonur Cosmodrome, permitting a Russian Progress cargo spacecraft to deliver supplies to the *Mir* space station. Kazakhstan had agreed to permit the Progress launch after receiving Russia's promise to pay a US$115 million

---

[742] Sujata Rao for Reuters, "Kazakhstan Sets Terms for Russian Space Launch," 13 July 1999.
[743] Audrey Hudson, "Analyst Fears U.S. Helps Iran Develop Missile via Moscow: Russia Diverts Aid from Its Space Agency," *Washington Times*, 14 July 1999.
[744] The Boeing Company, "NASA, Boeing Sign X-37 Vehicle Agreement," news release, 14 July 1999, http://www.boeing.com/news/releases/1999/news_release_990714t.html (accessed 19 February 2008); John O'Dell, "California Southland Focus Boeing, NASA Agree on Space Plane Contract," *Los Angeles Times*, 15 July 1999.

fee for the use of the launch facility. Russia had agreed to pay US$50 million in cash by November 1999 and US$65 million in goods by 2000. Kazakhstan's launch ban on Proton-K rockets remained in effect.[745]

*16 July*

A Russian Progress M42 cargo spacecraft launched with a Soyuz-U rocket to deliver supplies to the *Mir* space station. Supplies included equipment necessary to keep an uninhabited station operational and equipment for conducting a controlled reentry, in the event that Russia could not obtain the resources necessary to continue operation. The launch was the first since Kazakhstan's partial relaxation of a launch ban at Baikonur Cosmodrome.[746]

*17 July*

A Russian rocket launched from Baikonur Cosmodrome, carrying a scientific satellite into orbit. It was the second launch from the facility since Kazakhstan partially lifted the launch ban imposed following the crash of a Proton-K rocket on 5 July.[747]

*19 July*

The United Nations opened a conference, scheduled to run through 30 July, on space technology and exploration, the first such conference held in 17 years. The primary goals of the Unispace III conference included 1) adopting the Vienna Declaration, which provided "international guidelines for the use and environmental protection of outer space"; 2) addressing the cleanup of debris in space, including satellites or parts of spacecraft no longer in use, but continuing to orbit; and 3) ensuring the accessibility of information from outer space to all people, regardless of a country's ability to support costly space exploration programs. In his opening remarks, UN Secretary General Kofi Annan "urged the peaceful use of outer space," suggesting the necessity of a legal instrument, such as the Vienna Declaration, to prevent space from "becoming another arena of military confrontation." Because joint development programs between the UN and private industries were necessary to achieve the conference's goals, businesses involved in space technology participated in Unispace for the first time.

*20 July*

The salvage team lifted astronaut Virgil I. "Gus" Grissom's Mercury capsule to the surface of the Atlantic Ocean just one day before the 38th anniversary of the astronaut's historic suborbital flight. The second American in space, Grissom had flown in his *Liberty Bell 7* Mercury capsule on 21 July 1961. After its ocean landing, the capsule's hatch had blown open prematurely, nearly drowning him. Both the location of the capsule and the cause of the accident had remained a mystery until 1 May 1999, when a salvage team led by Curt Newport and funded by the Discovery Channel had discovered the spacecraft 3 miles (4.8 kilometers) beneath the ocean. Even after the team had retrieved the capsule, the cause of the accident remained a mystery. The

---

[745] Reuters, "Kazakhstan Lifts Space Ban, Russian Launch Cleared," 14 July 1999; Associated Press, "Russia-Kazakhstan-Space," 14 July 1999.
[746] *Spacewarn Bulletin*, no. 549, 1 August 1999, *http://nssdc.gsfc.nasa.gov/spacewarn/spx549.html* (accessed 29 October 2008).
[747] *Washington Times*, "Russian Satellite Launched in Kazakhstan," 18 July 1999.

salvage team had yet to locate the capsule's hatch, which Newport believed had the greatest potential for determining what had caused the malfunction at the end of Grissom's flight.[748]

*23 July*
During a spacewalk lasting nearly 6 hours, cosmonauts Viktor M. Afanasyev and Sergei V. Avdeyev searched for the source of a leak that was slowly causing pressure loss in *Mir*; they also attempted to install an antenna needed for an experiment. Both efforts failed. The leak, first detected in late June 1999, was "above the allowable limit." Although it posed no immediate threat to the space station, if the leak continued for three more months, the station would become uninhabitable.[749]

After two delays, Space Shuttle *Columbia* launched on Mission STS-93 from Kennedy Space Center (KSC) in Cape Canaveral, Florida, under the command of Eileen M. Collins, NASA's first female Shuttle commander in 38 years of human spaceflight. Commander Collins's crew comprised Pilot Jeffrey S. Ashby and Mission Specialists Steven A. Hawley, Catherine G. Coleman, and Michel Tognini. *Columbia* carried on board the US$1.5 billion Chandra X-ray Observatory, previously known as the Advanced X-ray Astrophysics Facility (AXAF), as well as secondary payloads and experiments—the Southwest Ultraviolet Imaging System to capture ultraviolet imagery of the Earth, Moon, Mercury, Venus, and Jupiter; plant growth experiments; biological cell-culture studies; and the Treadmill Vibration Information System and High Definition Television System, which the crew planned to test. NASA had scrubbed both of the Shuttle's earlier launch attempts late in the countdown, halting the 20 July launch 6 seconds before blastoff because of a technical malfunction and the 22 July launch because of stormy weather. The U.S. Air Force and the Boeing Company postponed a commercial launch to permit NASA a third chance to deliver Chandra into orbit. However, the Shuttle continued to have problems. During the 8-minute climb into outer space, a 0.5-second-long short circuit shut down computers controlling two of the Shuttle's three main engines. Furthermore, *Columbia* was 4,000 pounds (1,800 kilograms) short of liquid oxygen fuel in its external fuel tank, leaving the craft in orbit 7 miles (11 kilometers) lower than planned. Fuel carried aboard the orbiter made up the shortfall. Seven hours into the flight, the crew deployed Chandra into orbit, accomplishing its primary mission.[750]

*27 July*
Commander Eileen M. Collins became the first woman to "land a spaceship of any kind, anywhere," when she guided Space Shuttle *Columbia* to a safe late night landing at KSC in Cape Canaveral, Florida, the 12th nighttime touchdown in the Shuttle program's 95-flight history. A few hours after the landing, NASA confirmed that hydrogen fuel had leaked from one of the

---

[748] *Los Angeles Times*, "Grissom's Mercury Capsule Recovered," 21 July 1999; Marcia Dunn for Associated Press, "Expedition Begins To Recover Mercury Capsule from Ocean," 2 July 1999.
[749] Nikolai Pavlov for Reuters, "Mir Crew Spacewalks To Seek Pressure Leak," 23 July 1999; *Kansas City Star* (MO), "Antenna Failure Prevents Mir Test: Two Cosmonauts Give Up After Trying To Fix It for Six Hours," 24 July 1999.
[750] *USA Today*, "Shuttle Lifts Off with Woman in Command," 23 July 1999; Brad Liston for Reuters, "Shuttle Roars into History After Launch Problems," 23 July 1999; Beth Dickey, "After Hiccup at Liftoff, Shuttle Puts Telescope into Space," *New York Times*, 25 July 1999; NASA, "Mission Archives STS-93, Chandra X-ray Observatory," http://www.nasa.gov/mission_pages/shuttle/shuttlemissions/archives/sts-93.html (accessed 29 October 2008).

craft's main engines during liftoff on 23 July. *Columbia*'s mission, lasting just five days, was the shortest planned flight in nine years.[751]

*28 July*
Two Russian cosmonauts carried out a successful 5½-hour spacewalk outside *Mir*, installing equipment and opening the antenna that they had failed to unfold properly in a previous spacewalk. Russian space officials said that the spacewalk was most likely the last to occur at *Mir*. This was the seventh spacewalk for *Mir* Commander Viktor M. Afanasyev and the eighth spacewalk for Flight Engineer Sergei V. Avdeyev, who had spent a record-making, cumulative total of 717 days in space.[752]

NASA announced that, after Space Shuttle *Columbia*'s return to Earth on the night of 27 July, engineers had found "three little holes" in its right engine nozzle, confirming that the Shuttle had leaked hydrogen fuel during liftoff. The holes, 0.25 inches (0.64 centimeters) in size, had caused the craft to lose up to 5 pounds (2.3 kilograms) of fuel during each second of its 8½-minute climb to orbit.[753]

*29 July*
NASA scientists announced that, because the Deep Space 1 spacecraft had aimed its camera incorrectly when it flew within 15 miles (24 kilometers) of a small asteroid named Braille, the craft had failed to capture an image. However, the craft had succeeded in its primary mission to test 12 new technologies during the flyby, thereby validating the instruments' future use in solar-system probes. Deep Space 1 was equipped with "a mind of its own," enabling it to navigate through space without much assistance from ground control. The craft had traveled at 35,000 miles (56,000 kilometers) per hour relative to the asteroid, which was 117 million miles (188 million kilometers) from Earth. The flyby of the Braille asteroid, the final test of Deep Space 1's Autonomous Navigation System, had been the closest any spacecraft had come to an object in our solar system without landing.[754]

*31 July*
NASA's Lunar Prospector crashed into a frozen crater on the Moon, but detected no water upon impact. Scientists, believing that the crater held frozen water, had hoped that the "fire and violence of the collision would vaporize ice," sending a "wet plume, detectable by special instruments, spiraling into the lunar sky." Telescopes equipped with ultraviolet detectors recorded hours of data after the impact, searching for the "chemical signature" of water. The Prospector mission, which had lasted 18 months following its 6 January 1998 launch, had used five instruments to "map the magnetic, chemical and gravitational character of the [M]oon."[755]

---

[751] *Atlanta Journal* (GA), "Shuttle Columbia: Woman Commander Kept Cool During Snafus," 28 July 1999; *Los Angeles Times*, "1st Shuttle Led by Woman Lands Safely in Florida," 28 July 1999.
[752] Reuters, "Mir Cosmonauts Make Final Spacewalk," 28 July 1999.
[753] Marcia Dunn for Associated Press, "NASA Confirms Hydrogen Leak in Columbia Shuttle," 29 July 1999.
[754] *Washington Times*, "Spacecraft Misses Asteroid Snapshot: Camera Can't See Rock Called Braille," 30 July 1999.
[755] Associated Press, "Spacecraft Smashes into Moon: Scientists Hoping for Evidence of Ice Vapor in Lunar Sky," *Washington Post*, 1 August 1999.

## AUGUST 1999

*2 August*

NASA announced the student winners of the 1998–1999 National General Aviation Design Competition, sponsored by NASA and the Federal Aviation Administration (FAA). A team of 33 students at Embry-Riddle Aeronautical University in Daytona Beach, Florida, had produced the winning design, "aimed at attracting customers who want to move from propeller-driven craft to jets without needing a significant increase in pilot skill." A 13-member team from Pennsylvania State University in University Park, Pennsylvania, won second place for their "Baracuda," an acronym for Boldly Advanced and Refined Aircraft Concept Under Development for AGATE, a national, general aviation revitalization program. The team's "conventional-layout, modern-composite airplane featuring advanced aerodynamics, systems, and avionics" was a four-place, single-engine, jet-powered aircraft. A three-university team—the University of Virginia, Old Dominion University in Hampton, Virginia, and Pratt Institute in Brooklyn, New York—won third place for a "highly innovative design known as the 'Yeah Man'," a craft with two tail booms, each with vertical tails. The aircraft, which NASA scientists had tested in the Full Scale Wind Tunnel at Langley Research Center, showed good aerodynamic characteristics.[756]

*5 August*

NASA selected two software programs for the 1999 NASA Software of the Year award from 50 entries representing more than 150 corporations, universities, and government laboratories. In the 1970s, NASA's Glenn Research Center at Lewis Field (GRC)—at that time, the Lewis Research Center—had begun to develop the first award-winning program, the Genoa/Progressive Failure Analysis Software System, to simulate and predict "aging and failure in all sorts of structural materials, including high-tech alloys and ceramics used in airplanes, cars, engines, and bridges." In 1998 NASA's Small Business Innovation Research program had commercialized the software, primarily for the use of aircraft manufacturers. Ames Research Center and Jet Propulsion Laboratory (JPL) had developed the other winning software, the Remote Agent program, a "precursor to self-aware, self-controlled robots, exploring rovers and intelligent machines." NASA had used the Remote Agent software to plan three days of activities for Deep Space 1, and the spacecraft had "carried out the plan without ground intervention."[757]

*9 August*

SPACEHAB Inc. and the Canadian Space Agency (CSA) signed a contract granting SPACEHAB the use of one of the CSA's experiment lockers on the International Space Station (ISS), beginning in May 2001. According to the agreement between the two organizations, the CSA had committed to commercializing half of its allocated space aboard the space station, a global first in space commercialization.[758]

---

[756] NASA, "NASA and FAA Pick Student Aircraft-Design Contest Winners," news release 99-87, 2 August 1999, *http://www.nasa.gov/home/hqnews/1999/99-087.txt* (accessed 15 February 2008).

[757] NASA, "NASA's 1999 Software of the Year Makes Cars Safer and Spacecraft Cheaper," news release 99-90, 5 August 1999, *http://www.nasa.gov/home/hqnews/1999/99-090.txt* (accessed 15 February 2008); NASA, "NASA SBIR Success-Genoa," *http://sbir.gsfc.nasa.gov/SBIR/successes/ss/3-048text.html* (accessed 22 February 2008).

[758] SPACEHAB Inc., "Ground Breaking Deal Opens New Space Frontier: SPACEHAB Buys First Commercial Rights Aboard International Space Station," news release, 9 August 1999, *http://www.spacehab.com/news/1999/99_08_09.htm* (accessed 25 February 2008).

*10 August*
NASA and the National Research Council of Canada signed a protocol at the Aerospace North America conference in Vancouver, British Columbia, founding the Aircraft Icing Research Alliance and drafting the Aircraft Icing Research Strategic Plan, to "provide a framework for collaboration to develop critical aircraft icing technologies." The Alliance planned to add more partners, such as the United States' FAA, Environment Canada, Transport Canada, and other government agencies, universities, industrial firms, and organizations interested in aviation-icing research.[759]

*11 August*
Robert Thomas "R.T." Jones, who had "led the development of the swept-back jet wing design that revolutionized air travel and enabled planes to break the sound barrier," died at the age of 89. Jones had begun designing the swept-back wings in 1944, when airplane wings were perpendicular to the fuselage, while working at NACA (National Advisory Committee on Aeronautics), NASA's predecessor agency. In 1945 Jones had conducted airflow studies, showing that a cone-shaped shock wave cut across the tips of straight wings, causing deterioration at Mach 1, the speed of sound. Jones had concluded that a swept-back design would preserve the wings and reduce drag. Although his idea was initially ignored, "virtually every commercial and military jet uses the design today." Jones's later research had included work on the oblique wing, "mounted on a pivot on top of the fuselage." The oblique wing maintained a right angle to the fuselage for maximum lift during takeoff, but the pilot could manipulate it so that, at cruising altitude, one wing tip pointed forward and one backward, saving fuel, generating less engine noise, and eliminating the sonic boom. Although Jones's oblique wing design had never advanced beyond the testing stage, some aeronautical designers continued to study its possibilities.[760]

NASA's largest balloon—39 million cubic feet (1.1 million cubic meters) in volume and 60 stories high—lifted off from Lynn Lake in Manitoba, Canada, to carry out a 38-hour flight more than 20 miles (32 kilometers) above Earth. The balloon carried a Japanese-built instrument, the Superconducting Solenoidal Magnet, to collect particles of antimatter. Shuji Orito of the University of Tokyo led the project, called BESS (Balloon Borne Experiment with a Superconducting Solenoidal Magnet); NASA sponsored the United States' portion of the mission and Monbusho sponsored the Japanese portion.[761]

Kazakhstan's Finance Minister, Uraz Dzhandosov, announced that Russia had paid US$12.5 million as a first installment of its debt to Kazakhstan, fulfilling a condition of the agreement between the two countries. Kazakhstan had demanded the payment as part of the terms of lifting the launch ban it had imposed on Russia when a Russian Proton rocket exploded after liftoff from Baikonur Cosmodrome in Kazakhstan.[762]

---

[759] NASA, "NASA and Canada Join Forces To Combat Aircraft Icing," news release 99-92, 11 August 1999, *http://www.nasa.gov/home/hqnews/1999/99-092.txt* (accessed 15 February 2008).
[760] *Washington Post*, "Robert Jones Dies; Led Breakthrough on Wing Design," 15 August 1999.
[761] NASA, "Balloon-Borne Instrument Collects Antimatter," news release 99-93, 16 August 1999, *http://www.nasa.gov/home/hqnews/1999/99-093.txt* (accessed 15 February 2008).
[762] *Russia Today*, "Russia Pays First Part of Baikonur Debt, 12 August 1999.

China confirmed that it had fired its new Dong Feng-31 (DF-31) missile during a test launch on 2 August, publicizing details of the successful launch of the long-range, ground-to-ground, intercontinental, strategic ballistic missile. The Chinese *Guangzhou Daily* reported that the "three-stage, solid-fuel rocket was launched from northern Shanxi province and crashed down in the western territory of Xinjiang." *Jane's Defence Weekly* estimated that the DF-31 had an 8,000-kilometer (5,000-mile) range and was capable of carrying a 700-kilogram (1,500-pound) nuclear warhead. Robert Karniol, a foreign correspondent covering Asia for the defense industry publication *Jane's*, remarked that "the DF-31 should be operationally deployed by China in 2000," and that the country expected to build 10 to 20 of the missiles, some replacing 1960s-era missiles with half the range of the DF-31. The *Guangzhou Daily* also reported that China had developed the technology for piloted spaceflight. Wang Xinqing, head of the China Carrier-Rocket Research Institute, which designs military- and civilian-use rockets, stated that the core of the program was the development of a new series of carrier rockets, Long March 5. Wang Xinqing also refuted a report by the U.S. Congress alleging that China had stolen the United States' missile and nuclear warhead technology, insisting that China had perfected the rocketry before the United States had achieved it.[763]

The U.S. General Accounting Office (GAO) published a report critical of the X-33 program jointly sponsored by NASA and Lockheed Martin, concluding that the program was unlikely to meet its original cost, schedule, and performance goals because of increased costs, delayed testing, and revised objectives. However, NASA and Lockheed Martin responded that they expected the program to "achieve technical requirements, such as demonstrating the feasibility of building large liquid hydrogen fuel tanks made of graphite composite material." The purpose of the X-33 Program was to develop and demonstrate advanced technologies for use on future reusable launch vehicles, such as the company's VentureStar™.[764]

*12 August*
NASA's GRC announced that, for the first time, researchers had suspended particles of frozen hydrogen in liquid helium, the first step toward creating new rocket fuels that could revolutionize rocket-propulsion technology. Researchers had poured small amounts of liquid hydrogen at a temperature of 14 K (-435°F or -259°C), just above freezing point, onto the surface of liquid helium of a temperature just above absolute zero, at 4 K (-452°F or -269°C). The liquid hydrogen had formed small, solid hydrogen particles, which floated on the surface of the helium. Scientists planned to use the suspension to create "futuristic atomic fuels," making it possible to develop rockets with "liftoff weights one-fifth that of today's [rockets] or with payloads three to four times more massive." In addition, the suspension could "reduce or eliminate on-orbit assembly of large space vehicles," thereby contributing to the exploration of the entire solar system.[765]

---

[763] Matt Pottinger for Reuters, "China Details Missile Test, Planned Space Flight," 12 August 1999.
[764] U. S. General Accounting Office, "Space Transportation: Status of the X-33 Reusable Launch Vehicle Program" (report no. NIASD-99-176, Washington, DC, 11 August 1999), *http://www.gao.gov/archive/1999/ns99176.pdf* (accessed 4 March 2008).
[765] NASA Glenn Research Center, "Rocket Fuels Researchers Suspend Frozen Hydrogen Particles in Helium," news release 99-65, 12 August 1999, *http://www.nasa.gov/centers/glenn/news/pressrel/1999/99_65.html* (accessed 25 February 2008).

A burst of radiation from Jupiter hit NASA's Galileo spacecraft as it gathered data about Jupiter's moon Callisto, causing computer malfunctions. However, NASA reported, "recently loaded software automatically restarted command sequences, minimizing the loss of science data." James K. Erickson of NASA's JPL, Project Manager of the Galileo Program, remarked on the highly variable nature of the radiation, explaining that, previously, the spacecraft had entered safe modes during radiation bursts, resulting in the loss of data. Galileo's instruments had only lost about 1 hour of data during the radiation burst, successfully recording data about magnetic fields and particles. Project scientists expected that the data would help them study the event. To emphasize the value of the software, Erickson stated, "If we didn't have the automatic restart software, we wouldn't be getting any data at the encounters."[766]

*16 August*
A team of astrophysicists at NASA's Goddard Space Flight Center announced that they had found the first direct evidence of a black hole pulling in matter. Using the Advanced Satellite for Cosmology and Astrophysics, an x-ray satellite belonging to Japan and the United States and launched in 1993, the team observed superheated gas in the accretion disc of the black hole. Previously, scientists had only "seen" the phenomenon by observing these accretion discs, or "the swirling matter circling around as it is being pulled into the black hole." However, this time, the Goddard team had observed a "strange feature" buried in the x-rays that the gas emitted. Energy had been redshifted, an occurrence described as an "astronomical Doppler effect," because, just as the compression of sound waves, for example, causes the sound of a siren to rise and fall as it passes an observer, light stretches as it speeds away from Earth. The team observed light that was being stretched—redshifted—moving at about 6.5 million miles (10.5 million kilometers) per hour, toward a black hole in galaxy NGC 3516.[767]

Swedish state prosecutors charged Charlie Malm and Joel Soederberg of Stockholm, who had broken into NASA computer systems between October and December 1996, with violating Sweden's computer laws and with buying stolen equipment. In the trial, scheduled for sometime in the fall, NASA intended to demonstrate that the pair had "caused NASA great economic loss." Malm and Soederberg had also broken into the computer systems of the U.S. Air Force, U.S. Army, and U.S. Marines, as well as the system of a British Internet company.[768]

*19 August*
NASA announced its selection of three very small satellites, each about 16 inches (41 centimeters) across and 8 inches (20 centimeters) high, to conduct the Nanosat Constellation Trailblazer mission, the fifth mission in NASA's New Millennium program. NASA had developed the New Millennium program to test technology for future missions, with the goal of dramatically reducing the weight, size, and costs of missions, while simultaneously increasing science capabilities. NASA scheduled the US$28 million Nanosat Constellation mission to launch in 2003 as a secondary payload on an expendable launch vehicle.[769]

---

[766] Michael A. Dornheim, "Unexpected Jovian Radiation Hits Galileo," *Aviation Week and Space Technology* 151, no. 8 (23 August 1999): 42.
[767] Reuters, "Evidence of Super Gravity Reported: Backs Black Hole Astronomy Theory," *Washington Times*, 17 August 1999.
[768] Susanna Loof for Associated Press, "Swedes Charged with U.S. Hacking," 16 August 1999.
[769] NASA, "NASA Selects Miniature Spacecraft To Test Space Technology," news release 99-95, 19 August 1999, *http://www.nasa.gov/home/hqnews/1999/99-095.txt* (accessed 15 February 2008).

*24 August*
NASA announced that it had completed negotiations for a contract worth up to US$1.73 billion, to purchase 73 Space Shuttle reusable solid rocket motors from the Thiokol Propulsion company. Each Shuttle flight used two of these motors as the primary component of the Shuttle solid rocket boosters, providing 6.6 million pounds (2.9 million kilograms) of thrust, 71.4 percent of the thrust needed for liftoff. The contract covered the manufacture and delivery of the new motor components through September 2004, as well as postflight review of the last motors flown through 2005.[770]

SPACEHAB Inc., the first company to "commercially develop, own, and operate habitable modules that provide laboratory facilities and logistics re-supply aboard NASA's Space Shuttles," and leading global provider of commercial payload-processing services, participated in a Brazilian industry conference. Brazsat, a Brazilian commercial space company, hosted the Third Commercial Space Workshop, in Rio de Janeiro on 23 and 24 August. North American, South American, and European industry and government leaders attended the workshop, where SPACEHAB's Chief Executive Officer Shelley A. Harrison delivered one of two keynote addresses. Participants at the workshop, which had the theme "Commercial Space Technologies and its Benefits in the New Millenium," presented "state-of-the-art space technologies, research spin-offs and applications in areas of microgravity research, remote sensing, telecommunications satellites, distance learning, telemedicine and other disciplines" of human spaceflight.[771]

*25 August*
Ball Aerospace and Technologies Corporation announced that NASA had selected the company to build two spacecraft for the Space Technology 3 (ST-3) Mission, part of NASA's New Millenium Program. The contract, valued at US$50 million, covered the development and manufacture of the two spacecraft as well as the integration and testing of completed systems. NASA planned for the two spacecraft, flying 0.5 miles (0.8 kilometers) apart, to demonstrate interferometry, simulating a single large telescope. NASA's JPL was responsible for developing the interferometer and formation-flying sensing technologies, as well as for managing the mission.[772]

*26 August*
NASA released two initial images from its Chandra X-ray Observatory. One image traced "the aftermath of a gigantic stellar explosion in such stunning detail" that scientists believed they had detected evidence of a neutron star or black hole near the center. The second image showed "a powerful x-ray jet blasting 200,000 light-years into intergalactic space from a distant quasar."

---

[770] NASA, "NASA, Thiokol Complete $1.7 Billion Shuttle Motor Agreement," news release c99-c, 24 August 1999, http://www.nasa.gov/home/hqnews/contract/1999/c99-c.txt (accessed 15 February 2008).
[771] SPACEHAB Inc., "SPACEHAB Participates in Key Brazilian Industry Conference: SPACEHAB CEO Addresses Industry Officials on Essential Space Initiatives," news release, 31 August 1999, http://www.spacehab.com/news/1999/99_08_31.htm (accessed 3 March 2008).
[772] Ball Aerospace & Technologies Corporation, "BATC Wins NASA Contract for Two Spacecraft To Test Space Technology," news release, 25 August 1999, http://www.ballaerospace.com/page.jsp?page=30&id=36 (accessed 25 February 2008).

Chandra was still in its orbital "check-out and calibration phase" when it captured the images, confirming that the observatory was performing up to expectations.[773]

*27 August*
Cosmonauts Viktor M. Afanasyev and Sergei V. Avdeyev departed *Mir* with their French colleague Jean-Pierre Haigneré, leaving the station empty for the first time in 13½ years. On 2 August, in preparation for the departure, the crew had installed a new computer, which provided orientation and docking management systems to keep the station in orbit during the absence of a crew. The crew had also installed a ground-controlled backup system, which would prevent the station from crashing to Earth if a malfunction occurred on board the station.[774]

*30 August*
Lockheed Martin cleared its Atlas II rocket to resume the 1999 launch schedule, after having placed its schedule on hold when a Delta III launch failed in early May. Delta's RL10B-2, upper-stage engine was similar to the RL10A-4 used on Atlas's Centaur upper stage. The Boeing Company and Pratt & Whitney had led the investigation of the Delta III launch failure, with the support of Lockheed Martin. Meanwhile, Lockheed Martin Astronautics had formed two panels "to ensure that no aspects of commonality between the RL10 engine versions were overlooked." The Senior Engineering Review Panel and the Senior Management Review Team had addressed potential causes of launch failure identified in the Delta III investigation, ruling out those factors that did not apply to Atlas flights. When the investigators had determined that "an engine-brazing process that left voids in combustion chamber structural jacket splice joints" was common to all RL10 engines, Lockheed Martin had instituted a new set of inspection and analysis measures for accepting or rejecting each engine. Lockheed Martin intended to use the inspection techniques on new production chambers; Pratt & Whitney planned to continue using its standard procedure, with its Flight Certification Board certifying each set of RL10 engines for all Atlas missions.[775]

GAO published a report, "Space Station: Russian Commitment and Cost Control Problems," reiterating its previous recommendation to NASA to plan for a scenario in which Russia was unable to fulfill its commitments to the ISS. Although the report recognized that NASA was in the process of drafting such a plan, GAO's sole recommendation to NASA was to ensure the contingency plan's completion before the launch of the Russian-built Service Module.[776]

---

[773] NASA, "NASA Unveils First Images from Chandra X-ray Observatory," news release 99-98, 26 August 1999, *http://www.nasa.gov/home/hqnews/1999/99-098.txt* (accessed 15 February 2008).
[774] David Hoffman, "Cosmonauts Abandon Mir: Russian Space Station Faces Demise After 13 Years Aloft," *Washington Post*, 28 August 1999; Andrei Shukshin for Reuters, "Russia's Mir Has New Computer in Control," 2 August 1999; Reuters, "Russia's Mir Crew Fit Equipment To Leave Station," 6 August 1999.
[775] Lockheed Martin, "Lockheed Martin's Centaur RL10 Engines Cleared for Flight," news release, 30 August 1999, *http://www.lockheedmartin.com/news/press_releases/1999/LOCKHEEDMARTINSCENTAURRL10ENGINESCL. html* (25 February 2008).
[776] Jonathan Lipman, "GAO Warns NASA About the Russians," *Space.com*, 30 August 1999, *http://www.space. com/news/spacestation/gao_30.html* (accessed 9 April 2008); U. S. General Accounting Office, "Space Station: Russian Commitment and Cost Control Problems" (report no. GAO/NSIAD-99-175, Washington, DC, August 1999), *http://www.gao.gov/archive/1999/ns99175.pdf* (accessed 25 February 2008).

The Independent Assessment Team on Mission Success, which Lockheed Martin had commissioned in May 1999 after a series of Titan IV launch failures, met its 1 September deadline, reporting its findings and recommendations to Lockheed's senior management. Lockheed Martin had tasked the investigating team with assessing program management, engineering and manufacturing processes, and quality-control procedures within the company's Space and Strategic Missiles Sector. Having discovered problems in accountability, quality, subcontract and supplier management, and cost emphasis, the Team recommended that the company develop a "fly-out plan" for Titan IV's remaining missions, improve its quality control, and improve its management of suppliers and subcontractors. Chairperson of the review panel A. Thomas Young remarked that regardless of the problems the team had uncovered, the panel had also found "enormous fundamental technical strengths" throughout the sector, as well as "highly capable and dedicated" staff. Young also commented that the new leadership team was "off to a positive start." Vance D. Coffman, Chairperson and Chief Executive Officer for Lockheed Martin, "endorsed the panel's findings and pledged Lockheed Martin to implementing its recommendations."[777]

*31 August*
Kazakhstan lifted its ban on launches of Russian Proton rockets after Moscow paid US$270,000 in compensation for the 5 July crash of one of the booster rockets and completed an agreed-upon joint investigation. The lifting of the ban cleared the way for two Russian *Yamal-100* communications satellites to launch aboard a Proton rocket on 6 September. Although initially the country had banned all launches from Baikonur Cosmodrome, which Russia had rented since the collapse of the Soviet Union, Kazakhstan had modified the ban to bar only Proton rocket launches, pending a joint investigation of the accident.[778]

## SEPTEMBER 1999

*3 September*
NASA announced its decision to ground the entire Space Shuttle fleet while technicians continued detailed inspections of the fleet's wiring. During the first month of inspections, following the discovery that faulty wiring had caused a short circuit during *Columbia*'s 23 July launch, technicians had located dozens of nicked or exposed wires. Shuttle managers had decided to extend the inspections to allow technicians to examine additional parts of each Shuttle. Technicians had replaced damaged wiring, encased some wires in plastic tubing, added Teflon wrapping to others, and replaced some connectors, to protect the wiring systems against future damage. Manager of the Space Shuttle Program Ronald D. Dittemore explained that the wiring problems "appeared to have been caused by work-related mechanical damage"—from being rubbed or stepped on or from having "heavy objects set down on them." The age of the wiring, normal wear, or vibrations from Shuttle operations did not appear to have caused the damage.[779]

---

[777] Lockheed Martin, "Lockheed Martin Implements Space Panel Recommendations," news release, 8 September 1999, *http://www.lockheedmartin.com/news/press_releases/1999/LOCKHEEDMARTINIMPLEMENTSSPACE PANELR.html* (accessed 27 February 2008).
[778] *Russia Today*, "Kazakhstan Ends Russia Proton Rocket Ban," 1 September 1999.
[779] Warren E. Leary, "Shuttle Fleet Is Grounded by Damage to Wiring," *New York Times*, 4 September 1999.

NASA announced that its new orbiting SeaWinds radar instrument, flying aboard the QuikSCAT satellite, was successfully tracking Iceberg B10A, which had broken off the Thwaites glacier of Antarctica in 1992, drifting into a shipping lane. Earlier in 1999, Iceberg B10A had disappeared when conventional methods of tracking sea-surface ice—ships' radar, shipping reports, optical images from satellites, and microwave sensor data—were unable to track it. David G. Long of the SeaWinds science team at Utah's Brigham Young University remarked that, although "a ship was dispatched to the iceberg's last known position, we were unable to find it until we started receiving data from the SeaWinds instrument in July." During its first pass over Antarctica, SeaWinds had spotted the iceberg, and the National Ice Center in Suitland, Maryland, had confirmed that the iceberg was B10A. Scientists had continued to track it as it moved through the Drake Passage and headed northeast between Tierra del Fuego, at the southern tip of South America, and the Antarctic Peninsula. At that time, the National Ice Center had issued an iceberg navigation warning to Argentina.[780]

*8 September*
NASA announced that scientists at Ames Research Center (ARC), who were developing an autonomous robot called the Personal Satellite Assistant (PSA), had completed a key test of the robot's components. NASA planned to use the robot to support future space missions, equipping it with a variety of sensors to monitor environmental conditions inside a spacecraft, such as the amounts of oxygen, carbon dioxide, and other gases in the air; the amount of bacterial growth in the environment; the air temperature; and the air pressure. The development of the PSA was the next phase in developing advanced information technologies, following the success of the Wireless Network Experiment, which NASA's ARC had conducted for the International Space Station (ISS) in 1995. *Atlantis* astronauts had discovered that wireless computer network systems worked well in a space environment during Mission STS-76, when they tested these networks and found that radio signals from wireless computers did not interfere with the electronic equipment of the Shuttle or of *Mir*. The experiment's success had prompted the astronauts to recommend using handheld, wireless, portable data assistants to support mission operations on the future ISS. ARC scientists had taken the idea a step further, designing autonomous intelligent robots that would free the astronauts' hands. ARC scientists had also designed the PSA to handle routine "housekeeping chores," such as monitoring inventory and performing environmental sensor-calibration checks, so that astronauts would have more time to focus on research tasks.[781]

*9 September*
A Russian Soyuz booster launched from Plesetsk Cosmodrome 450 miles (724 kilometers) north of Moscow, carrying German, French, and Swedish research devices. In selecting the launch location, Russia was attempting to increase its use of the Plesetsk launch facility because of its problems continuing to launch from Baikonur Cosmodrome. Since the collapse of the Soviet Union, Russia had rented the Baikonur launch facility from Kazakhstan.[782]

*10 September*

---

[780] NASA, "New NASA Ocean Radar Watches for Breakup of Giant Iceberg," news release 99-102, 3 September 1999, *http://www.nasa.gov/home/hqnews/1999/99-102.txt* (accessed 15 February 2008).
[781] NASA Ames Research Center, "NASA Developing Autonomous Robot for Future Space Missions," news release 99-53AR, 8 September 1999.
[782] Reuters, "Russian Rocket Blasts Off from Plesetsk Cosmodrome," 9 September 1999.

NASA's Inspector General Roberta L. Gross published findings and recommendations on behalf of NASA's Office of Inspector General, regarding Vice President Albert A. Gore Jr.'s proposed *Triana* satellite project, named for Rodrigo de Triana, the sailor on Christopher Columbus's 1492 voyage who had first spotted North America. The Triana project would call on NASA to provide "continuous, satellite-generated images of the Earth for posting on the Internet." Gross suggested that it was "ill-advised for the financially strapped [A]gency to fund a new system to collect pictures of Earth when many images" were already widely available. Gross said the Triana project would cost four times as much as the Vice President's US$50 million estimate, noting that NASA had already spent US$41 million on the project, even though Congress had not yet endorsed it. She criticized NASA for "pressing forward with the project without fully assessing the scientific value of photographing the Earth." NASA had scheduled the satellite for launch aboard a Space Shuttle in December 2000.[783]

*20 September*
AlliedSignal Technical Services, based in Columbia, Maryland, named as its president former astronaut and Space Shuttle Program Office executive James C. Adamson, who had logged 334 hours in orbit during Missions STS-28 and STS-43. AlliedSignal Technical Services had a staff of 1,800 in Maryland and responsibility for oversight of most of the "spacecraft ground and flight control at NASA's Goddard Space Flight Center in Greenbelt [Maryland]." In accepting the appointment, Adamson was leaving his position as Chief Operating Officer of United Space Alliance (USA), which operated the Shuttle program for NASA.[784]

Lockheed Martin announced a "sweeping reorganization" of key leadership positions, in response to recommendations of the review panel chaired by A. Thomas Young. The Young Panel had urged the company to make structural changes to correct quality-control problems at Astronautics, the unit responsible for Atlas, Titan, and Athena rocket development. In response to the panel's recommendation to divide the two tasks of product assurance and mission success, Astronautics President G. Thomas Marsh appointed John P. Mari as Vice President of Product Assurance and Roman Matherne as Vice President for Mission Success. Marsh named John Parker, President of the former Mission Success and Product Assurance Section, to head the operations post in Lockheed Martin's VentureStar™ Program, which the company hoped to develop into a future reusable launch vehicle with the potential to replace the Space Shuttle. Marsh appointed Joel S. Porter as Vice President for Business Development and Advanced Programs for Astronautics, succeeding Matthew B. Foster, who resigned. Other changes at Astronautics included the appointment of Claude McAnally as Vice President of Operations; the appointment of Grover W. Hall Jr. as Vice President of Technical Operations; and the separation of Technical Operations from Manufacturing Operations to better delineate responsibility and accountability, according to the recommendations of the Young Panel. Lockheed tasked Gareth D. Flora, under Marsh's supervision, with developing, managing, and improving subcontract management. In addition, Flora became responsible for ensuring that the company properly

---

[783] NASA Office of Inspector General, "Assessment of the Triana Mission" (final report no. G-99-013, Washington, DC, 10 September 1999), *http://oig.nasa.gov/old/inspections_assessments/g-99-013.pdf* (accessed 3 March 2008); Steve Lash, "Gore's Pet Project Not Worth Its Cost, NASA Auditor Says," *Houston Chronicle*, 17 September 1999.
[784] Greg Schneider, "AlliedSignal Technical Gets a New President; Adamson Is Former Astronaut, Shuttle Executive, Engineer," *Baltimore Sun* (MD), 21 September 1999; NASA Johnson Space Center, "Biographical Data: James C. Adamson," *http://www.jsc.nasa.gov/Bios/htmlbios/adamson-jc.html* (accessed 28 October 2008).

implemented the recommendations of the investigating team, as well as those of a separate U.S. Air Force review of the aerospace industry.[785]

NASA announced the establishment of a review panel to examine the overall safety of Shuttle maintenance and refurbishment practices following the discovery of maintenance-related damage to the Shuttle fleet's electrical wiring. NASA had named Director of ARC Henry McDonald to chair the panel. NASA Associate Administrator for Space Flight Joseph H. Rothenberg indicated that the other team members, not yet named, would include top maintenance experts from NASA, the military, the commercial aerospace industry, and the commercial aircraft industry. NASA required the team to assess NASA's standard practices and recommend improvements, with preliminary findings due in October.[786]

*21 September*
Space.com, an Internet site devoted to covering the news and science of space, appointed Sally K. Ride as its President. Lou Dobbs had begun Space.com in July 1999, after resigning as president of CNN's Financial News. Ride, the first American woman in space, had worked closely with Dobbs as he launched the site. The position was her "first foray into the cyberspace after a long career in space research, policy, and education."[787]

*23 September*
Former astronauts Michael J. McCulley, Vice President and Deputy Program Manager for USA, and William F. Readdy, NASA's Deputy Associate Administrator for Space Flight, testified before the U.S. House Science Subcommittee on Space and Aeronautics regarding wiring damage found in Space Shuttles *Columbia*, *Discovery*, *Endeavour*, and *Atlantis*. In its first official report since discovering that frayed wires had caused the short circuit 5 seconds into *Columbia*'s launch on 23 July, NASA informed committee members that wiring inspections and maintenance had cost US$350,000. NASA had not yet calculated the cost of the ongoing safety "stand down." NASA also indicated that it did not intend to pay USA the US$2.5 million the company would have earned if it had punctually delivered the next Shuttle scheduled for launch. McCulley had indicated that his company accepted "full responsibility for wiring damage" found in all four Shuttles, despite the fact that the wiring problems might have stemmed from maintenance conducted before USA took over operational responsibility. Committee members expressed their concern that NASA planned to request more money for the repairs, but Readdy replied that such a scenario was unlikely, because USA would be responsible for making the repairs. Committee members also were worried about whether Shuttle launches and flights could maintain adequate safety levels when the Shuttle team had only a few opportunities to practice their skills. NASA Associate Administrator for Safety and Mission Assurance Frederick D. Gregory responded that simulation launches, in combination with a flight rate of two launches per year, were sufficient to keep a launch team's skills sharp.[788]

---

[785] Jeffrey Leib, "Lockheed Replaces Key Staff in Astronautics Unit," *Denver Post* (CO), 21 September 1999.
[786] NASA, "NASA Forms Independent Industry-Government Team To Review Shuttle Maintenance and Refurbishment Practices," news release 99-104, 20 September 1999, *http://www.nasa.gov/home/hqnews/1999/99-104.txt* (accessed 15 February 2008).
[787] Andrea Orr for Reuters, "Space.com Names Space Woman as Company President," 21 September 1999.
[788] Larry Wheeler, "Shuttle Wiring Inspections, Maintenance Cost $350,000," *Florida Today* (Brevard, FL), 24 September 1999; Tamara Lytle, "NASA Officials Bring Wiring Problems to House Panel's Attention," *Orlando Sentinel* (FL), 24 September 1999.

NASA scientists lost contact with the Mars Climate Orbiter spacecraft as it entered its Martian orbit and disappeared behind the planet, a little less than a year after the launch of its Mars mission in December 1998. Mars Surveyor Project Manager John B. McNamee said that the orbit-insertion burn, a 16-minute-long burn to slow the spacecraft down as it enters orbit, had begun on time. The team of NASA scientists observed 5 minutes of the burn before the craft went behind Mars. At that point, the craft stopped communicating. Communication did not resume at the time the team expected the craft to reappear from the far side of the planet. McNamee reported that the latest navigation results showed that the craft had entered its orbit lower than his team had planned, indicating that NASA's navigation predictions at the tracking station may have been incorrect. Carl B. Pilcher, Science Director for Solar System Exploration at NASA Headquarters, commented that, although losing the spacecraft was very serious, the loss was "not devastating to the Mars Surveyor Program as a whole," because the mission of Mars Polar Lander, scheduled to arrive at Mars two months after Mars Climate Orbiter, was independent of that of Orbiter.[789]

*25 September*
Space Shuttle *Columbia*, the oldest of NASA's four orbiters, landed at a U.S. Air Force plant in California for a "long-overdue tune-up." The orbiter's overhaul, expected to last until July 2000, came months late because of delays related to the launch of the Chandra X-ray Observatory. *Columbia* astronauts had deployed Chandra during a mission in July 1999. The inspection and overhaul was *Columbia*'s fourth since entering service in 1981.[790]

*28 September*
NASA announced that after just two months in space, the Chandra X-ray Observatory had taken "a stunning image of the Crab Nebula," revealing for the first time "a brilliant ring about the nebula's heart." The Crab Nebula, located 6,000 light-years from Earth, in the constellation Taurus, is the remnant of a star that Chinese astronomers had observed exploding in 1054 A.D., when it "appeared suddenly and remained visible for weeks, even during daytime." Astronomers had used "virtually every astronomical instrument" to study the Crab's area of the sky, discovering that unlocking its mysteries led to "insight after insight" into how the universe works. Astronomers had linked the origin of pulsars to supernovas and confirmed the cosmic origin of chemical elements through the study of the "expanding cloud of filaments" in the Crab. The new image showed the Crab pulsar "accelerating particles up to the speed of light and flinging them out into interstellar space at an incredible rate." Using Chandra's "exceptional resolution," scientists had traced the jet all the way to the neutron star, where the ring pattern clearly appeared. Previously, Hubble Space Telescope images had revealed "moving knots and wisps around the neutron star," and other x-ray images had shown the outer portions of the jet, only hinting at the ring structure.[791]

---

[789] NASA Jet Propulsion Laboratory, "NASA's Mars Climate Orbiter Believed To Be Lost," news release, 23 September 1999, *http://www.jpl.nasa.gov/releases/99/mcolost.html* (accessed 3 March 2008); Reuters, "NASA Loses Contact with Mars Orbiter," 23 September 1999.
[790] Associated Press, "Shuttle Tuneup," 25 September 1999.
[791] NASA, "Chandra Discovers X-ray Ring Around Cosmic Powerhouse in Crab Nebula," news release 99-109, 28 September 1999, *http://www.nasa.gov/home/hqnews/1999/99-109.txt* (accessed 15 February 2008); Paul Hoversten, "Photos Get to the Source of Pulsar," *USA Today*, 29 September 1999.

*30 September*
NASA released preliminary findings of an internal peer review conducted at Jet Propulsion Laboratory (JPL) after the apparent loss of the Mars Climate Orbiter spacecraft as it entered its orbit on 23 September. The results of the review indicated that "a failure to recognize and correct an error in a transfer of information between the Mars Climate Orbiter spacecraft team in Colorado and the mission navigation team in California led to the loss of the spacecraft." The peer review discovered that "one team had used English units while the other used metric units for a key spacecraft operation." NASA's systems engineers had failed to detect the error. In addition to the JPL committee, NASA had formed a separate review committee of experts from within and outside of JPL and planned to establish an independent failure-review board shortly.[792]

NASA announced that new findings from the Galileo spacecraft indicated the presence of sulfuric acid on the frozen surface of Jupiter's moon Europa. Sulfuric acid is a corrosive chemical found in car batteries. Robert W. Carlson of NASA's JPL, the lead author of a paper published in the journal *Science*, was Principal Investigator for Galileo's Near Infrared Mapping Spectrometer, which located the chemical. The new findings corroborated Galileo's earlier spectrometer data analyses, suggesting the presence of sulfate salts on Europa. Despite indications that a liquid ocean may exist under Europa's crust, Carlson had initially thought the presence of sulfuric acid would end debate over whether it was possible for Europa to contain biological life. However, his colleague Kenneth H. Nealson suggested the opposite, remarking that sulfur and sulfuric acid are oxidants—energy sources known to us on Earth. Therefore, Nealson proposed that the findings should encourage scientists to search for "any possible links between the sulfur oxidants on Europa's surface and natural fuels produced from Europa's hot interior."[793]

## OCTOBER 1999

*1 October*
V. Philip Rasmussen Jr., Head of the Department of Plants, Soils, and Biometeorology at Utah State University, began work as NASA's first and only agricultural extension agent, with the task of disseminating to American farmers satellite data that could increase their productivity. In 1917 Congress had established the U.S. Department of Agriculture's Cooperative Extension Service, operating out of land-grant universities and serving every county in the country. Through this program, the federal government had made research-based agricultural information available to the public in exchange for federal resource support at the universities. Initially, Cooperative Extension Service agents had focused on farm and ranch questions, but the program had evolved to assist home gardeners, greenhouse users, and businesses as well. After studying the land-grant university system, NASA had decided to create a space-grant consortium system to provide the public with access to space science, such as the sciences of remote sensing and crop management. These scientific fields had developed significantly since the 1970s, when

---

[792] NASA, "Mars Climate Orbiter Team Finds Likely Cause of Loss," news release 99-113, 30 September 1999, *http://www.nasa.gov/home/hqnews/1999/99-113.txt* (accessed 15 February 2008).
[793] NASA, "Battery Acid Chemical Found on Jupiter's Moon Europa," news release 99-112, 30 September 1999, *http://www.nasa.gov/home/hqnews/1999/99-112.txt* (accessed 15 February 2008); R. W. Carlson et al., "Sulfuric Acid on Europa and the Radiolytic Sulfur Cycle," *Science* 286, no. 5437 (1 October 1999): 97–99.

Rasmussen had used Landsat data strategically to predict Russian crop yields. At that time, the U.S. government was not using Landsat data to assist American farmers, but under NASA's new space-grant program, Rasmussen's primary task would be to make such satellite data available, through the existing county extension service, so that county agents could teach farmers how to use the data to improve crop yield. Rasmussen described his extension work—using the global positioning satellite (GPS) program in tandem with the geographic information system—as "precision agriculture," because farmers could use the data to make informed adjustments concerning methods of watering and fertilizing fields of various soil types.[794]

The Perseus B research aircraft, built by Aurora Flight Sciences Inc., sustained moderate damage when it crashed on a California highway during a flight from NASA's Dryden Flight Research Center (DFRC) in Edwards, California. The craft was a "developmental vehicle designed to operate at high altitudes for extended periods on scientific sampling missions," one of several aircraft that NASA was evaluating as part of its Environmental Research Aircraft and Sensor Technology program. Controllers could operate the craft remotely from a ground station, or the Perseus B could navigate autonomously along a preprogrammed flight path. At the time of the accident, ground-based Aurora Flight Sciences mission controllers at Edwards Air Force Base were guiding the Perseus B. DFRC flight safety officials, with assistance from Aurora Flight Sciences operations staff, formed an accident investigation team to determine the exact cause of the crash.[795]

NASA announced that scientists studying high-resolution images from Mars Global Surveyor had detected no evidence that ocean shorelines once existed on Mars. Previously, researchers had interpreted features in images from NASA's Viking missions of the 1970s as remnants of ancient coastlines. However, Surveyor images taken in 1998, with a resolution 5 to 10 times sharper than images that Viking had produced had not provided evidence that water in a coastal environment had formed those features. Michael C. Malin of Malin Space Science Systems, Principal Investigator for the Mars Orbiter Camera aboard Surveyor, remarked that although scientists could not rule out the presence of oceans on Mars at one time, the evidence in the new images appeared to undermine the "foundation for the 'ocean hypothesis' developed in the 1980s on the basis of suspected shorelines." Nevertheless, Malin acknowledged "significant other evidence of water on Mars in the past," suggesting that as the Mars Orbiter Camera continued to acquire new high-resolution images, scientists might have new clues to the role of water in the evolution of Mars.[796]

*2 October*
The U.S. Department of Defense (DOD) successfully conducted the first of a series of tests of a proposed missile defense system. An unarmed Minuteman missile, carrying a dummy warhead and decoy balloon, launched over the Pacific Ocean from Vandenberg Air Force Base; 20 minutes later an Exoatmospheric Kill Vehicle, a 55-inch-long (1.4-meter-long), 120-pound (54.4-kilogram) device built by Raytheon Corporation, launched aboard a booster rocket from the Marshall Islands. The two missiles collided 10 minutes later, 3,000 miles (4,800 kilometers)

---

[794] Joe Bauman, "NASA Agent Tackles Down-to-Earth Work," *Deseret News* (Salt Lake City, UT), 23 October 1999.
[795] NASA, "Perseus B Damaged in Crash on California Highway," news release 99-115, 1 October 1999.
[796] NASA, "New Mars Images: No Evidence of Ancient Ocean Shorelines," news release 99-114, 1 October 1999.

from California and 140 miles (225 kilometers) above the ocean. According to the CIA, when DOD conducted the test, North Korea, Iran, and Syria were developing long-range missiles that could hit the United States.[797]

*4 October*
NASA announced its selection of three graduate students to receive Michelson fellowships offered by NASA's Origins Program and its Space Interferometry Mission. NASA awarded a fellowship to Philip M. Hinz of the University of Arizona for his work building a new type of nulling interferometer designed to null the glare from nearby stars, thereby enabling scientists to observe, in infrared wavelengths, dust and giant planets orbiting the stars. NASA chose Erin M. Sabatke, also of the University of Arizona, to work on creating models of large, stretched, flat, plastic membranes, for collecting light from several telescopes placed on separate spacecraft flying in formation. Sabatke planned to explore the use of this technique to photograph planets around other stars. NASA selected Benjamin F. Lane, a student at California Institute of Technology, for his work advancing the "technique of using two stars with a narrow angle separating them to measure relative motion of one with respect to the other, utilizing a ground-based interferometer." NASA had named the fellowship program for the first American to win a Nobel Prize in physics, Albert Michelson, known as the father of interferometry. Interferometry is a technique of combing and processing light from multiple telescopes to obtain a clear image of distant objects. The fellowship covered tuition, a student stipend, and a small budget for travel and other research expenses, for three years of graduate research at the student's host institution.[798]

Vladimir Petrovsky, a member of the Russian Academy of Military Sciences, condemned the United States' 2 October testing of its proposed missile defense system, noting that testing could "aggravate relations not only between the United States and the Asian-Pacific region, but also between Japan—a U.S. ally in the development of the system—and other countries." Russia had refused to amend the 1972 Anti-Ballistic Missile Treaty, which would have permitted the United States to develop a full-fledged missile defense system. Preventing either country from developing missile defense systems, the Cold War–era treaty acted as a deterrent by keeping both the Soviet Union and the United States vulnerable to attack. Officials in Washington, DC, stated that DOD was testing a missile defense system intended to destroy lone missiles launched by rogue states or terrorists, not a system meant to protect against a multi-missile attack from a major nuclear power.[799]

*5 October*
Analyzing data from NASA's Tropical Rainfall Measuring Mission (TRMM) spacecraft, researchers proved for the first time that smoke from forest fires inhibits rainfall. TRMM, a mission of the United States and Japan and part of NASA's Earth Science Enterprise, had been producing continuous data since December 1997. Daniel Rosenfeld of the Institute of Earth Sciences at Hebrew University in Jerusalem had studied the data, finding that clouds polluted with heavy smoke significantly inhibit warm rain processes. When smoke pollutes them, cloud

---

[797] Associated Press, "Missile Test Sparks Russian Gripes," 4 October 1999.
[798] NASA Jet Propulsion Laboratory, "Three Planet-Hunters Earn NASA's Michelson Fellowships," news release 99-086, 4 October 1999, *http://www.jpl.nasa.gov/releases/99/fellowships.html* (accessed 7 March 2008).
[799] Associated Press, "Missile Test."

tops must "grow considerably above the freezing level" to produce rain through a different process. Rosenfeld remarked that the results of his research had validated earlier studies, which showed that urban air pollution inhibits rainfall. Christian D. Kummerow, a TRMM scientist at NASA's Goddard Space Flight Center (GSFC) commented that such findings were "making the first inroads into the difficult problem of understanding humanity's impacts on global precipitation."[800]

*6 October*
NASA Administrator Daniel S. Goldin named Arthur G. Stephenson, Director of NASA's Marshall Space Flight Center (MSFC) in Huntsville, Alabama, to head the Mars Climate Orbiter Mission Failure Investigation Board. Mars Climate Orbiter had disappeared on 23 September as it entered orbit around Mars. Goldin tasked the board with independently reviewing all aspects of the mission's failure, to verify whether it had been the result of an inadequacy in NASA's interplanetary navigation systems. Preliminary findings from an internal peer review at NASA's Jet Propulsion Laboratory indicated that "a failure to recognize and correct an error in a transfer of information between the spacecraft team in Colorado and the mission navigation team in California" had caused the loss of the spacecraft. Goldin instructed the board to report its initial findings by 3 November.[801]

NASA researchers at MSFC unveiled an experimental 50-foot (15-meter) track using magnetic levitation (maglev) technology to propel a 30-pound (13.6-kilogram) model of a spacecraft, at a speed of 60 miles (97 kilometers) per hour, in 0.5 seconds. MSFC researchers had sought to reduce launch costs, as well as to reduce space-vehicle size with the development of a maglev launch-assist system. The maglev track was theoretically capable of propelling and releasing a spacecraft at 600 miles (965 kilometers) per hour; wing design would then lift the craft and "give it a kick start" before the ignition of the rocket's engines. MSFC Manager of Launch Technologies Sherry Buschmann commented that the researchers believed the use of maglev technology could cut launch vehicle size by 20 percent. NASA's industry partner in the experimental technology was PRT Advanced Maglev Systems Inc. of Park Forest, Illinois.[802]

NASA released new Hubble Space Telescope (HST) images of the cores of spiral galaxies like the Milky Way, and astronomers studying the images for clues about galactic development discussed their findings at a briefing at NASA Headquarters. Two complementary surveys by independent research teams had concluded that the Hubble images confirmed the evolutionary link between a disc-like cloud of dust and gas at the center of a galaxy and a bulge of millions of stars at the center of the disc-like cloud. The central bulge stabilizes a galaxy's development, controls the birth of stars in the galaxy's core, and "holds secrets as to how and when a galaxy formed." Reynier F. Peletier of the University of Nottingham in the United Kingdom confirmed that central bulges of "more tightly wound spirals" formed at approximately the same time in the early universe. C. Marcello Carollo of Columbia University, who had led a team studying

---

[800] NASA, "Spacecraft Provides First Direct Evidence: Smoke in the Atmosphere Inhibits Rainfall," news release 99-110, 5 October 1999.
[801] NASA, "Head of Mars Climate Orbiter Investigation Board Named," news release 99-117, 6 October 1999.
[802] NASA Marshall Space Flight Center, "New NASA Track Races Toward Cheaper Trips to Space," news release 99-260, 4 October 1999; Kent Faulk, "NASA Spacecraft Float to Future: Researchers Test Levitation as Way To Boost Launches," *Birmingham News* (AL), 7 October 1999.

galaxies with small bulges and "bar-like structures that bisect the nucleus," had discovered that such galaxies developed more recently. Both teams had used HST's visible-light and infrared cameras to determine the stars' true colors, a measure of the age of the stars inside the galaxies' core.[803]

*7 October*
A Delta II rocket launched from Cape Canaveral, Florida, carrying a NAVSTAR Block 2R GPS satellite, the third in a new series of GPS spacecraft; Lockheed Martin was under contract with the U.S. Air Force to launch 21 GPS craft. The NAVSTAR Block 2R military navigational satellite became the 28th operational spacecraft in the GPS constellation, to replace an older satellite in the network after completing a one-month test period. The U.S. Air Force had originally planned to launch the replacement craft in May 1999, but after a rainstorm had damaged the satellite while it was sitting on the launchpad, the Air Force had returned it to Lockheed Martin Missile and Space for repairs. Two hurricanes, a tropical storm, and other rainstorms had caused additional weather delays, preventing the launch of an alternate replacement satellite.[804]

*10 October*
NASA's Galileo probe survived intense radiation emanating from Jupiter during a "do-or-die effort," when the spacecraft came within 380 miles (611 kilometers) of Jupiter's volcanic moon Io. The craft experienced computer problems when passing through the densest part of Jupiter's radiation belt, but 1 hour after Galileo's closest encounter with Io, all of its systems had resumed normal functioning. Duane Bindschadler, Manager of Science Operations for the Galileo Program, explained that close-up study of Io could help scientists learn about the behavior of volcanoes, thereby helping them predict the behavior of volcanoes on Earth. However, because Jupiter's radiation field could damage or destroy the spacecraft, NASA had planned the close flyby for the end of the two-year extended mission, following Galileo's original two-year mission.[805]

*12 October*
Space Imaging Inc., a private company owned by Lockheed Martin and based in Thornton, Colorado, released a photograph taken from space of the intersection of 14th Street, NW, and Constitution Avenue in Washington, DC. President William J. Clinton had approved the development of private space cameras in 1994, but Space Imaging had been the first company to successfully build and launch a spacecraft with such a camera, launching its Eastman Kodak–built camera on 24 September and testing it on 30 September. The commercial image rivaled the products of military spy technology, showing objects as small as 3 feet (0.9 meter) wide. Space Imaging's test image showed the Washington Monument, part of the Ellipse, and the buildings of the U.S. Department of Commerce and the Museum of American History.[806]

---

[803] NASA, "Starry Bulges Yield Secrets to Galaxy Growth," news release 99-107, 6 October 1999; Deborah Zabarenko for Reuters, "Hubble Tackles Questions About Cosmic Bulges," 6 October 1999.
[804] Irene Brown, "Delta Rocket Soars into Space," *Space.com*, 7 October 1999.
[805] NASA Jet Propulsion Laboratory, "Galileo Succeeds in Historic Flyby of Jupiter's Volcanic Moon," news release, 11 October 1999; Reuters, "Galileo Probe Risks Ruin for a Closer Look at Io," *Washington Post*, 12 October 1999.
[806] William J. Broad, "Giant Leap for Private Industry: Spies in Space," *New York Times*, 13 October 1999.

*13 October*
At the annual meeting of the American Astronomical Society's Division for Planetary Sciences, research group leaders from the University of Texas at Austin announced that scientists had sifted through the data from Earth-based observatories and from spacecraft such as the HST, obtained from the controlled crash of NASA's Lunar Prospector into a crater near the south pole of the Moon on 31 July. The researchers reported that they had identified "no observable signature of water" in the Moon's crater. The announcement confirmed the conclusion reached at the time of the crash. The Lunar Prospector spacecraft had launched on 6 January 1998. In March 1998, mission scientists had announced the first tentative findings of the presence of water ice on both of the Moon's poles. The controlled crash, proposed by engineers and astronomers at the University of Texas, was "a low-budget attempt to wring one last bit of productivity from the low-cost" mission. NASA had accepted the proposal because of the successful peer review of the idea, and because the craft's useful life was nearing its end.[807]

*14 October*
NASA announced its selection of the next two missions of its Medium-Class Explorer (MIDEX) Program. The first mission, headed by Neil Gehrels of NASA's GSFC and planned for launch in 2003, was the Swift Gamma Ray Burst Explorer, a three-telescope space observatory designed with the "unique ability to rotate in orbit and point its gamma-ray telescope, x-ray telescope, and ultraviolet/optical telescope at gamma-ray bursts [GRBs] within minutes of the burst's first appearance." The second mission, led by Kenneth J. Johnston of the U.S. Naval Observatory, was the Full-Sky Astrometric Mapping Explorer (FAME), scheduled to launch in 2004, a space telescope designed to "obtain highly precise position and brightness measurements of 40 million stars." From a group of five missions selected in January 1999, NASA had chosen these two missions for detailed four-month feasibility studies, examining cost, management, and technical plans, such as small business involvement and educational outreach. NASA had originally received 31 full proposals in August 1998. Associate Administrator for Space Science Edward J. Weiler remarked that the selection of the two missions had been the most difficult he had made during his 21 years at NASA, because, over the years, the space science community had been submitting to NASA a steadily increasing number of first-class concepts for smaller missions.[808]

Two satellites, the China-Brazil Earth Resources Satellite (CBERS-1), sometimes called ZY-1 in Chinese reports, and the Brazilian *Saci-1* satellite, launched atop a Chinese Long March 3B rocket from Taiyuan Launch Center in central China. Brazil and China had jointly financed the 1,500-kilogram (3,307-pound) CBERS-1, which carried three high-resolution cameras for monitoring environmental and vegetation conditions in Brazil and China, as well as in other, unspecified locations. Brazil had designed the 60-kilogram (132-pound) *Saci-1* microsatellite to monitor cosmic rays, the magnetic field, and plasma. Shortly after launch, communications with the craft failed.[809]

---

[807] NASA, "No Water Ice Detected from Lunar Prospector Impact," news release 99-119, 13 October 1999; Associated Press, "Craft's Crash into Moon Finds No Water," *Los Angeles Times*, 13 October 1999.
[808] NASA, "NASA Selects Missions To Search for Planetary Systems and Observe Cosmic Explosions," news release 99-120, 14 October 1999.
[809] *Spacewarn Bulletin*, no. 552, 1 November 1999, *http://nssdc.gsfc.nasa.gov/spacewarn/spx552.html* (accessed 6 August 2008).

*18 October*
NASA released its newest and most accurate map of the continent of Antarctica, created from data that the Canadian Space Agency's (CSA's) RADARSAT-1 satellite had collected over an 18-day period during the spring of 1997. Vexel Corporation of Boulder, Colorado, had developed software to generate a mosaic of the continent from "many small images made from different angles and orientations of the satellite." Project officials and scientists remarked that the most important discoveries resulting from the radar images concerned the network of ice streams; the new satellite data revealed that the streams traveled "enormous distances at speeds up to 3,000 feet (914 meters) per year—100 times faster than the flow of surrounding ice." The CSA had begun planning RADARSAT in 1980, because of Canada's interest in using a high-resolution radar satellite to monitor shipping channels in the Arctic. NASA's negotiations with the CSA had resulted in an agreement: NASA would launch the satellite and provide software and data analysis in exchange for access to some of the data the satellite produced. Additionally, the CSA would make at least two imaging scans of Antarctica. RADARSAT had launched from Vandenberg Air Force Base on 4 November 1995, but, until 1997, Canadian controllers had not completed a difficult rotation of the satellite, necessary to place Antarctica in full view. Once correctly oriented, RADARSAT had been able to complete the imaging in 18 days because it could collect data anytime of the day or night and in any weather. The last satellite map of Antarctica had used images from five different satellites, spanning the years 1980 to 1994.[810]

*20 October*
President William J. Clinton signed into law the FY 2000 appropriations bill for the Department of Veterans Administration, Housing and Urban Development, and Other Agencies, which included the US$13.65 billion NASA budget. Congress had approved NASA's budget on 7 October. In keeping with the President's original request, the final appropriation included full funding for the International Space Station and the Space Shuttle program; US$80 million for Spaceliner 100, an MSFC program to find new propulsion technology; US$25 million for Shuttle upgrades; US$5 million for the National Center for Space Research and Technology, a joint venture including MSFC, the University of Alabama in Huntsville, and private industry; and US$3 million to continue research into tether-guided satellites. The U.S. House of Representatives' version of the bill had cut US$900 million from NASA's funding, but the U.S. Senate had approved a budget equal to NASA's FY 1999 budget: US$13.6 billion. House Republicans had voted to eliminate funding for *Triana*, a controversial Earth-observing satellite first envisioned by Vice President Albert A. Gore Jr., but the House and Senate conference committee had voted on a compromise, providing for the National Academy of Sciences to review the program and forbidding NASA from launching *Triana* until 1 January 2001.[811]

*21 October*
British astronomers Martin J. Ward of the University of Leicester and Keith Mason of Mullard Space Laboratory at University College, London, announced that NASA had selected them to participate in its Swift Gamma Ray Burst Explorer mission. NASA's mission called for the two

---

[810] NASA, "NASA Unveils New, Most Accurate Map of Antarctic Continent," news release 99-122, 18 October 1999; Malcolm W. Browne, "Under Antarctica, Clues to an Icecap's Fate: Radar Uncovers a Network of Ice Streams Larger and Faster than Expected, and More Ominous," *New York Times*, 26 October 1999.
[811] Brett Davis, "Congress Approves Full NASA Budget," *Huntsville Times* (AL), 8 October 1999; Stephen Koff and Tom Diemer, "Cleveland NASA Unit, CMHA Get Funding," *Cleveland Plain Dealer* (OH), 21 October 1999.

scientists to use a pair of specialized telescopes, one measuring ultraviolet and visible light and the other measuring x-ray light, to study GRBs, the "unpredictable powerful explosions in space that sometimes last less than one second." A collaboration of Italian, British, and U.S. scientists, the Swift Gamma Ray Burst Explorer was one of NASA's medium-class explorer missions, scheduled for launch in 2003.[812]

Brazilian space officials, having lost contact with their country's *Saci-1* satellite days after its launch, asked NASA to help them rescue the US$4.6 million satellite. *Saci-1* had launched on 14 October from China along with another satellite, the CBERS-1, which the two countries had built jointly. Brazil asked NASA to photograph the satellite, check its condition, and devise a plan for its recovery. President of the Brazilian Space Agency Luis Meira Filho remarked that Brazil had not given "the satellite up for lost," because space officials knew its location. The CBERS-1 satellite was functioning normally and had returned images of Brazil.[813]

*22 October*
NASA released an image of Jupiter's volcanic moon Io, captured at the closest range ever accomplished. The image, which NASA's spacecraft Galileo took during a flyby of the moon on 10 October, showed a lava field near the center of an erupting volcano named Pillan, with visible new lava flows from its volcanic center. To capture most of the new images of Io, Galileo had used a fast camera mode, in which the camera preprocesses the image, averaging the brightness in its adjacent parts. However, Io's radiation had disrupted the process, degrading the quality of the images. Radiation had not affected images, including the newly released image of Pillan, taken in other camera modes. Galileo Project Manager James K. Erickson remarked that Galileo would not use the fast camera mode during the next Io flyby.[814]

*27 October*
New York Governor George E. Pataki awarded NASA astronaut Eileen M. Collins of Elmira, New York, the state's highest award, the Jackie Robinson Empire State Freedom Medal, for her achievement as the first female Space Shuttle Commander. Collins had been Commander of the historic Mission STS-93. During that mission, she and her crew had deployed the Chandra X-ray Observatory, the heaviest, largest, and most powerful x-ray telescope ever launched into space. Governor Pataki described Collins as "a bold pioneer of the reaches of space" and remarked that "her talent, intelligence and courage set an example that every woman and every man can hope to duplicate in their lives." Every year New York presents the Empire State Freedom Medal, established in 1997, to "those who best demonstrate the qualities of determination, dignity, fairness, and honor that were exemplified by Jackie Robinson, who broke major league baseball's color barrier."[815]

A Russian Proton rocket carrying a communications satellite manufactured by Lockheed Martin exploded 6 minutes after launching from Baikonur Cosmodrome in Kazakhstan, the second

---

[812] Reuters, "British Scientists Selected for New NASA Project," 21 October 1999.
[813] Reuters, "Brazil Asks NASA Help To Rescue Lost Satellite," 21 October 1999.
[814] NASA Jet Propulsion Laboratory, "Closest-Ever Picture of Volcanic Moon Io Released," news release, 22 October 1999.
[815] NASA, "First Female Shuttle Commander Eileen Collins Receives Jackie Robinson Medal," news release 99-125, 27 October 1999.

failure of a Proton rocket in three months. The Russian Space Agency formed a commission to investigate the crash. Kazakhstan had temporarily banned Proton launches from Baikonur following a launch failure on 5 July, but Proton rockets had launched from the facility successfully on 6 and 26 September.[816]

*28 October*
Kazakhstan imposed a new ban on Russian launches from Baikonur Cosmodrome following the previous day's crash of a Proton booster rocket, which had malfunctioned shortly after launch. Without naming a figure, Kazakhstan announced that it would require Russia to pay compensation in an amount significantly higher than the fee that Kazakhstan had charged Russia after the 5 July Proton crash. In addition, the Kazakhstan government formed a special commission headed by Deputy Prime Minister Alexander Pavlov to investigate the accident.[817]

**NOVEMBER 1999**

*2 November*
NASA announced that Administrator Daniel S. Goldin would present Simon Ramo and Bernard A. Schriever with NASA's Distinguished Public Service Medal on 5 November at the start of a daylong conference. NASA, the U.S. Air Force Space Command, the National Air and Space Museum, the Space Policy Institute at George Washington University, and the National Space Society were cosponsoring the conference, "Developing U.S. Launch Capability: The Role of Civil-Military Cooperation." The award recognized the leadership of Ramo and Schriever during the early years of the U.S. space program. Ramo had been the chief scientist and leading civilian in the Air Force program to build the first U.S. intercontinental ballistic missile system. His work had fostered cooperation between the U.S. Air Force and aeronautics industry, the necessary foundation of the space program. Schriever had pioneered the development of the first ballistic missile for the United States; the resulting rocket technology had led to NASA's successes in its early human spaceflight programs.[818]

*3 November*
A large section of the outer wall of one of the Lockheed Martin X-33 rocket plane's two liquid hydrogen fuel tanks separated 2 hours after completing pressure and structural tests at NASA's Marshall Space Flight Center. Lockheed intended the X-33 prototype to lead to the development of its VentureStar™ Reusable Launch Vehicle (RLV), and NASA planned to replace its fleet of Space Shuttles with a fleet of Lockheed's VentureStar™ RLVs after 2012. The latest problem potentially delayed the X-33's first test flight at least an additional six months, essentially guaranteeing that Lockheed Martin would be unable to produce an operational VentureStar™ vehicle by 2004. NASA had "placed a huge wager" on the VentureStar™, committing US$941 million to the X-33 program, which had been threatened by technical problems and cost overruns for more than a year. U.S. General Accounting Office (GAO) examiners had criticized the

---

[816] Miles O'Brien, "Russian Rocket Explodes After Liftoff: Incident Likely To Delay International Space Station," *CNN.com*, 27 October 1999, http://www.cnn.com/TECH/space/9910/27/proton.explodes/ (accessed 10 April 2008).
[817] United Press International, "Kazakhstan Bans Russian Space Launches," 28 October 1999.
[818] NASA, "NASA Honors Missile Pioneers at Launch Vehicle Conference," news release N99-56, 2 November 1999, http://www.nasa.gov/home/hqnews/note2edt/1999/n99-056.txt (accessed 12 March 2008).

project for its rising costs, warning that "delays in the prototype would harm NASA's plans for a full-scale reusable vehicle."[819]

GAO reported that NASA had "inadvertently exported radiation-hardened, microcircuit, optical coupler parts to Russia," relying on the manufacturer's statement that the so-called rad-hard parts were not on the U.S. Department of State's Munitions List and that, therefore, NASA was not required to obtain an export license. GAO also reported that NASA was "preparing to export sensitive encryption technology to Japan and Europe," technology transfers related to building the International Space Station (ISS). Under international agreements related to the space station, NASA was obligated to deliver, disclose, or transfer certain technology, data, and commodities to other nations involved in building the ISS. GAO had investigated NASA's export-control procedures after members of Congress expressed concern about safeguards to protect technology and information exported to support the ISS. GAO evaluators' conclusions were similar to those of NASA's Inspector General, who had reported earlier that NASA's export-control procedures were likely inadequate for the technology exchanges necessary under the ISS program.[820]

*5 November*
NASA announced that its scientists had developed a new Digital Tectonic Activity Map (DTAM) of the Earth, pinpointing geologically and volcanically active features of the planet over the last 1 million years. Whereas most global geological maps were "plate maps," emphasizing the definition of plate boundaries of the planet's crust and current seismic or volcanic activity, the new tectonic map portrayed the broad architecture of Earth's crust, identifying all current and past geological activity. NASA had primarily generated the DTAM using publicly available data and commercial software, but had also used spacecraft data, including Landsat images, hand-held astronaut photography, and radar altimetry of sea-surface measurements gathered by NASA spacecraft.[821]

*8 November*
Engineers discovered that exposure to 11 months of cold temperatures in outer space could cause the descent engine on Mars Polar Lander to malfunction. NASA had scheduled Mars Polar Lander to land on Mars on 3 December. Lander's companion craft, Mars Climate Orbiter had burned up as it entered its Martian orbit on 23 September. The investigative panel studying the cause of Orbiter's loss had uncovered the potential problem with Lander's engines. Because of the discovery, scientists at NASA's Jet Propulsion Laboratory (JPL) altered their plans, deciding

---

[819] Jeff Leeds, "Test Mishap Delays Development of Lockheed's X-33 Rocket Plane," *Los Angeles Times*, 6 November 1999.
[820] U. S. General Accounting Office, "Export Controls: International Space Station Technology Transfers" (report no. NSIAD-00-14, Washington, DC, 3 November 1999), *http://www.gao.gov/archive/2000/ns00014.pdf* (accessed 21 March 2008); *Aerospace Daily*, "NASA Illegally Exported Technology for Space Station, GAO reports," 22 November 1999.
[821] NASA Goddard Space Flight Center, "New Global Digital Tectonic Activity Map of the Earth Produced," news release 99-116, 5 November 1999, *http://www.gsfc.nasa.gov/news-release/releases/1999/99-116.htm* (accessed 24 March 2008).

to turn on the craft's descent engine heaters earlier than originally scheduled, to warm the engines to 46.4°F (8°C) before Lander's descent.[822]

*10 November*
The Mars Climate Orbiter Mission Failure Investigation Board released its first report, identifying eight contributing factors that had led to the 23 September loss of Orbiter as it entered its Martian orbit. Arthur Stephenson, chairperson of the board investigating the failure, agreed with NASA on the causes of the mission's failure. The board reported that "the failed translation of English units into metric units in a segment of ground-based, navigation-related mission software" was the primary cause of the loss, but that other significant factors had set the stage for the mission's failure. Moreover, the team had failed to identify and correct multiple mistakes, leading to "a major error in our understanding of the spacecraft's path as it approached Mars." The failure board faulted the Mars Climate Orbiter team with "inadequate consideration of the entire mission and its postlaunch operation as a total system, inconsistent communications and training within the project, and lack of complete end-to-end verification of navigation software." Although engineers working with Lockheed Martin Astronautics' mission operations team had failed to convert English units into metric for entering data into ground-based navigation software, the investigating board focused most of its attention on the navigation team at NASA's JPL, which had overall management authority for the mission. The board remarked that, because it had navigated interplanetary spacecraft successfully for 30 years, JPL had developed a "widespread perception that 'Orbiting Mars is routine'," leading it to pay inadequate attention to the risk of faulty navigation.[823]

*12 November*
NASA's Office of Earth Science Enterprise selected Ball Aerospace of Boulder, Colorado, and Aerojet General Corporation of Azusa, California, to study the next-generation, spaceborne, microwave, atmospheric instrument for weather-forecasting and climate-change research, awarding each company a one-year, US$4 million, fixed-price contract through Goddard Space Flight Center. The new instrument, the Advanced Technology Microwave Sounder (ATMS), was one-third the size and weight of existing microwave-sounding instruments on board the Polar-Orbiting Environmental Satellite (POES) and the Earth Observing System-PM spacecraft. The application of new technologies, particularly in microwave electronics, had made the significantly smaller size of the ATMS possible. NASA intended to fly the first ATMS unit on the National Polar-Orbiting Operational Environment Satellite System (NPOESS) Preparatory Project (NPP) mission, a joint effort between NASA and the NPOESS Program Office. NPOESS, a tri-agency program consisting of parts of NASA, the National Oceanic and

---

[822] Larry Wheeler, "NASA Finds Flaw in Mars Lander Engine Turning On Descent Heaters Earlier Should Prevent Failure of Mission," *Florida Today*, 9 November 1999; Mark Carreau, "Mars Probe To Get Brake Check: NASA Engineers To Warm Descent Rockets of Polar Lander," *Houston Chronicle*, 9 November 1999.
[823] NASA, "Mars Climate Orbiter Failure Board Releases Report, Numerous NASA Actions Underway in Response," news release 99-134, 10 November 1999, http://www.nasa.gov/home/hqnews/1999/99-134.txt (accessed 12 March 2008); Earl Lane, "Report Details NASA Failings: Assigns Blame for Orbiter's Loss," *Newsday* (Long Island, NY), 11 November 1999.

Atmospheric Administration, and the U.S. Air Force, merged civilian and military polar-orbiting weather satellite systems into a single system.[824]

*13 November*

The Hubble Space Telescope (HST) entered safe mode when the fourth of six gyroscopes failed, halting its astronomical observations. Requiring a minimum of three operating gyroscopes to conduct observations, HST would remain in safe mode with a reduced power load, until Space Shuttle astronauts arrived with six new gyroscopes, another data recorder, a radio transmitter, and an improved computer. The telescope was not at risk. NASA had divided the servicing mission, originally scheduled for June 2000, into two missions, when the third gyroscope failed in February. NASA had scheduled the first part of the mission, Servicing Mission-3A, for the fall but rescheduled it for 6 December, because the Shuttle had wiring problems and needed its engine replaced. NASA had scheduled Servicing Mission-3B for mid-2001.[825]

*15 November*

NASA announced the winner of an essay contest to name the Deep Space 2 microprobes, which the Mars Polar Lander spacecraft was carrying to Mars. NASA had scheduled the two probes to crash into Mars's south pole on 3 December. Paul Withers, a graduate student at the University of Arizona in Tucson, studying the thin upper atmosphere of Mars, had suggested naming the pair Amundsen and Scott, in honor of Roald Amundsen and Robert F. Scott, the first explorers to reach Earth's South Pole. In his winning essay, Withers recalled that one century ago, Antarctica was Earth's only unexplored continent. Withers wrote, "Scott perished in Antarctica. His memorial's inscription reads: 'To strive, to seek, to find, not to yield.' These are the aims of the Deep Space 2." NASA had designed the two probes with a dual purpose: to test advanced technology for future planetary-surface microlanders and to search for water ice 3 feet (0.9 meters) below the Martian surface. Deep Space 2 Project Manager Sarah A. Gavit remarked that the names of the Antarctic explorers were appropriate for the probes, because "like Amundsen and Scott, Deep Space 2 will have to survive great odds, including not only braving the elements but also crashing into the terrain with unbelievable force."[826]

The main engine of Japan's H-2 rocket malfunctioned 4 minutes after launching from Tanegashima Space Center, failing to put the *Mtsat* satellite into orbit. Fearing they could lose control of its trajectory, officials ordered the rocket's destruction 8 minutes into its flight, marking the first time the National Space Development Agency of Japan had destroyed a rocket in flight. It was the second launch failure for the H-2 rocket during 1999; another H-2 had failed to place its payload in orbit in February. However, before the February malfunction, Japan had

---

[824] NASA Goddard Space Flight Center, "Ball and Aerojet—General Picked To Study New Weather Sensor," news release 99-117, 12 November 1999, *http://www.gsfc.nasa.gov/news-release/releases/1999/99-117.htm* (accessed 24 March 2008).

[825] NASA, "Hubble Telescope Placed into Safe Hold as Gyroscope Fails," news release 99-136, 15 November 1999, *http://www.nasa.gov/home/hqnews/1999/99-136.txt* (accessed 12 March 2008); Marcia Dunn for Associated Press, "Hubble Trouble," 15 November 1999.

[826] NASA, "Mars Penetrator Probes Named for Pioneering Polar Explorers," news release 99-135, 15 November 1999, *http://www.nasa.gov/home/hqnews/1999/99-135.txt* (accessed 12 March 2008).

launched five H-2 rockets successfully. The *Mtsat* satellite, intended to replace the *Himawari 5* satellite, would have observed weather patterns and monitored aircraft.[827]

*16 November*
Officials at NASA's Langley Research Center (LARC) officially opened the Center's new Experimental Test Range (ETR), described as an "electromagnetic wind tunnel." LARC engineers had designed the ETR to determine how low-frequency electromagnetic radiation affects aircraft, to help the military improve stealth technology. The ETR also would assist in testing technologies to enable airline passengers to watch television signals from satellites or to help Internet users make wireless connections. The first military customer that had signed up to use the ETR, the U.S. Army's National Ground Intelligence Center, planned to use models of enemy targets in the test chamber, to see how they appeared on radar screens.[828]

*17 November*
NASA announced that new analysis of data from the Galileo spacecraft suggested that Jupiter is possibly much older and colder than previously thought. Upon reaching the planet on 7 December 1995, Galileo had dropped a probe carrying a mass spectrometer into Jupiter's atmosphere to measure its chemical composition. The spectrometer had "detected surprisingly high concentrations of argon, krypton, and xenon," raising questions about the noble gases' provenance. Because Jupiter would have trapped the gases through condensation or freezing, scientists did not believe Jupiter's atmosphere had trapped the gases at its present site. As positioned at present, Jupiter is too close to the Sun and too warm to have trapped the gases. Tobias Owen, an astronomy professor at the Institute for Astronomy of the University of Hawaii and a member of the Galileo probe Neutral Mass Spectrometer team, suggested three hypotheses to explain how the gases had been trapped within the Jovian atmosphere: 1) Jupiter had formed in the area around the Kuiper Belt[829] and was dragged inward to its present location; 2) the solar nebula, the cloud of gas and dust that formed the solar system, had been much colder than scientists had previously believed; and 3) the solid materials that had brought the gases to Jupiter had begun to form in the original interstellar cloud of gas and dust, before the cloud had collapsed to form the solar nebula. Owen remarked further that, if either of the last two hypotheses were correct, then giant planets might be able to form closer to their stars than current theories had predicted. Such a finding could help explain "the new observations of planetary systems around other stars, in which such close-in giant planets are relatively common."[830]

*18 November*

---

[827] Associated Press, "Japan Destroys Rocket in Midair as It Lifted $95 Million Satellite," *New York Times*, 16 November 1999; Dow Jones Newswire, "Japan Satellite Fails To Reach Orbit, Rocket Destroyed," 15 November 1999.
[828] Jeff Long for Knight-Ridder Tribune Business News, "Hampton, Va., NASA Research Center Tests Plane Design with Wind Tunnel," 17 November 1999.
[829] The Kuiper Belt, a region of the solar system beyond the planets, is composed of the remnants of the solar system's formation.
[830] NASA Jet Propulsion Laboratory, "Galileo Probe Results Suggest Jupiter Had an Ancient, Chilly Past," news release 99-096, 17 November 1999, *http://www.jpl.nasa.gov/releases/99/glprobeargon.html* (accessed 12 March 2008); Tobias Owen et al., "A Low-Temperature Origin for the Planetesimals that Formed Jupiter," *Nature* 402, no. 6759 (18 November 1999): 269–270.

Interfax news agency reported that Russia had agreed to pay Kazakhstan US$400,000 in damages following the 27 October crash of a Proton rocket during its launch from Baikonur Cosmodrome. After reaching an agreement on compensation, Kazakhstan had lifted its launch ban for all Russian launches except the Proton rocket.[831]

*19 November*
NASA announced that new data and images, which the Galileo spacecraft had collected on its closest-ever flyby of Jupiter's moon Io on 11 October 1999, revealed that Io, the most volcanic body in the solar system, was "even more active than previously suspected, with more than 100 erupting volcanoes." Data results from the flyby, which brought the spacecraft within 380 miles (611 kilometers) of the moon's surface, had focused on Pele, Loki, and Prometheus, three of Io's most active volcanoes. A close-up image of Pele showed part of the volcano glowing in the dark, with Pele's hot lava behaving similarly to active lava lakes in Hawaii. Galileo's Photopolarimeter Radiometer and Near Infrared Mapping Spectrometer had provided detailed temperature maps of Loki, indicating the presence of an enormous caldera repeatedly flooded by lava. New data had clarified the location of lava from Prometheus, erupting, advancing, and producing plumes. NASA's 1979 Voyager mission had observed a plume, and early Galileo images had shown a new lava flow and plume, erupting from a location 60 miles (97 kilometers) west of the area observed during Voyager's mission.[832]

*21 November*
The People's Republic of China announced it had launched into orbit its first piloted spacecraft, a non-reusable capsule named *Shenzhou*, which carried a mannequin for test purposes but was capable of carrying a crew of four. Xinhua news agency reported that the capsule, launched from Jiuquan satellite launch center aboard the new model Long March 2F, had spent 21 hours in space, orbiting Earth 14 times. China had built a new land- and sea-based, space-monitoring-and-control network for the launch. China heralded the launch as a breakthrough in the Chinese government's effort to "join the United States and Russia in the elite club of manned space flight." The Chinese had abandoned the pursuit of human spaceflight in the late 1960s, because Premier Zhou Enlai thought it was too costly, but in 1992, President Jiang Zemin had decided to renew Chinese efforts, with the goal of putting a human in space by the end of the century. The head of China's Manned Spaceflight Program told Xinhua news agency that the successful test flight of the *Shenzhou* capsule demonstrated that the Chinese spacecraft and the new Long March rocket performed superbly. During the mission, the Chinese had conducted experiments in remote sensing, environmental monitoring, space materials, astronomy, and physics. *Shenzhou* had landed in Inner Mongolia using a parachute.[833]

*22 November*
Lockheed Martin launched a Navy communications satellite aboard an Atlas 2A rocket, a badly needed success in a year of launch failures, delays, investigations, and diminishing business. The

---
[831] Associated Press, "Russia-Kazakhstan-Rockets," 19 November 1999.
[832] NASA Jet Propulsion Laboratory, "Jupiter's Moon Io: A Flashback to Earth's Volcanic Past," news release 99-138, *http://www.jpl.nasa.gov/releases/99/ioishot.html* (accessed 12 March 2008).
[833] Michael Laris, "Chinese Test Craft for Manned Orbits: Space Launch Boosts National Pride," *Washington Post*, 22 November 1999; *Washington Times*, "Chinese Successfully Launch First Spacecraft," 22 November 1999; *Spacewarn Bulletin*, no. 553, 1 December 1999, *http://nssdc.gsfc.nasa.gov/spacewarn/spx553.html* (accessed 6 August 2008).

company had failed to put three satellites in orbit in April, sustaining a total loss of US$1.5 billion. Investigations into the launch failure of a rocket using similar parts had delayed Lockheed International Launch Services' maiden flight of the Atlas IIIA, causing its first customer, Space Systems/Loral, to launch its *Telstar 7* satellite on a European Ariane booster, instead. NASA had lost contact with its US$125 million Lockheed Martin–built Mars Climate Orbiter in September and, in the subsequent investigation, had discovered that Lockheed had used English units of measurement instead of metric, resulting in miscalculations in navigational data. Although for many years Lockheed Martin had supplied satellites to the National Reconnaissance Office (NRO), the U.S. government agency responsible for spy satellites, in September the NRO had awarded a major contract to the Boeing Company to develop the next generation of imaging-reconnaissance satellites. Lockheed Martin also faced increased competition from Russia's new Dnepr rocket, Europe's Eurockot, and the U.S.-based Orbital Sciences Corporation's Pegasus.[834]

A Russian Soyuz rocket launched from Baikonur Cosmodrome carrying four Globalstar satellites into orbit. The launch was Russia's first from the space center in Kazakhstan, since a Proton rocket exploded after launch on 27 October, prompting the Kazakhstan government to reimpose its launch ban. Kazakhstan had lifted the ban partially after Russia agreed to pay US$400,000 in compensation. Media reports indicated that Russia had earned US$70 million from commercial launches such as that conducted for Globalstar Inc., an important income source for the Russian space program.[835]

*23 November*
Oleksandr Serdiuk, head of the International Relations Department of Ukraine's Space Agency, announced that Ukraine and Brazil had signed an agreement the previous week, allowing Ukraine to use Brazil's Alcântara space complex to launch a new generation of Cyclone-4 booster rockets, medium-weight rockets capable of putting 4 tons (3,600 kilograms or 3.6 tonnes) of payload into a geostationary orbit. In 1992, when the USSR dissolved, Ukraine had retained part of the former Soviet Union's space program, using several former Soviet factories to build rockets and selling rocket space to commercial satellite companies. Before the agreement with Brazil, Ukraine had only used launchpads in other former Soviet countries. However, Ukraine wanted to take advantage of the location of the Alcântara space complex near the equator, where spacecraft reach orbit more easily and commercial launches a

*24 November*
Secretary of the Smithsonian Institution I. Michael Heyman announced the appointment of General John R. Dailey, U.S. Marine Corps (Retired), Associate Deputy Administrator of NASA, as Director of the Smithsonian's National Air and Space Museum, effective January 2000. Dailey succeeded Vice Admiral Donald D. Engen, U.S. Navy (Retired), who had died suddenly in a glider accident in July after serving as the museum's director for three years. Dailey had begun work at NASA in 1992, after retiring from 36 years of highly decorated service in the U.S. Marine Corps. NASA Administrator Daniel S. Goldin remarked that Dailey's leadership at NASA had been unparalleled, and that he had "shaped and strengthened the [A]gency, and was responsible for developing an infrastructure that [would] carry NASA into the

---

[834] United Press International, "Lockheed Martin Struggles Through Year," 23 November 1999.
[835] Associated Press, "Russia Launches First Rocket Since October Explosion," 22 November 1999.

new millennium." Secretary Heyman commented on Dailey's selection from a strong field of candidates, describing him as "a most impressive individual," who had earned confidence and admiration throughout the air and space community and would continue the strong, dedicated leadership expected at the museum. In addition to managing 260 employees and a US$25 million budget, Dailey's appointment placed him in charge of construction of the museum's new 710,000-square-foot (66,000-square-meter), hangar-style facility, expected to open at Dulles International Airport in December 2003. The Smithsonian had conceived the idea of the new facility to display its vast aeronautics and space holdings; the original National Air and Space Museum building was too small to house the additional collection of more than 180 aircraft and 100 spacecraft.[836]

*27 November*
NASA's Galileo spacecraft was unable to record its closest encounter with Jupiter's moon Io, as it passed within 186 miles (299 kilometers) of the volcanic moon's surface. Although radiation shut down the craft's instruments, NASA was able to restart the instruments 4 minutes later, permitting Galileo to complete more than half of its planned observations. The craft had encountered a similar problem when it had flown within 380 miles (611 kilometers) of Io in October. NASA had added the close flybys of Io, located deep within the Jovian system's radiation belts, to the end of Galileo's two-year extended mission. The spacecraft had launched from Space Shuttle *Atlantis* on 18 October 1989, had begun orbiting Jupiter in December 1995, and had completed its primary mission in December 1997, thereafter, continuing an extended mission focused on Jupiter's moon Europa, with flybys of the planet's moons Callisto and Io, as well.[837]

## DECEMBER 1999

*3 December*
NASA officials and Mars Polar Lander Mission scientists were unable to communicate with Lander following its scheduled landing. The team had expected the first signal to arrive from the spacecraft at a tracking station in the Mojave Desert in California, approximately 30 minutes after the scheduled touchdown at 3:01 p.m. (EST). A second attempt to communicate with the craft had produced nothing. Before the craft's landing, tracking data indicated that Lander had veered slightly off course. Flight controllers had fired its thrusters briefly to alter its approach angle slightly. Flight Operations Manager Sam Thurman had expressed his satisfaction with the new tracking data, which showed the craft heading for a point within a mile of its target.[838]

*7 December*
Mission controllers indicated that chances of ever contacting the US$165 million Mars Polar Lander were "remote at best," after they failed to detect a signal during their "last best chance"

---

[836] Smithsonian National Air and Space Museum, "New Director of the National Air and Space Museum To Take Office in January," news release, 24 November 1999, *http://www.nasm.si.edu/events/pressroom/releaseDetail.cfm?releaseID=97* (accessed 19 March 2008); Peter Carlson, "New Air & Space Chief: General Succeeds Director Killed in Crash," *Washington Post*, 24 November 1999.
[837] John Antozak for the Associated Press, "Galileo's Moon Shots Impeded by Radiation," *Washington Times*, 27 November 1999; John Antozak for the Associated Press, "Galileo Spacecraft Faces Huge Doses of Radiation," *USA Today*, 26 November 1999.
[838] John Noble Wilford, "Probe to Mars Becomes Silent, Its Fate Unclear," *New York Times*, 4 December 1999.

of communicating with the probe. NASA officials announced that the loss of Lander meant that NASA would need to drop the timetable for future Mars missions and rethink the entire Mars exploration program, including the scheduled launch of the next lander in 2001. NASA had already spent US$193 million on the planned 2001 launch. Edward J. Weiler, head of NASA's Office of Space Science, indicated that NASA would convene a panel of experts to diagnose the Mars Polar Lander failure and to suggest a "new architecture" for its Mars program. Although they had set no dates, members of Congress said that they planned to hold hearings about NASA's procedures and budget.[839]

*9 December*
The National Research Council (NRC) published a NASA-funded study, which found that NASA needed more safeguards to protect astronauts from potentially cancer-causing radiation during spacewalks they would make while building the International Space Station (ISS). The study recommended that NASA change flight rules, improve space-weather forecasting, and install a radiation-monitoring device outside the ISS. The NRC recommended the extra precautions because the station's assembly schedule, consisting of more than three dozen flights and 1,500 hours of spacewalks over four years, would coincide with the most active part of the 11-year solar cycle. Moreover, the report estimated a nearly 100 percent chance that at least two assembly missions would overlap with a serious solar storm and a 50 percent chance that five flights would overlap with such a storm. Additionally, the NRC predicted that a change in the station's planned orbit, which the Russian Space Agency had requested to accommodate Russian launches to help build and resupply the station, would increase the level of radiation exposure. George L. Siscoe, a Boston University physics professor and chairperson of the study, remarked that, although the radiation levels were not life threatening, they exceeded the 30-day and 60-day limits set to protect the skin and eyes. To safeguard crew from exposure to the increased risk of developing cancer later in life, countries participating in the ISS program would have to pay careful attention to flight schedules and crew rotation. Furthermore, the amount of radiation exposure sustained during ISS missions might affect an astronaut's chances of assignment to additional spacewalks and their selection for future missions.[840]

*10 December*
Europe's new 746-ton (677,000-kilogram or 677-tonne) Ariane 5 rocket launched smoothly on its first mission, carrying the European Space Agency's 3.7-ton (3,400-kilogram or 3.4-tonne) X-ray Multi-Mirror (XMM) scientific satellite into orbit. In what France called an industry milestone and "hailed as proof of Europe's superiority in the commercial launch market," the powerful rocket lifted off at the opening of its launch window from Europe's facility in Kourou, French Guiana, depositing the satellite in orbit exactly on schedule, 29 minutes later. Jean-Marie Luton, chairperson of Arianespace, the France-based commercial space company that dominated the world's commercial satellite-launching market, remarked that, although the United States remained its most serious competitor and China was catching up quickly, "Europeans were still

---

[839] Down Jones Newswire, "NASA Fails To Detect Signal from Mars Polar Lander," 7 December 1999; Peter Kendall and Vincent J. Schodolski for the Chicago Tribune News Service, "Mars Exploration Put on Hold: NASA Scraps Its Timetable, and the Launch of a $193 Million Lander Set for 2001 Is Uncertain," *Oregonian* (Portland), 8 December 1999.
[840] Michael Cabbage for Knight-Ridder Tribune Business News, "Study Urges NASA To Increase Radiation Safeguards," *Orlando Sentinel* (FL), 10 December 1999.

one step ahead." Arianespace intended the Ariane 5, twice as heavy and twice as powerful as the Ariane 4, to carry 10- to 12-ton (9,100- to 10,900-kilogram or 9.1- to 10.9-tonne) payloads by 2005, transporting space-exploration modules and serving the ISS. The XMM satellite was the largest scientific satellite ever built in Europe, equipped with three powerful telescopes, with 58 mirrors each, and "capable of observing everything from supernovas to the remains of exploded stars."[841]

A six-member team of Brown University scientists used data from NASA's Mars Global Surveyor to examine six categories of evidence supporting the hypothesis that oceans once existed on Mars. The team, led by planetary geologist James W. Head III, published their findings in the journal *Science*. Primarily using altimetry data and images captured by Surveyor, the researchers had found that the border between two geologically dissimilar areas was nearly level in elevation, suggesting an ancient coastline. Furthermore, they found that the topography below the possible coastline was consistent with the effects of sedimentation and the volume of the hypothetical sea was within range of previous estimates of Martian water. A series of terraces running parallel to the supposed shoreline gave credence to the idea of receding water; low areas contained the appearance of possible mud cracks; and scars from impact craters suggested groundwater or ice near the surface.[842]

At a scheduled 13 December ribbon-cutting ceremony, NASA announced the opening of FutureFlight Central (FFC) at Ames Research Center's Moffett Field in Mountain View, California. NASA had designed the two-story facility—the world's first full-scale, virtual, airport control tower—to test methods of solving, under realistic conditions and configurations, potential air and ground traffic problems at commercial airports. NASA and the Federal Aviation Administration had jointly funded the US$10 million project. The design of FFC permitted "integration of tomorrow's technologies in a risk-free simulation of any airport, airfield, and tower-cab environment."[843]

*12 December*
The U.S. Air Force successfully launched a Titan II rocket from Vandenberg Air Force Base carrying the first of a new generation of military weather satellites, a Defense Meteorological Satellite Program (DMSP) Block 5D-3 spacecraft. Lockheed Martin Missiles and Space had built the DMSP Block 5D-3 craft under contract with the U.S. Air Force, and Lockheed Martin Astronautics had supplied the launch vehicle. U.S. Air Force Space and Missile Systems Center at Los Angeles Air Force Base managed the DMSP and Titan programs. The first launch of a Titan II since 19 June 1999, it was the ninth consecutive successful launch of the space launch vehicle. The Titan II, a type of booster previously used as an intercontinental ballistic missile (ICBM) and a fundamental element of the United States' strategic deterrent for two decades, was one of 14 two-stage, liquid-fueled former ICBMs that Lockheed Martin Astronautics had

---

[841] Jean-Marie Godard for Associated Press, "Mission Accomplished for Powerful New European Launcher," 10 December 1998.
[842] James W. Head III et al., "Possible Ancient Oceans on Mars: Evidence from Mars Orbiter Laser Altimeter Data," *Science* 286, no. 5447 (10 December 1999): 2134–2137; William J. Broad, "Tantalizing Signs of Ancient Martian Ocean," *New York Times*, 10 December 1999.
[843] NASA, "NASA Opens New, Virtual Airport Control Tower at Ames," news release 99-143, 10 December 1999, *http://www.nasa.gov/home/hqnews/1999/99-143.txt* (accessed 13 March 2008).

refurbished for Air Force space launches. NASA had also used Titan II rockets during its Gemini program in the 1960s, launching 10 piloted and 2 unpiloted missions. The DMSP Block 5D-3 series accommodated larger sensor payloads than previous models, featuring a larger power supply; a more powerful on-board computer with increased memory, giving the craft greater autonomy; and increased battery power, intended to extend the mission's duration. The National Oceanic and Atmospheric Administration operated DMSP, a program for strategic and tactical weather prediction to aid the U.S. military in planning sea, land, and air operations.[844]

*14 December*
NASA Administrator Daniel S. Goldin selected NASA Chief Engineer Daniel R. Mulville to replace General John R. Dailey as Associate Deputy Administrator, effective 1 January 2000. Dailey was leaving to become Director of the National Air and Space Museum. NASA had appointed Mulville as Chief Engineer in 1995, responsible for overall review of technical readiness and for execution of all NASA programs. From 1990 to 1995, Mulville had been Director of the Engineering and Quality Management Division in NASA's Office of Safety and Mission Assurance, and, from 1986 to 1990, he was Deputy Director of the Materials and Structures Division in NASA's Office of Aeronautics and Space Technology. Before that appointment, Mulville had served as the Structures Technology Manager at the Naval Air Systems Command since 1979, where he led the development of structural design, testing, and certification methods, as Program Manager for the Development of Composites for the AV-8B and F/A-18 aircraft, and advanced aircraft and missile programs. Mulville had received NASA's Distinguished Service Medal, NASA's Outstanding Leadership Medal, and NASA's Exceptional Service Medal, among other awards. Goldin expressed his pleasure at Mulville's acceptance of the position, remarking that, as Associate Deputy Administrator, Mulville would be his most senior advisor on NASA operations and commenting on Mulville's outstanding work as NASA's Chief Engineer.[845]

*17 December*
NASA Administrator Daniel S. Goldin named A. Thomas Young to chair the Mars Program Independent Assessment Team, charged with reviewing NASA's approach to robotic exploration of Mars following the recent loss of the Mars Polar Lander Mission. Among the review team's tasks were the evaluation of several recent NASA missions to deep space, including Mars Pathfinder, Mars Global Surveyor, Mars Climate Orbiter, Mars Polar Lander, Deep Space 1, and Deep Space 2. The team would analyze the projects' budgets, content, schedule, management structure, and scientific organization, assessing how roles and responsibilities of those missions related to the missions' safety, reliability, and success.[846]

At the American Geophysical Union's fall meeting in San Francisco, NASA presented images from the Galileo spacecraft's close flyby of Jupiter's moon Io on 25 November, showing "a curtain of lava erupting within a giant volcanic crater." Because of the intense heat and height of

---

[844] Lockheed Martin, "Lockheed Martin Companies Launch Military Weather Satellite," news release, 12 December 1999, http://www.lockheedmartin.com/news/press_releases/1999/LOCKHEEDMARTINCOMPANIESLAUNCH MILITA.html (accessed 13 March 2008).
[845] NASA, "Mulville Named Associate Deputy Administrator," news release 99-146, 14 December 1999, http://www.nasa.gov/home/hqnews/1999/99-146.txt (accessed 13 March 2008).
[846] NASA, "Young To Lead Mars Program Assessment Team," news release 99-147, 17 December 1999, http://www.nasa.gov/home/hqnews/1999/99-147.txt (accessed 13 March 2008).

the lava fountains, NASA's Infrared Telescope on Mauna Kea, Hawaii, was also able to observe them. NASA scientists hoped to determine the temperature of the extremely hot lava on Io, by combining data from the telescope and from Galileo observations. Galileo scientist Alfred McEwen of the University of Arizona in Tucson remarked that capturing images of the fountains was a 1-in-500-chance observation. NASA announced that engineers at Jet Propulsion Laboratory's (JPL's) Measurement Technology Center had been able to repair images damaged by radiation during Galileo's 10 October flyby using LabVIEW software from National Instruments. JPL's Torrence V. Johnson compared the work to unscrambling a television cable signal: "JPL engineers had to break the code that was inadvertently introduced by the radiation near Io."[847]

*18 December*
NASA launched its *Terra* satellite, carrying instruments made in the United States, Japan, and Canada, from Vandenberg Air Force Base atop a Lockheed Martin Atlas IIAS rocket. NASA had originally scheduled the mission to launch on 14 December, but a computer had aborted the launch moments before ignition. *Terra* carried five sophisticated instruments for observing the interactions among Earth's landmasses, atmosphere, ocean, and biosphere. NASA considered the craft the flagship of its Earth Observing System because of its instruments' measurement capabilities and accuracy. The five instruments were 1) the Advanced Spaceborne Thermal Emission and Reflection Radiometer (ASTER), designed to capture high-resolution images of Earth in visible, near-infrared, shortwave-infrared, and thermal-infrared areas of the spectrum; 2) the Clouds and the Earth's Radiant Energy System (CERES), a set of two broadband scanning radiometers for measuring Earth's radiation; 3) the Multi-Angle Imaging Spectroradiometer (MISR), an instrument using nine cameras to measure the amount of sunlight scattered in different directions; 4) the Moderate-Resolution Imaging Spectroradiometer (MODIS), designed to view the entire surface of the Earth every one to two days and to make observations in 36 spectral bands; and 5) the Measurement of Pollution in the Troposphere (MOPITT), designed to observe the interaction of the lower atmosphere with the land and oceans.[848]

*19 December*
After an unprecedented nine delays, Space Shuttle *Discovery* launched from Kennedy Space Center in Cape Canaveral, Florida. During Mission STS-103, the crew planned to restore the Hubble Space Telescope (HST) to working order and to upgrade its systems. *Discovery*'s crew comprised Commander Curtis L. Brown Jr., Pilot Scott J. Kelly, and Mission Specialists Steven L. Smith, C. Michael Foale, John M. Grunsfeld, Claude Nicollier, and Jean-François Clervoy.[849]

*20 December*
A Taurus rocket launched from Vandenberg Air Force Base, carrying satellites for NASA and the Korean Aerospace Institute, as well as 36 capsules holding 7 ounces of cremated remains for Celestis Inc., a Houston-based company that arranged to send human remains to space. NASA's 253-pound (114.8-kilogram) Active Cavity Radiometer Irradiance Monitor (EOS ACRIM III),

---

[847] NASA, "Galileo Sees Dazzling Lava Fountain on Io," news release 99-148, 17 December 1999, *http://www.nasa.gov/home/hqnews/1999/99-148.txt* (accessed 13 March 2008).
[848] Associated Press, "NASA Launched Earth Observing Satellite on $1.3 Billion Mission," 20 December 1999.
[849] NASA, "Mission Archives STS-103, Third Hubble Space Telescope Servicing Mission," *http://www.nasa.gov/mission_pages/shuttle/shuttlemissions/archives/sts-103.html* (accessed 29 October 2008).

expected to return data for a minimum of five years, was deployed to measure sunlight reaching Earth's atmosphere, oceans, and land, data needed for scientists to study whether slight changes in solar output affected global warming and cooling. ACRIM III was the third in a series of long-term, solar-monitoring tolls that NASA's JPL had built. ACRIM III's data would expand the database begun with the ACRIM I mission in 1980, aboard the Solar Maximum Mission spacecraft, and continued with the ACRIM II mission in 1991, aboard the Upper Atmosphere Research Satellite. ACRIM I was the first instrument to demonstrate clearly that the total radiant energy from the Sun is not a constant. The Korean satellite, built by TRW Inc. for a three-year mission, carried three instruments for creating digital elevation maps of Korea, studying the biology of the ocean, and performing physics experiments on the effects of radiation on electronics. The Celestis mission carried 36 capsules of human remains, which would orbit for 45 years before burning up in Earth's atmosphere.[850]

*21 December*
Space Shuttle *Discovery* astronauts captured the HST using the Shuttle's robotic arm. *Discovery* had to alter its approach to the space telescope, because the telescope's failed gyroscopes had "left the telescope a little shaky," but Commander Curtis L. Brown Jr. and his crew had prepared for such a scenario. A fully operational HST would have had its aft pointed at the center of Earth; Brown would have carried out a straightforward approach, rising up from below until the robotic arm simply locked on. Without its gyroscopes functioning, the HST slowly rotated at one revolution per hour. Therefore, controllers activated backup gyroscopes to eliminate the rotation as much as possible, permitting Brown to fly around the HST and properly position the Shuttle's robotic arm. As both spacecraft traveled around the Earth at 17,500 miles (28,000 kilometers) per hour with Brown at the controls, *Discovery* moved closer to the HST until French astronaut Jean-François Clervoy had removed the 43-foot (13-meter), 25,000-pound (11,300-kilogram) telescope from orbit and anchored it in the Shuttle cargo bay.[851]

*22 December*
Robert S. Ruggeri, a retired NASA engineer and pioneer in de-icing research, died at the age of 75. Ruggeri had begun working for NASA's predecessor agency in 1944, retiring from NASA in 1980. NASA had applied his findings from studying ice buildup on airplane wings in the 1940s and 1950s, to the development of rockets and Space Shuttles. When the United States sought to speed up its space program following the Soviet Union's launch of *Sputnik*, Ruggeri had also researched the effects of cavitation—partial vacuums in flowing liquids—on cryogenic rocket pumps, looking for ways to make the pumps more efficient for rockets. Later in his career at NASA, Ruggeri's work had focused on the design of compressors for advanced aircraft engines, as he sought ways to increase fuel efficiency, reduce air pollution, and minimize the noise of jet engines.[852]

---

[850] NASA, "Item 1—ACRIMSAT Launch—HQ," *NASA Daily News Summary*, 21 December 1999, http://www.nasa.gov/home/hqnews/media/1999/m99-262.txt (accessed 13 March 2008); Associated Press, "NASA Satellite, Korea Probe Rocket into Space," 21 December 1999.
[851] Associated Press, "Discovery Astronauts Capture Hubble Telescope for Repairs," *Washington Times*, 22 December 1999; William Harwood, "Discovery Catches Up with Hubble for a Service Call in Space," *Washington Post*, 22 December 1999.
[852] "Robert Ruggeri, NASA Engineer Was a Pioneer in Aircraft De-Icing," *Cleveland Plain Dealer* (OH), 26 December 1999.

In the second-longest spacewalk in NASA's history, lasting 8 hours and 15 minutes, Steven L. Smith and John M. Grunsfeld successfully replaced all six of HST's gyroscopes and had "just enough time to equip each of the telescope's six batteries with a voltage regulator to prevent overheating." NASA believed corroded wires had caused the gyroscope failures. To avoid repeating the failure, engineers used pressurized nitrogen rather than air to force fluid into the new gyroscopes. Smith and Grunsfeld alternated working inside the tight space housing the gyroscopes. Because of his long arms, NASA assigned the 6-foot-3.5-inch-tall (1.9-meter-tall) Smith, who had worked on HST during its last service call in 1997, the task of replacing the two sets of gyroscopes on the sides, which were more difficult to reach. The spacewalk lasted 2 hours longer than scheduled. Although Smith and Grunsfeld went to work an hour early and quickly organizing their tools in the cargo bay, the pair needed extra time to open coolant-line valves on a disabled infrared camera and encountered difficulty latching the doors of the cargo bay holding the equipment.[853]

*23 December*
British-born U.S. astronaut C. Michael Foale and Swiss astronaut Claude Nicollier installed a new central computer on HST, 20 times faster than the telescope's previous one, and a refurbished, 550-pound (250-kilogram), fine guidance sensor. At 8 hours and 10 minutes, the pair's spacewalk was the third longest in NASA's history.[854]

*24 December*
Astronauts Steven L. Smith and John M. Grunsfeld completed the HST servicing mission, replacing a nonfunctioning radio transmitter and an outdated tape recorder. The spacewalk again exceeded 8 hours and, for the third time, ran behind schedule. As the pair prepared to begin their spacewalk, Grunsfeld encountered a mechanical problem with his suit and had to exchange it before exiting *Discovery*'s airlock. Grunsfeld also faced the most difficult task, replacing the transmitter, which was not easy to service in space. The job required him to disconnect and reconnect a series of thin coaxial cables while wearing bulky pressurized gloves. NASA had designed a special tool for the job, but even so, the task required more than an hour, and Grunsfeld needed to stop to rest his hands. The eight-day repair flight had accomplished all of the mission's major objectives, although NASA cancelled a fourth spacewalk, preventing the astronauts from installing all six steel sunshades that protect HST from solar damage.[855]

*25 December*
*Discovery* astronauts placed the HST back in orbit using the Shuttle's robotic arm. Astronaut Jean-François Clervoy operated the arm, grabbing a handle on the telescope's side. After releasing the latches securing the observatory inside the Shuttle's bay, the robotic arm lifted the telescope above the spacecraft, and ground controllers opened the door covering HST's

---

[853] Marcia Dunn for the Associated Press, "Hubble Gets a Transplant: Initial Check Says New Parts Are Working," *Los Angeles Times*, 23 December 1999, Associated Press, "Crew Replaces All 6 Hubble Gyroscopes," *Washington Times*, 23 December 1999.
[854] NASA, "Mission Archives: STS-103," *http://www.nasa.gov/mission_pages/shuttle/shuttlemissions/archives/sts-103.html* (accessed 25 March 2008).
[855] Michael Cabbage, "Good Job: Hubble Looks Fit as Fiddle at the End of a Happy Day in Space, the Astronauts Sent Holiday Greetings to Earth," *Orlando Sentinel* (FL), 25 December 1999; Reuters, "Astronauts Wrap Up Work on Hubble: Christmas Eve Mission Replaces Bad Transmitter," *Washington Times*, 25 December 1999; Associated Press, "Repairs Done, Shuttle Crew Makes Plans for Landing," *New York Times*, 27 December 1999.

telescopic eye. Once ground controllers gave the command, *Discovery*'s robotic arm released the telescope. Commander Curtis L. Brown Jr. and Pilot Scott J. Kelly fired the Shuttle's thrusters to move away from the telescope slowly.[856]

*27 December*
Although crosswinds caused a late touchdown, Space Shuttle *Discovery* landed safely after its eight-day servicing mission to HST. NASA managers had instructed *Discovery* to fly an extra orbit before landing, while Mission Control monitored the weather on the ground. NASA had shortened the mission to eight days from the scheduled 10 because of numerous launch delays, and because *Discovery* must return to Earth before New Year's Eve to avoid potential Y2K computer problems.[857]

## JANUARY 2000

*4 January*
After detailing only a few computer glitches, which had occurred during the dreaded Y2K transition, *Federal Computer Week* reported that NASA had narrowly avoided a Y2K data flaw that might have redirected the orbit of its Upper Atmosphere Research Satellite. According to NASA's Chief Information Officer, the software that transmitted the commands directing satellite orbits had automatically reverted to 1 January 1999 commands, when it should have used 1 January 2000 commands. Attentive operators caught the potential problem before the software had sent the faulty data to any satellites. Although the episode accentuated the seriousness of the Y2K computer transition, it was notable that, in spite of the risks, very few computer crashes actually occurred during the changeover.[858]

*7 January*
In the wake of the failure of the Mars Polar Lander Mission, an independent assessment team appointed by NASA Administrator Daniel S. Goldin assembled for the first time to begin a review of NASA's approach to the exploration of Mars. The 16-member team included aeronautics experts from NASA, aerospace industries, the U.S. Air Force, and academia. Although NASA officials held out hope that Lander would eventually respond to signals from Earth, Goldin tasked the team with evaluating NASA's successful and unsuccessful missions to Mars. The independent investigation continued NASA's pattern of scrutinizing all failures to improve its proficiency in the exploration of space.[859]

*10 January*
NASA released recently obtained images that further supported the possibility that Jupiter's moon Europa might be home to a liquid ocean. The Galileo spacecraft had captured the images when passing Europa's north pole. Using a magnetometer instrument, Galileo had recorded

---

[856] Robyn Suriano, "Hubble Set Free: Discovery's Crew Ready for Monday Landing at KSC," *Florida Today* (Brevard, FL), 26 December 1999.
[857] *Washington Times*, "Shuttle Back on Earth after Hubble Repairs," 28 December 1999; Robyn Suriano, "'Welcome Back to Earth', Discovery: Shuttle Astronauts Come in for Late but Safe Landing," *USA Today*, 28 December 1999.
[858] Paula Shaki Trimble, "NASA Discovered Last-Minute Y2K Glitch," *Federal Computer Week*, 4 January 2000.
[859] NASA, "Mars Program Independent Assessment Team Begins Work," news release 00-6, 7 January 2000.

changing currents in the moon's magnetic field—a finding consistent with the presence of a conducting material such as a watery liquid.[860]

*11 January*
The U.S. Department of Justice filed a lawsuit against Rockwell International Inc., Boeing North American Inc., and United Space Alliance for concealing millions of dollars in fraud committed by a subcontractor. According to the suit, the subcontractor had used funds from the Shuttle program to purchase homes, jewelry, and vacations. The subcontractor accused of the fraud had pled guilty to all charges five years earlier, and the culpability of the major aerospace companies in concealing the theft was the last significant legal matter to come to trial. The three companies claimed that they had cooperated with government investigations into the 180 felony fraud violations and were not liable for the actions of the rogue subcontractor.[861]

*12 January*
Using NASA's Far Ultraviolet Spectroscopic Explorer (FUSE) spacecraft, astronomers found significant proof that thousands of exploding stars had generated the halo of gas around Earth's Milky Way, possibly originating as the galaxy evolved. The findings were among the first generated from the FUSE project, and scientists hoped that they represented the first of many significant discoveries. The halo surrounding the Milky Way had intrigued scientists for decades, but researchers had been unable to shed light on how or why the gas formed as it did. A team of astronomers presented the findings at the American Astronomical Society conference, proclaiming the FUSE observatory "open for business." H. Warren Moos of Johns Hopkins University, Principal Investigator of the FUSE project, stated that the "debugging period" had concluded, and that project managers had moved from fine-tuning the instrument to performing space observations. The FUSE spectrograph had more than 100 times the power of previous instruments, and many researchers hoped that it would allow more extensive investigation of the formation and collapse of stars—a research agenda that might eventually uncover the sequence of events that resulted in the formation of the Earth[862]

A National Research Council panel made up of climate and environment experts announced that it had found significant evidence to support the theory of global warming. The panel found that the Earth's surface had warmed at a rate substantially greater than average for the past millennium. Panel Chairperson John M. Wallace of the University of Washington's Environment Program clarified the group's findings: "The surface of the temperature is rising and has risen substantially in the past 20 years." Wallace also carefully pointed out that the group had not addressed the issue of causation: "We are not saying that the rise is due to greenhouse gases nor are we saying that it is going to continue." The findings added to the ongoing debate over global

---

[860] NASA, "Galileo Findings Boost Idea of Other-Worldly Ocean," news release 00-7, 10 January 2000.
[861] Associated Press, "Government Sues Three Companies, Including Boeing North American, over Space Shuttle Program Fraud," 12 January 2000.
[862] NASA, "FUSE Spacecraft Observes Interstellar Lifeblood of Galaxies," news release 00-5, 12 January 2000; John Noble Wilford, "Scientists Are Gaining a New Understanding of the Dynamics of Galaxies," *New York Times*, 12 January 2000; Reuters, "USA: Milky Way's Halo Caused by Exploding Stars—NASA," 12 January 2000.

warming. Not surprisingly, environmental advocates praised the panel's conclusions, and global-warming critics dismissed them.[863]

The Russian Space Agency announced another delay in the launch of the crew compartment for the International Space Station (ISS), after two Proton rockets exploded when metal and mineral particles contaminated their engines. The Russian Space Agency had planned to use a Proton rocket to launch the ISS module. The S. P. Korolev Rocket and Space Corporation Energia, which built the boosters, promised to replace the Proton rocket's faulty parts. Funding problems had previously slowed construction of the crew compartment, disrupting the timeline for the international project.[864]

As a part of its "Celebrate the Century" program, the U.S. Postal Service unveiled a stamp commemorating the Space Shuttle Program. The stamp with the Shuttle's image joined 12 other stamps with images recalling significant American memories of the past century, such as video games, the fall of the Berlin Wall, personal computers, and Cabbage Patch Dolls. The Space Shuttle stamp was part of the celebration of the 1980s portion of the millennium. Kennedy Space Center Director Roy D. Bridges Jr. issued a statement thanking the Postal Service for the honor.[865]

*13 January*
Astronomers using images from the Hubble Space Telescope (HST) discovered what they believed to be the first examples of "isolated stellar-mass black holes adrift among the stars of our galaxy." They announced their discovery at a convention of the American Astronomical Society. The findings supported a long-standing theory that black holes could form with the collapse of just one massive star, although an opposing theory had concluded that multiple stars must collapse in tandem. Moreover, the researchers' contribution was the latest in a string of recent findings illuminating the relatively commonplace nature of black holes.[866]

*14 January*
Science and business leaders from the United States and Singapore signed an agreement to collaborate on biotechnology research conducted in space. The pact, focused on improving drugs and crops, outlined experiments planned for a 2001 Shuttle *Columbia* mission. U.S. Senator Christopher Bond (R-MO) called the agreement an "important step toward increasing scientific knowledge."[867]

*17 January*
NASA declared an end to its attempts to contact Mars Polar Lander, which had descended to the Red Planet's surface on 3 December 1999. The planned 90-day mission had derailed when

---

[863] Robert Lee Hotz, "Global Warming Real, Says National Panel of Climate Experts," *Los Angeles Times*, 13 January 2000; Associated Press, "Scientists Report Warmer Earth, Cause Uncertain," 12 January 2000; Arthur B. Robinson, "Global Warming Is 300-Year-Old News," *Wall Street Journal*, 18 January 2000.

[864] Associated Press, "Russia Again Delays Launch of Space Station Unit," 12 January 2000; Associated Press, "Space Station Launch Delayed," 12 January 2000.

[865] NASA, "Space Shuttle Program Joins American Icons in Commemorative Stamp Collection," news release 00-12, 12 January 2000.

[866] NASA, "Lone Black Holes Discovered Adrift in the Galaxy," news release 00-4, 13 January 2000.

[867] Associated Press, "Singapore, U.S. Scientists Sign Accord on Outer Space Research," 14 January 2000.

mission controllers at NASA's Jet Propulsion Laboratory (JPL) were unable to communicate with the probe after it landed. Investigators had methodically tried different commands to reach Lander. Meanwhile, scientists and amateur space enthusiasts alike had offered a wide range of hypotheses concerning the cause of Lander's silence, speculating that Lander had exploded before actually reaching Mars's surface or that it had simply sunk in the Martian dust. After exhausting all means of contacting the probe, NASA had declared the US$165 million mission concluded, indicating that future missions would attempt to make up for the loss of Polar Lander. Project Manager Richard A. Cook expressed his sense of closure about the decision to stop investigating the matter: "we feel somewhat complete in the sense that we did go through the things we thought were reasonable . . . we gave it a good shot."[868]

*24 January*
NASA released the first HST images taken after the Shuttle's Mission STS-103, in which the crew of *Discovery* realigned and refurbished the telescope. The stunning clarity of the new images suggested that the mission specialists had succeeded in their sensitive work. According to Steven Beckwith, Director of the Space Telescope Science Institute, the work returned HST to "a condition that was better than it was even before the fourth gyroscope failed." Astronomers had focused on targets more than 5,000 light-years away to test the upgraded telescope. During the servicing mission, astronauts had also installed a Wide Field and Planetary Camera 2 and a Space Telescope Imaging Spectrograph, new instruments that would contribute to the already formidable strength of the HST.[869]

*28 January*
NASA announced that the Food and Drug Administration had cleared for diagnostic use a new technology to fight breast cancer, originally developed at JPL. OmniCorder Technologies, Inc. had built the device, called the BioScan System, using a sensor developed by JPL researcher Sarath D. Gunapala. Gunapala had originally developed the sensor, named the Quantum Well Infrared Photodetector, to locate hot spots in fires and volcanoes. When researchers had learned that the increased blood flow surrounding cancerous tissues often raises skin temperatures slightly, medical scientists realized that doctors could use the technology to conduct noninvasive screenings, helping to detect breast cancer. The new device had charted temperature changes as slight as 0.027°F (-18°C) from one area of the body to another, making it a promising tool for doctors and cancer researchers.[870]

NASA released the research results of astronaut John H. Glenn Jr.'s historic 1998 voyage aboard Shuttle *Discovery*. The report detailed the 88 experiments conducted on Mission STS-95, the most of any Shuttle mission. NASA responded to some critics, who called the mission a publicity-driven flight with little scientific value, disclosing that Glenn himself had participated in 10 experiments during the mission. After becoming the first American to orbit the Earth in

---

[868] Usha Lee McFarling, "NASA Gives Up Hope of Finding Mars Probe," *Los Angeles Times*, 18 January 2000; Associated Press, "JPL To End Search Today for Mars Polar Lander," 17 January 2000; *USA Today*, "NASA Quits Trying To Contact Missing Mars Polar Lander," *USA Today*, 18 January 2000; Associated Press, "Report: Mars Lander May Have Touched Down in Steep Canyon," 6 January 2000.
[869] NASA, "Hubble Reopens Eye on the Universe," news release 00-16, 24 January 2000.
[870] NASA, "Breast Cancer Screening Aid Cleared for Diagnostic Use," news release 00-17, 28 January 2000.

1962, Glenn had returned to space 36 years later to achieve another record—the oldest man to fly in space.[871]

## FEBRUARY 2000

*2 February*
The publication *Florida Today* announced that astronaut Janice E. Voss would carry aboard the next scheduled *Endeavour* flight a commemorative stuffed bear, manufactured to raise money for victims of the shooting at the Columbine High School in Colorado. Voss had agreed to the plan when students from Columbine, visiting Kennedy Space Center as NASA's special guests, explained their fund-raising goals. NASA had invited the Columbine students to Cape Canaveral, Florida, to watch a Space Shuttle launch. NASA also announced that at the conclusion of Voss's mission, it would donate to the still-recovering high school an official *Endeavour* flight kit. Of the class's trip to NASA to watch *Endeavour*'s launch, a parent of one of the Columbine students said simply that it was "a once-in-a-lifetime opportunity."[872]

*3 February*
In another escalation of the increasingly tense debate over how to handle Russia's repeated delays in building and launching the service module of the International Space Station (ISS), NASA Administrator Daniel S. Goldin announced that, if Russia did not meet a July 2000 deadline, the United States would provide a substitute module. Russia was already more than two years behind schedule, affecting the ISS's timeline and hampering the cooperative international effort. Goldin's decision followed Russia's announcement of its plan to direct more resources to the failing *Mir* space station. Some space experts perceived Russia's designation of scarce financial funds to *Mir*, rather than to the ISS, as a signal of Russia's reluctance to participate fully in the international space consortium. In his strongest criticism to date, Goldin said of Russia's ISS involvement: "To say we are frustrated and disappointed is an understatement . . . We want the Russians to be there, but we felt we had to be responsible custodians to keep the program on track."[873]

*7 February*
President William J. Clinton's proposed budget for NASA was unveiled at NASA Headquarters in Washington, DC. The proposal contained none of the cutbacks that had driven the Clinton administration's previous NASA budgets, instead calling for steady increases in funding for space exploration over a period of five years. The key caveat, however, was that most of the increases were scheduled to be implemented after Clinton left office. Most noteworthy, the proposal set aside US$6 billion to begin research into designing a replacement for the Space Shuttle fleet. Officials hoped that NASA's plan to build a reusable launch vehicle to replace the aging Shuttles would come to fruition around the year 2005.[874]

---

[871] Associated Press, "Scientific Results of Glenn's Mission Released by NASA," 28 January 2000.
[872] Billy Cox, "Endeavour To Launch with Columbine Bear," *Florida Today* (Brevard, FL), 2 February 2000.
[873] Warren E. Leary, "NASA Dictates Date for Russia on Station Port," *New York Times*, 4 February 2000; Associated Press, "NASA: We'll Launch if Russia Fails," 4 February 2000.
[874] Joseph C. Anselmo, "NASA Funds Research for Shuttle Successor," *Aviation Week and Space Technology* 152, no. 7 (14 February 2000): 11–12.

NASA Administrator Daniel S. Goldin announced that NASA would merge the Office of the Chief Technologist and the Office of Aero-Space Technology, to better facilitate NASA's technological advancements. According to Goldin, placing a range of issues, from IT operation to technology-based research, under the domain of a single leader would promote IT compatibility. Goldin appointed Chief Technologist Samuel L. Venneri to head the new Office, giving Venneri the responsibility of developing a long-term strategy to keep NASA at the forefront of technological development. Goldin also tasked Venneri with forging new relationships with companies using technologies similar to the type that NASA developed and used. Venneri had been NASA's Chief Technologist since 1996.[875]

*8 February*
The Solar and Heliospheric Observatory (SOHO) spacecraft set an astronomical record by facilitating the discovery of 102 comets. The spacecraft featured a Large Angle and Spectrometric Coronagraph (LASCO) instrument that allowed astronomers to observe comets that previously would have been invisible. Douglas A. Biesecker, an astronomer at NASA's Goddard Space Flight Center, was responsible for making 45 discoveries on his own. According to the researchers working on the project, SOHO had revealed far more suicidal comets (those plunging into the Sun's atmosphere) and sungrazers (those that pass by the Sun) than scientists had previously expected to find. The findings had practical implications as well. By observing the patterns of comets as they circled, collided, and split, scientists hoped that they might be better prepared to predict the behavior of a comet headed for Earth.[876]

*9 February*
Russia successfully launched a Soyuz test rocket equipped with a Fregat accelerator unit able to carry payloads into high orbit, but lost the Fregat unit upon its reentry to Earth. The rocket took off from Baikonur Cosmodrome in Kazakhstan, carrying only a dummy payload into space, in a test that was part of Russia's ongoing research into developing vehicles and technologies that would eventually serve the orbiting ISS.[877]

More than 10,000 white-collar workers walked off the job at the Boeing Company, in protest over failed contract negotiations. Most of the strikers came from the aerospace giant's work site in Seattle, Washington.[878]

*10 February*
In a much-anticipated launch, the *Astro-E* satellite lifted off aboard an M-5 rocket from the Kagoshima Space Center on the Japanese island of Kyushu. However, soon after launch, *Astro-E* was lost when the rocket suffered a "control system breakdown," propelling the satellite into a lower orbit than planned. U.S. officials monitoring the launch determined that the satellite had most likely burned up in Earth's atmosphere. *Astro-E*, the product of a joint venture between the United States and Japan, carried new x-ray–sensing equipment, which scientists had hoped would further illuminate the material surrounding black holes. Scientists had designed the new

---
[875] NASA, "Goldin Names Venneri To Head Merged Technology, Aero-Space Office; New Chief Engineer Appointed," news release 00-21, 7 February 2000.
[876] NASA, "SOHO Spacecraft Bags 102 Comets," news release 00-23, 8 February 2000.
[877] Dow Jones Newswire, "Russia Tests, Then Loses Rocket with Union for Higher Orbit," 9 February 2000.
[878] *Orlando Sentinel* (FL), "Thousands Walk Out on Boeing in Seattle," 10 February 2000.

equipment to sense minute changes in the heat of single protons. The specialists overseeing the launch and mission had hoped that *Astro-E* would orbit and gather data for five years. The satellite had cost more than US$100 million to develop.[879]

*11 February*
Shuttle *Endeavour* lifted off in Mission STS-99, planning to use new radar technology to map Earth's terrain. Crew included astronauts Kevin R. Kregel, Dominic L. P. Gorie, Janet L. Kavandi, Janice E. Voss, Mamoru Mohri, and Gerhard P. J. Thiele. The astronauts planned to extend the Shuttle Radar Topography Mission antenna from *Endeavour*'s cargo bay. The 197-foot (60-meter) antenna would gather mapping data necessary to create the most comprehensive map of Earth ever drawn. NASA collaborated with the National Imagery and Mapping Agency in planning the mission and hoped to cover more than 70 percent of Earth's surface during the 11-day mission.[880]

*14 February*
The Near Earth Asteroid Rendezvous (NEAR) spacecraft successfully orbited the asteroid Eros, 160 million miles (258 million kilometers) from Earth, gathering close-up images of the craggy rock, in the first successful orbit of a spacecraft around an asteroid. The unique images revealed large boulders and craters on Eros. Scientists noted with amusement that on the Valentine's Day encounter, NEAR obtained images of a heart-shaped chasm on Eros. NASA planned for NEAR to orbit Eros for about one year.[881]

*18 February*
NASA announced that a team of astronomers had discovered what they believed to be the earliest known structure ever to form in the universe. The researchers had used the Hale Telescope at Palomar Observatory in California and the National Science Foundation's Mayall Telescope in Kitt Peake, Arizona, to find an ancient quasar. By measuring the quasar's redshift to determine how fast the quasar was moving away from the galaxy, scientists had been able to calculate the cosmic distance separating Earth and the quasar. The odds of discovering such a distant and relatively fast-moving quasar were remote, especially since scientists could monitor only a fraction of the sky at any one time. The discovery had the broader value of providing a reference point by which to assess those bodies between Earth and the quasar. Daniel Stern of NASA described the finding's utility: "Finding a quasar at this distance is like turning on a flashlight at the edge of the universe."[882]

At a ceremony at Peterson Air Force Base in Colorado Springs, Colorado, U.S. Air Force General Ralph E. Eberhart succeeded General Richard B. Myers as the nation's top military space official, overseeing the U.S. Space Command, the U.S. Air Force Space Command, and the North American Aerospace Defense Command, in addition to thousands of troops and more

---

[879] NASA, "Japanese-U.S. Satellite Ushers in Golden Era of X-ray Astronomy," news release 00-18, 2 February 2000; Associated Press, "Japan Suffers Another Setback When Satellite Fails To Reach Orbit," 10 February 2000; Justin Ray, "Astro-E Believed Lost Following Botched Launch," *Space Flight Now*, 10 February 2000.
[880] Associated Press, "Endeavour on Mission To Get Best Map of Earth," 12 February 2000.
[881] NASA, "NEAR Begins Looking Closely at Eros," news release 00-28, 17 February 2000; Peter Kendall, "Cameras Get Close-Up View of Asteroid," *Chicago Tribune*, 15 February 2000; Dow Jones Newswire, "Robot Spacecraft Successfully Orbits Asteroid Eros," 14 February 2000.
[882] NASA, "Newfound Quasar Wins Title 'Most Distant in the Universe'," news release 00-30, 18 February 2000.

than 100 satellites. Eberhart took over the U.S. space operations post after having led the Air Force's Air Combat Command at Langley Air Force Base in Virginia.[883]

*19 February*
The American Museum of Natural History's new Hayden Planetarium, which had used information gathered from NASA and the European Space Agency to compile a database containing billions of stars, opened to the public. From 1935 to 1997, the New York City museum had hosted thousands of visitors annually with a less elaborate planetarium. The new planetarium was only one part of the sparkling US$210 million Rose Center for Earth and Space, which museum curator Michael M. Shara hoped would inspire future scientists. The facility, resembling a spacecraft on the outside, featured informative programming hosted by celebrities, such as actors Tom Hanks and Jodie Foster. As they walked through the new facility, visitors could observe displays demonstrating how scientists believed the universe had evolved over a period of 13 billion years.[884]

*20 February*
By conserving fuel throughout their mission, the astronauts aboard Shuttle *Endeavour* were able to prolong their time in space by one extra day, allowing the crew to map the Earth's terrain, covering several million square miles more than they had originally estimated possible. The extra 9 hours and 10 minutes of mapping allowed the radar flown from *Endeavour* to survey Australia in detail, making up for time lost when a Shuttle thruster had failed.[885]

*22 February*
Shuttle *Endeavour* returned safely to Earth after a successful 11-day voyage. The scientific community and the public greeted the Shuttle's return with even more fanfare than usual, because the crew had obtained stunning images of Earth. *Endeavour*'s six-person crew had worked nearly around the clock, keeping radar antennas running to compile the data for three-dimensional digital maps of Earth's surface. The crew had gathered more than 300 digital tapes of radar data during the mission, which NASA Administrator Daniel S. Goldin called "one of the most challenging, difficult missions we ever undertook." The mission had successfully mapped 43.5 million square miles (113 million square kilometers) of Earth's terrain. NASA and the National Imagery and Mapping Agency expected that transferring the data into readable maps would take between one and two years. NASA had obtained the images primarily for the U.S. Department of Defense, to improve the accuracy of missiles and to assist in navigation for military planes and troops.[886]

*23 February*
NASA Inspector General Roberta L. Gross reported that the Boeing Company would have to forfeit most of its profit from the ISS contract because of its nearly US$1 billion in cost overruns. NASA's incentive-laden contract with Boeing allowed for significant bonuses for meeting

---

[883] Tom Breen, "U.S. Gets New Military Space Chief," *Florida Today* (Brevard, FL), 19 February 2000.
[884] Associated Press, "Museum Unveils Planetarium for the 21 Century," 2 February 2000.
[885] Associated Press, "Endeavour Squeezes in Extra Mapping Day," 21 February 2000.
[886] Associated Press, "Endeavour Returns with Dazzling Images of the Home Planet," 22 February 2000; *Los Angeles Times*, "Endeavour's Astronauts End Earth-Mapping Work," 22 February 2000; *Chicago Tribune*, "Mission: Map the Earth," 22 February 2000.

timeline and cost limits. NASA Administrator Daniel S. Goldin had ordered an audit of the NASA-Boeing contract in 1999, because projected cost overages had continued to rise. Nonetheless, as Boeing neared completion of the construction of its portion of the ISS, both NASA and the aerospace giant were optimistic that cost overruns had stabilized, and that the space station hardware would perform well in space.[887]

## MARCH 2000

*3 March*
NASA's Glenn Research Center (GRC) in Cleveland, Ohio, opened a new laboratory dedicated to the study of ballistics. The laboratory, housed in the complex's Building 49, featured a 40-foot (12-meter) gas gun that could shoot projectiles at speeds of up to 1,500 feet (457.2 meters) per second. A camera with the ability to capture 2.5 million images per second allowed researchers to observe the behavior of the projectiles and chart the results of their impact on aviation equipment. "The whole idea," Team Leader Dale A. Hopkins explained, "is to watch the impact and see how materials struck by the projectiles behave . . . not just whether they survive, but how they deform and fail." GRC had designed the new facility to test materials used for aircraft engine housings and for flywheel containment. Researchers hoped that the new tools would help engineer aircraft that could withstand the pressures of high-speed and high-altitude flight better than those already available could. GRC had been testing ballistic materials since 1980.[888]

*7 March*
Chester M. Lee, who had served as Mission Director for six Apollo Missions to the Moon, including Apollo 13, died from complications of open-heart surgery. Lee had begun his 23-year career with NASA in 1965 after retiring from the U.S. Navy, specializing primarily on the Apollo Missions. He had also served as Program Director for the 1973–1975 Apollo-Soyuz Test Project, the first joint venture between the United States and the former Soviet Union.[889]

*8 March*
A National Academy of Sciences task force issued a report supporting the development and deployment of the *Triana* research satellite proposed by Vice President Albert A. Gore Jr. Congress had authorized the task force in 1999, to investigate the feasibility of Gore's plan, which called for a satellite to orbit Earth continuously, capturing live pictures of the planet. Gore believed that obtaining the images and making them available on the Internet would increase public awareness of environmental issues. The proposal had sparked a political controversy between Gore and the Republican Congress, prompting Congress to demand that a task force investigate the value of the project. The task force reported that the satellite had scientific merit. White House science advisor Neal Lane praised the study's results, urging House members to act with "bipartisan Congressional support." The process of investigation and approval of the project had drawn public attention to the political debate about space exploration.[890]

---

[887] Tony Capaccio, "Boeing Overruns on Space Station Near $1Billion," *Los Angeles Times*, 23 February 2000.
[888] NASA, "New Lab Ready To Test Sudden Impact," news release 00-34, 3 March 2000.
[889] Associated Press, "Chet Lee, Former NASA Worker, Dies at 80," 7 March 2000; *Los Angeles Times*, "Obituaries Chester Lee; Navy Captain, Apollo Mission Director," 9 March 2000.
[890] Warren E. Leary, "Science Panel Supports Gore Satellite Plan," *New York Times*, 9 March 2000; Associated Press, "Panel Finds Merit in Space Mission Proposed by Gore," 8 March 2000.

*9 March*
An independent study, commissioned by NASA in the wake of the highly public failures of Mars Climate Orbiter and Mars Polar Lander, found that cost-cutting measures had created significant risks for NASA's Shuttle program. Although the report maintained that the overall safety of NASA's operations remained within an acceptable range, the investigators warned that the use of contractors and the lack of oversight compromised the quality of NASA's missions. NASA's chief of the Office of Space Flight accepted the report's recommendations to improve NASA's operations, also noting, however, that NASA had already investigated and dealt with many of the weaknesses highlighted in the report. With budgetary matters always a pressing concern, NASA had made various necessary reductions in personnel, relying on companies such as Boeing and Lockheed Martin more than ever. NASA had initiated the investigation to understand more fully the impact of the changes on NASA's highest priority—the safety of its astronauts.[891]

*10 March*
In an article in the journal *Science*, a team of researchers led by Maria T. Zuber of the Massachusetts Institute of Technology reported that they had found compelling evidence that Mars once had a wet climate. According to Zuber, "Evidence is building of more water on the surface of Mars at one time." The researchers had used the laser measurements taken by Mars Global Surveyor to reach their conclusions. Although Zuber indicated that scientists did not yet have enough evidence to determine whether Mars once had a surface ocean, the study did find large, buried channels beneath the planet's crust, suggesting that an enormous flow of water once deluged the landscape. The researchers had also found further evidence of two distinct "crustal provinces" on Mars, similar to those on Earth. The team estimated that Mars's crust was 50 miles (80 kilometers) thick beneath the planet's southern highlands, but only 22 miles (35 kilometers) thick beneath the northernmost areas of the planet. By using Surveyor's data to analyze thoroughly the topography of the planet, the research team had also drawn the conclusion that Mars had once been home to more water than ice. The variety of landmasses suggested that, for a period, the planet's core had released heat, turning ice trapped beneath Mars's surface into water. As one newspaper described the formative events estimated to have taken place approximately 4 billion years ago, "It steamed and streamed."[892]

*13 March*
The Sea Launch Company suffered a setback in its bid to become a major participant in the satellite launch industry, when one of its Russian-Ukrainian Zenit 3SL rockets carrying a British communications satellite crashed into the Pacific Ocean. The launch was only the third undertaken by the international company, which had planned to use a converted ocean oil rig known as Odyssey as its permanent launchpad. The Sea Launch Company comprised the Boeing Company, together with companies from Russia, Ukraine, and Norway. The company had designed its innovative Sea Launch System to use Earth's rotation and the faster movement of

---

[891] Associated Press, "NASA Says Shuttle Cuts Have Led to Higher Risk," 10 March 2000; *Los Angeles Times*, "Study Raises Concerns About Shuttle Safety," 13 March 2000.
[892] Marie T. Zuber et al., "Internal Structure and Early Thermal Evolution of Mars from Mars Global Surveyor Topography and Gravity," *Science* 287, no. 5459 (10 March 2000): 1788–1893; NASA, "View Inside Mars Reveals Rapid Cooling and Buried Channels," news release 00-36, 9 March 2000; Dan Vergano, "Northern Mars Once Was Drenched," *USA Today*, 13 March 2000.

Earth's surface at the equator to make it easier to launch rockets with heavy payloads. The failed satellite had cost London-based ICO Global Communications more than US$100 million to develop.[893]

*14 March*
NASA announced that it had renamed the Near Earth Asteroid Rendezvous (NEAR) spacecraft, already orbiting asteroid 433 (Eros), after Eugene M. Shoemaker, an expert in the study of asteroid craters and their origins who died in 1997. Shoemaker's work on Meteor Crater in Arizona, during the 1960s, had set the standard for crater investigation. Shoemaker had worked for NASA teaching astronauts about meteors and craters. He died in the Australian outback while researching asteroid-impact craters.[894]

*16 March*
NASA dedicated a new aerospace laboratory at Anne Beers Elementary School in Washington, DC, featuring a simulated Shuttle cockpit and Mission Control Center. NASA had named the facility, designed to encourage student interest in space exploration, for U.S. Representative Louis Stokes (D-OH), who had proposed legislation to fund the endeavor in 1998. NASA's GRC and Goddard Space Flight Center had worked with the Orchard Glenn School of Cleveland, Ohio, and the University of the District of Columbia to construct the innovative facility.[895]

*17 March*
The U.S. General Accounting Office (GAO) issued a report stating that the Russian components for the International Space Station (ISS) did not meet NASA's standard safety requirements. GAO's findings were the latest in a long series of controversies, particularly between the United States and Russia, in the international effort to build the space station. Most significant of its safety deficits, the Russian crew module did not have adequate shielding to protect the ISS crew from orbiting space debris. At the time of the report, the Russians were already more than two years behind in delivering the crew module.[896]

*21 March*
During routine testing in preparation for a July 2000 launch, excessive vibration damaged NASA's High Energy Solar Spectroscopic Images (HESSI) spacecraft. The vibration test device was accidentally set at a level 10 times higher than appropriate, causing structural damage and cracks in two of the satellite's four solar arrays. Officials estimated that the damage would delay the planned launch by at least six months. NASA had designated about US$75 million for the project. The faulty test, occurring at NASA's Jet Propulsion Laboratory, revealed the high degree of precision required of engineers when creating and testing a space-bound craft. Although the satellite had vibrated at the incorrect setting for only about 200 milliseconds, it still sustained

---

[893] Associated Press, "Sea Launch Satellite Falls into Ocean; Boeing-Led Venture Out $100 million," 13 March 2000.
[894] NASA, "NASA Renames NEAR Spacecraft for Planetary Science Pioneer Gene Shoemaker," news release 14, March 2000.
[895] NASA, "Aerospace Laboratory Encourages Students To Reach for the Stars with NASA Help," news release m00-051, 15 March 2000.
[896] Patty Reinhert, "Space Station Safety at Center of Debate," *Houston Chronicle*, 17 March 2000; Associated Press, "General Accounting Office Slams Russian Modules," 17 March 2000.

damage. However, the project engineers were optimistic that the mission would resume eventually.[897]

*29 March*
Planet sleuths Geoffrey W. Marcy of the University of California at Berkeley and Steven S. Vogt of the University of California at Santa Cruz announced the discovery of two very small planets outside Earth's solar system. Smaller in mass than Saturn, the planets resided approximately 100 light-years from Earth. Although Marcy and Vogt had concluded that the planets were very hot, and that neither planet was capable of supporting life, their discovery reinforced the long-standing theory that planets form by "snowball effect," growing from smaller to larger, as well as suggesting that many other stars in the galaxy might harbor small planets. To make their discovery, the astronomers had used the W. M. Keck Observatory in Hawaii in one phase of a multiyear project to observe stars within 300 light-years of Earth.[898]

*30 March*
Engineers at NASA's Dryden Flight Research Center (DFRC) successfully tested a prototype for the "flying lifeboat" that would transport astronauts to and from the ISS. DFRC officials launched a robotic X-38 from under the wing of a B-52 flying at 39,000 feet (12,000 meters). It free fell for more than 40 seconds before deploying a 5,500-square-foot (511-square-meter) parachute, which provided a soft landing for the spacecraft. The test was a significant step in evaluating the technology planned for the crew return vehicle, a low-cost ISS initiative.[899]

*31 March*
The National Partnership for Reinventing Government (NPR) released the results of a survey indicating that NASA employees had the highest level of job satisfaction among federal government employees. NASA employees rated very favorably in the categories of employee job satisfaction, customer orientation, making reinvention a priority, management communication, and employee participation in cross-functional teams. NASA Administrator Daniel S. Goldin announced the survey's findings, stating, "I am incredibly proud of these results. They represent a strong statement of the top-to-bottom excellence of the NASA team." NPR and the Office of Personnel Management administered the survey to understand employees' perspectives on "reinvention and workplace issues." When informed of NASA's outstanding score, Keith Cowing, the editor of Nasawatch.com remarked, "I'm not at all surprised. I used to work there. I loved my job."[900]

## APRIL 2000

*4 April*

---

[897] Associated Press, "NASA Engineers Damage Satellite During Shake Test," 24 March 2000; NASA, "HESSI Sustains Damage During Vibration Testing," news release 00-45, 24 March 2000.
[898] NASA, "Planet Hunters on Trail of Worlds Smaller than Saturn," news release 00-47, 29 March 2000; John Noble Wilford, "2 Relatively Small Planets Are Found," *New York Times*, 30 March 2000; Dan Vergano, "Other Worldly Discovery," *USA Today*, 30 March 2000.
[899] Associated Press, "Space Station 'Lifeboat' Prototype Successfully Tested," 30 March 2000.
[900] NASA, "Survey Shows NASA Employees Among Most Satisfied Federal Workers," news release 00-52, 31 March 2000.

Russian cosmonauts Sergei V. Zalyotin and Alexander Y. Kaleri launched from Earth aboard *Soyuz TM-30*, headed for the *Mir* space station. In their planned two-month mission, more than seven months after the last crew members had left the high-maintenance spacecraft, the men aimed to restore *Mir* to working order. Many experts had assumed that *Mir*'s days as an active space research center were finished when the previous crew of cosmonauts left the station. Although unable to pay to keep cosmonauts aboard *Mir*, Russia was still hoping to obtain funding to keep the space station in operation and had not taken it out of orbit. When the Netherlands-based MirCorp had agreed to pay US$20 million to lease the Russian-owned station, Russia had scrambled to put together a crew for a repair mission. The departure of *Soyuz* caused some concern in the international space community, once again raising speculation regarding Russia's commitment to the International Space Station.[901]

*5 April*
NASA awarded its Commercial and Government Inventions of the Year awards. Inventors Anne K. St. Clair, Terry L. St. Clair, and William P. Winfree, working as a research team at NASA's Langley Research Center, won the award for inventing a material that they had named Colorless and Low Dielectric Polyimide Thin Film. Commercial manufacturers could apply the thermoplastic material to many types of surfaces to protect against ultraviolet radiation and high temperatures. The researchers also suggested adding the material to liquids, such as paints and cosmetics, to provide ultraviolet protection. Douglas B. Leviton received the Government Inventor of the Year award for his Ultra-High Sensitivity, Incremental, and Absolute Optical Position Encoder. The encoder allowed scientists to calibrate space-bound instruments more accurately than they had done previously. Both awards recognized inventions completed during 1999. NASA honored the recipients at a ceremony, presenting them with plaques and cash awards.[902]

NASA released a comprehensive new study, which found that the levels of ozone in the Arctic stratosphere were declining rapidly. In some parts of the stratosphere, the atmosphere had lost 60 percent of its ozone over the course of one year. The report, compiled by hundreds of European and American scientists, also demonstrated that changes in the global climate (primarily global warming) threatened to delay the recovery of the ozone layer. The scientists were optimistic that the ozone layer would recover eventually, but perhaps not until the mid-21$^{st}$ century. The study had cost NASA more than US$20 million, and the European scientists contributed US$10 million. The researchers had gathered data using satellites, airplanes, and hundreds of weather balloons.[903]

*6 April*
*Soyuz* successfully docked with the unoccupied *Mir* space station. At the last minute, cosmonauts Sergei V. Zalyotin and Alexander Y. Kaleri docked *Soyuz* manually, because Russian space

---

[901] *Florida Today* (Brevard, FL), "Cosmonauts Head to Mir Today," 4 April 2000; *Des Moines Register* (IA), 5 April 2000.
[902] NASA, "NASA Selected Its Commercial and Government Inventions of the Year," news release 00-53, 5 April 2000.
[903] Knight-Ridder News Service, "Thinning Ozone over Arctic Raises Concerns," 6 April 2000; William K. Stevens, "New Survey Shows Growing Loss of Arctic Atmosphere's Ozone," *New York Times*, 6 April 2000.

officials feared the autopilot was about to malfunction. The cosmonauts planned to concentrate on plugging a small air leak and conducting several scientific experiments during a spacewalk.[904]

In two *Nature* articles, research teams announced that through independent research they had discovered the same result: the Ulysses spacecraft had passed through the longest comet tail ever discovered. Although the Ulysses vehicle had made its unintended discovery in 1996, when it passed through the tail of Comet Hyakutake, scientists had taken four years to confirm the discovery. The scientists concluded that Ulysses had discovered a comet tail approximately 300 million miles (483 million kilometers) long. "The discovery was made quite by accident, a bit like finding a needle in a haystack when you weren't even looking for it in the first place," was how one project member described the fortunate discovery. Ulysses, a joint mission of NASA and the European Space Agency, had launched in 1990 to study solar winds and rays, not comets. However, the spacecraft had picked up some unusual readings during its mission, which turned out to be signs of Hyakutake's tail, extending much farther than previously estimated.[905]

NASA announced that follow-up studies had determined that the distant mass known as TMR-1C, discovered in 1997, was probably not a protoplanet—a young, still-forming precursor of one of the giant planets. In 1998 astronomer Susan Terebey of the Extrasolar Research Corporation had published findings suggesting that the body was a protoplanet several times larger than Jupiter. Terebey's research had received widespread attention. However, although NASA had released the images of TMR-1C captured by the Hubble Space Telescope (HST), NASA had urged scientists to obtain further verification before identifying the body as a protoplanet. After observing the body over several months, astronomers had concluded that TMR-1C was a bright star, rather than a planet. Reflective dust had likely increased the brightness of the star, located approximately 135 billion miles (217 billion kilometers) from Earth, making it appear to be a planet. After her initial high-profile announcement, Terebey had continued to study the star using the W. M. Keck Observatory in Hawaii and had amended her own hypothesis, reporting, "the new data do not lend weight to the protoplanet interpretation and the results remain consistent with the explanation that TMR-1C may be a background star." Most newspapers gave as much coverage to the news of the changing assessment as they had given to the original discovery.[906]

*15 April*
Princeton University graduate student Xiaohui Fan discovered a distant quasar that scientists believed to be the oldest object ever observed by a human being. Fan had worked as part of a team using the 32-foot (10-meter) W. M. Keck Telescope. Estimating that the quasar was 12

---

[904] Associated Press, "Latest Space Crew Docks Flawlessly with Mir Space Station," 6 April 2000.

[905] Geraint H Jones et al., "Identification of Comet Hyakutake's Extremely Long Ion Tail from Magnetic Field Signatures," *Nature* 404, no. 6778 (6 April 2000): 574; G. Gloeckler et al., "Interception of Comet Hyakutake's Ion Tail at a Distance of 500 Million Kilometers," *Nature* 404, no. 6778 (6 April 2000): 576; NASA, "Strangers in the Night: Ulysses Spacecraft Meets a Comet," news release 00-55, 5 April 2000; Associated Press, "Solar Probe Finds Longest Comet Tail Ever Recorded," 10 April 2000.

[906] James Glanz, "Scientists Retract Claim of Planet Discovery," *New York Times*, 7 April 2000; Associated Press, "Presumed Protoplanet May Be a Background Star," 7 April 2000; NASA, "Suspected Protoplanet May Really Be a Distant Star," news release 00-58, 6 April 2000.

billion years old, astronomers expressed their excitement that the benchmark discovery would enable them to observe the universe in its formative stages.[907]

*18 April*
The United States and Brazil signed a treaty giving U.S. aerospace firms access to Brazil's Alcântara launch site, located just 3° south of the equator. The treaty was the first accord signed by the United States that allowed U.S. satellites to launch from foreign soil. The launch site's location near the center of the globe would save money for aerospace companies, because satellites could take a shorter path to equatorial orbit, requiring less fuel. Brazil benefited from the boost to its embryonic space program and the possibility of capturing a portion of the lucrative commercial-satellite business. At the time of the agreement, U.S. firms controlled two-thirds of the annual US$66 billion commercial satellite industry.[908]

*19 April*
After completing four months of on-orbit checkout and verification, *Terra*, the newest in NASA's series of Earth Observing System Satellites, became functional. NASA had developed and launched *Terra* to monitor Earth's atmosphere on a daily basis. Researchers hoped that gathering frequent data would allow a more comprehensive assessment of Earth's climate change. According to Yoram J. Kaufman, Project Scientist for *Terra* at Goddard Space Flight Center (GSFC), "*Terra* is measuring and documenting the Earth's vital signs, many of them for the first time." In the midst of public and scholarly discussions of global warming and climate change, the new research satellite was a step toward understanding the causes of the changes. "The data will help us understand our planet, aid in our distinguishing between natural and human-induced changes, and show us how the Earth's climate affects the quality of our lives," Kaufman clarified. The satellite's first transmitted images thrilled scientists. The satellite surveyed North America, creating images in many different layers, to reveal how different levels of population and types of vegetation affected the climate. NASA officials planned to make the images obtained by *Terra* available to business, scientists, and the public.[909]

*20 April*
NASA revealed what it called "the first detailed images of the early universe," captured by the Balloon Observations of Millimetric Extragalactic Radiation and Geophysics (BOOMERANG) experiment. For the BOOMERANG experiment, scientists had suspended a powerful telescope from a research balloon circumnavigating the Antarctic. Andrew E. Lange of California Institute of Technology called the experiment and the resulting images "an incredible triumph of modern cosmology." For the most part, the data, obtained by closely measuring variations in the microwave background radiation, confirmed scientists' existing theory that, in its earliest days, the universe was flatter than it is today. According to this theory, confirmed by other evidence besides these images, the universe is flat and forever expanding. In summarizing the significance

---

[907] Associated Press, "Quasar Discovered by Student May Be Most Distant Ever Seen," 15 April 2000; Kevin Coughlin, "Young Princeton Expert Finds Piece of a Fledgling Universe," *Star-Ledger* (Newark, NJ), 14 April 2000; *Washington Post*, "Most Distant Object Observed," 17 April 2000.
[908] Reuters, "Brazil: Joins Space Race with U.S. Satellite Treaty," 17 April 2000; Peter Fritsch, "Brazil To Sign U.S. Accord To Enter Satellite Business," *Wall Street Journal*, 18 April 2000.
[909] NASA, "Terra Spacecraft Open for Business," news release 00-62, 19 April 2000; NASA, "New Views of Earth from Terra Satellite Wednesday," news release N00-16, 18 April 2000.

of the experiment, one Massachusetts Institute of Technology researcher stated, "it's confirmation of the prediction of our best theory of what caused the structure of the universe."[910]

*24 April*
NASA and the science community celebrated the 10-year anniversary of the deployment of the HST. During its decade in space, the HST had enabled astronomers to make hundreds of discoveries. According to NASA, the HST's "rapid-fire rate of discoveries and cosmic images has profoundly changed the science of astronomy, and astounded and inspired people around the world." The U.S. Postal Service and NASA unveiled a series of five stamps designed to commemorate the HST's decade of contributions. The images on the stamps would not have been possible without the HST's power, which enabled it to capture images of the Eagle Nebula, Ring Nebula, Lagoon Nebula, Egg Nebula, and Galaxy NGC 1316. NASA held a series of events to celebrate the anniversary. Hubble's 10 years in space had not begun smoothly, as many critics of the program had been quick to point out. As the *Baltimore Sun* remembered, the HST was "once the butt of late-night talk-show jokes," because a flawed mirror initially had prevented the telescope from focusing. However, after a series of costly repairs, the HST had produced a litany of awe-inspiring images. The HST made more than 271,000 observations in its first 10 years, with another decade of usefulness anticipated. HST Project Scientist David S. Leckrone of NASA's GSFC summarized the HST's impact, "Not since Galileo aimed a small, 30-power telescope into the night sky in 1609 has humanity's vision of the universe been so revolutionized in such a short time span by a single instrument."[911]

*25 April*
NASA announced that a team of scientists had used the Chandra X-ray Observatory to solve one of astronomy's fundamental problems—determining the distance from Earth to a cosmic object. Peter Predehl, the lead researcher on the project, had found that by scattering x-rays between Earth and a given object, the dust grains and scattered materials between the two points could serve as measuring posts to determine the distance covered by the x-ray. Because of the Chandra Observatory's superior resolution and strength, particles that scientists could not otherwise have observed would serve as points of context. Researchers were confident that they could use the new approach to help determine the size of nearby galaxies. Furthermore, the newly devised means of measuring the distance between objects and Earth had implications for understanding the universe and its age.[912]

**MAY 2000**

*1 May*
In delivering the Louisiana State University Chancellor's Distinguished Lecture, NASA Administrator Daniel S. Goldin announced that "in no less than 10 and no more than 20 years," U.S. astronauts would land on Mars. Proclaiming that Mars was NASA's next frontier, Goldin

---
[910] Associated Press, "Scientists Reveal First Detailed Images of Early Universe," 27 April 2000; NASA, "The Universe in Its Infancy: New Findings Unveiled at News Briefing," news release N00-17, 20 April 2000.
[911] NASA, "April 24 Marks a Triumphant Ten Years in Space for Hubble Telescope," news release 00-59, 11 April 2000; NASA, "NASA, U.S. Postal Service To Celebrate Hubble Space Telescope 10th Anniversary with Stamp Unveiling and Other Events," news release N00-14, 4 April 2000; Frank D. Roylance, "Hubble Goes from Flop to Star; Once the Butt of Jokes, Telescope Now Thrills with Space Images," *Baltimore Sun* (MD), 21 April 2000.
[912] NASA, "Chandra Shows New Way To Measure Cosmic Distances," news release 00-66, 25 April 2000.

stated that the Red Planet held special promise for researchers. The possible presence of water and other signs of life on Mars intrigued scientists, making a Mars landing a priority for NASA. In addition, because scientists have estimated that the two planets are roughly the same age, the exploration of Mars could lead to new information about Earth. Goldin made his remarks in the aftermath of the high-profile crash of Mars Polar Lander.[913]

NASA announced the creation of a new office dedicated to improving heath and safety for its workforce. NASA Administrator Daniel S. Goldin appointed Dr. Arnauld E. Nicogossian as Chief Health and Medical Officer, giving him the responsibility for developing programs to research and implement the best medical procedures for astronauts and other NASA personnel. "On the ground and in space, rapid advances in medical knowledge and tools need to be adapted and incorporated into our planning and practices," Goldin said when announcing the new position. Nicogossian faced the task of establishing NASA's Health Council, in keeping with NASA's commitment to making heath and safety its number-one priority.[914]

*2 May*
The Carrier Test Pilot Hall of Honor, aboard the USS *Yorktown*, inducted U.S. Senator John H. Glenn Jr., honoring him for his distinguished aviation career before becoming an astronaut. Before Glenn became the first man to orbit Earth, he had already achieved a long record of flight exploits. Glenn's fellow test pilot Neil A. Armstrong, the first astronaut to walk on the Moon, attended the induction. Upon receiving the award, Glenn expressed his gratitude at joining the distinguished ranks of the Hall of Honor and spoke of his lifetime of flight experiences.[915]

*3 May*
The "Love Bug" computer virus infected the computer system of NASA, along with those of the Pentagon, the CIA, Microsoft Corporation, Ford Motor Company, and other major business and government offices. Considered one of the most damaging and costly computer viruses in history, "Love Bug" spread primarily through e-mail messages, corrupting files on millions of computers. Four of NASA's 10 space centers suffered from the attack, causing both Johnson Space Center (JSC) and Kennedy Space Center (KSC) to shut down e-mail and other communication between computers.[916]

News reached the United States that space debris had crashed to the ground near Johannesburg, South Africa, prompting NASA to respond. Several metal balls had fallen from the sky, leaving 8-inch (20-centimeter) dents in the ground. NASA scientists clarified that the debris probably came from a Delta rocket launched in 1996. Chief Scientist for Orbital Debris Nicholas L. Johnson explained that manufactured objects sent into space plummet back to Earth nearly every day, landing mostly in the ocean. Johnson also stated that NASA would compensate anyone hurt by the falling debris, remarking that no such injuries had occurred in more than 40 years.[917]

---

[913] Associated Press, "Red Planet Is Next Frontier," 2 May 2000.
[914] NASA, "NASA Creates New Office To Foster Heath and Safety," news release 00-69, 1 May 2000.
[915] Associated Press, "Glenn Honored for Test Pilot's Role as Armstrong Watches," 3 May 2000.
[916] *Los Angeles Times*, "Fast Moving Virus Hits Computers Worldwide," 5 May 2000.
[917] Associated Press, "Hail of Metal Balls Puzzles Farmers, Until NASA Steps In," 3 May 2000.

NASA successfully launched the GOES-L satellite aboard an Atlas 2A rocket from Cape Canaveral, Florida. The new weather satellite would provide the National Oceanic and Atmospheric Administration with detailed readings on severe weather systems affecting the United States. NASA deemed the GOES-L satellite a top priority because of the declining returns of the aging GOES-8 satellite. NASA had delayed the US$220 million satellite project because of engine problems, causing anxiety to weather forecasters who depended on the satellites to track the powerful hurricanes that affected the United States in 1999. As Director of the National Weather Service's Southern Region, William Proenza explained, "It is vitally important that we have continuity of data or we may be finding ourselves impaired in our ability of forecasting." NASA had kept the May 2000 launch of the GOES-L on track, despite some calls to postpone it so that NASA could launch Shuttle *Atlantis* on an urgent repair mission to the International Space Station (ISS). However, NASA officials had determined that the risk of inadequate weather forecasting was too great to take a chance on delaying the launch.[918]

*4 May*
*Nature* published a letter from physicist Leonard Reiffel revealing that, during the Cold War in the 1950s, the United States had funded a research program focused on detonating a nuclear bomb on the Moon, part of a plan to demonstrate the strength of the U.S. military. The Armour Research Foundation—now a part of Illinois Institute of Technology—had directed the project, known as A Study of Lunar Research Flights. Astronomer Carl Sagan, then a young graduate student, had worked on the project. As Reiffel explained, Cold War politics had dramatically affected science agendas during the period: "There was lots of talk on the part of the Air Force about the [M]oon being 'military high ground'." The scenario had called for the U.S. Air Force to launch a small nuclear device from an undisclosed location and detonate it on the Moon. Those planning the detonation had selected an atomic bomb, because a hydrogen explosive would have been too heavy for a rocket to carry 238,000 miles (383,000 kilometers) to the lunar landing spot. Military leaders eventually called off the plan because of concern that, during the building and launch of the bomb, a nuclear accident could occur on Earth.[919]

Astronomers using the telescope at the Arecibo Observatory in Puerto Rico captured the first-ever images of the so-called Metal Dog Bone asteroid, officially named 216 Kleopatra. Scientists had categorized the asteroid as a main-belt asteroid, theorizing that the New Jersey–sized rock was the result of an ancient, violent collision. Steven Ostro of NASA's Jet Propulsion Laboratory called the find "one of the most unusual asteroids we've seen in the [s]olar [s]ystem." Scientists made the discovery by bouncing radar signals off the asteroid and cataloging the signals' echoes. By gathering a comprehensive collection of echoes, the research team had been able to assemble a computer model of the asteroid's shape. The use of radar signals with a moderately powered telescope had made the long-distance discovery possible. Ostro marveled at the technology, describing the find as akin to "using a Los Angeles telescope the size of the human eye's lens to image a car in New York." Many of the astronomers working on the project spoke of their

---

[918] Associated Press, "NASA Launches Hurricane-Tracking Satellite," 3 May 2000; Knight-Ridder Tribune Business News, "Weather Satellite Has Successful Launch, Heads for Orbit," 3 May 2000.
[919] Leonard Reiffel, "Sagan Breached Security by Revealing U.S. Work on a Lunar Bomb Project," *Nature* 405, no. 6782 (4 May 2000): 13–14; Associated Press, "Physicist Says U.S. Contemplated Detonating an Atom Bomb on the Moon," 18 May 2000.

amazement at the size, shape, and metallic complexion of the asteroid. They also clarified that the asteroid posed no danger to Earth.[920]

*5 May*

Laurence Vico released a study on astronauts' loss of bone density because of time spent in space. Vico and his team of researchers had examined the bone-mineral density of 15 cosmonauts who had spent from one to six months aboard the *Mir* space station. They had found a "striking" loss of bone density, in some cases as serious as that experienced by paraplegic patients. The team of doctors suggested that physicians should examine astronauts more closely for susceptibility to bone-weakening conditions such as osteoporosis, before they traveled into space, and that postflight recovery periods should be much longer than previous standards dictated. The sample study found that astronauts experienced an average bone-density loss of more than 5 percent, and that the lost bone did not immediately regenerate after postflight recovery periods. Scientists had long been aware of the physical challenges posed by the lack of gravity in the space environment. Therefore, NASA and the space agencies of other countries had carefully dictated diet and exercise programs for astronauts. The study had special relevance, however, because the international space community was continuing work on the ISS and anticipating even longer stays in space.[921]

*12 May*

The board appointed to investigate the mishap that had damaged the High Energy Solar Spectroscopic Imager (HESSI) spacecraft in preflight testing announced its findings. The investigators had discovered that the computer controlling the vibration device had sensed an abnormally high level of static friction and had responded by delivering approximately 10 times the exertion of gravity suitable for the test. A slight misalignment of the shaker mechanism had proved just enough to throw off the preflight testing. The board had determined that the accident had been avoidable. Routine maintenance and a pretest of the testing device itself would have caught the problem before any damage had occurred. NASA planned for the HESSI satellite to undergo repairs at the University of California at Berkeley.[922]

Cosmonauts Sergei V. Zalyotin and Alexander Y. Kaleri conducted a 5-hour spacewalk to repair tiny cracks weakening the hull of the *Mir* space station. The cracks, the result of the June 1997 collision of a cargo craft with the station, had caused the module to lose air pressure steadily. The spacewalk was not the first attempt to solve the problem, but previous teams had been unable to locate the cracks. Zalyotin and Kaleri tested a new "cosmic version of super glue," designed to seal the fractures even in the harsh environment of space. The Russian Space Agency clarified that the cracks were not crippling to the space station, and that the crew could maintain acceptable levels of air pressure by manipulating oxygen outflow inside *Mir*. However, the agency hoped to solve the problem permanently, because Russia planned to use the aging space station for commercial purposes. The Netherlands-based MirCorp, which had committed nearly

---

[920] NASA, "Astronomers Catch Images of Giant Metal Dog Bone Asteroid," news release 00-74, 4 May 2000.
[921] *Birmingham Post* (AL), "Serious Bone Loss for Cosmonauts Not Mir Speculation," 5 May 2000.
[922] NASA, "Board Finds Cause of HESSI Mishap," news release 00-80, 12 May 2000; Associated Press, "NASA Identifies Causes of Test Accident that Broke Spacecraft," 12 May 2000.

US$20 million to leasing *Mir*, observed the cosmonauts' activities, calling the mission the first privately funded spacewalk in history.[923]

*19 May*
Shuttle *Atlantis* lifted off from KSC in Cape Canaveral, Florida, on Mission STS-101 bound for the ISS. The seven-astronaut crew comprised Americans—James D. Halsell Jr., Scott J. Horowitz, Susan J. Helms, James S. Voss, Mary Ellen Weber, and Jeffrey N. Williams—and Russian cosmonaut Yury V. Usachev. The crew planned to make repairs to the space station, focusing primarily on replacing four malfunctioning solar-powered batteries. The crew also planned to use the thrusters of the Shuttle to realign the ISS's orbit. Having been unoccupied and underpowered for nearly a year, the station had fallen below the optimal orbiting altitude. The station had gradually descended at a rate of about 1.5 miles (2.4 kilometers) per week. The Shuttle mission plan also called for *Atlantis*'s crew to repair a damaged radio antenna and stabilize a shaky construction crane mounted on the outside of the ISS. Because of volatile weather in April 2000 and a full docket of rocket launches, NASA had delayed *Atlantis*'s launch three times, pushing the Shuttle's liftoff into mid-May. Russian cosmonaut Yury V. Usachev, a veteran of two tenures aboard *Mir*, had spent more time in space than the rest of *Atlantis*'s crew combined.[924]

*22 May*
Astronauts Jeffrey N. Williams and James S. Voss conducted a 6-hour spacewalk to make repairs to the ISS. The pair successfully repaired the station's construction crane, replaced a failed antenna, and began installing a much larger, Russian-built crane, designed to aid workers in adding the final modules to the ISS. NASA officials highlighted the exercise as a precursor of the maintenance schedule to come: "This spacewalk was a good example of what future spacewalks are going to look like, where we will need to perform maintenance on the space station." The spacewalk occurred after *Atlantis* had docked at the space station, but while the hatch doors between the two spacecraft remained shut.[925]

*23 May*
Shuttle *Atlantis*, commanded by James D. Halsell Jr., began boosting the ISS into a higher orbit. By firing its steering jets 27 times during a 1-hour push, Shuttle *Atlantis* raised the space station about 9 miles (14 kilometers) farther from Earth than it had been. The effort was the first of three planned realignment maneuvers. NASA officials wanted to allow *Atlantis*'s thrusters to have a day to cool down before continuing with the project. In addition to changing the ISS's orbit, *Atlantis*'s astronauts serviced the exterior of the space station and replaced four of its failing

---

[923] Reuters, "Russia: Russian Cosmonauts Leave Mir and Start Spacewalk," 12 May 2000; Associated Press, "Mir Cosmonauts Experiment with Space Glue To Seal Cracks," 12 May 2000.
[924] Tom Breen, "NASA Goes for 4th Launch Try in a Month," *Florida Today* (Brevard, FL), 19 May 2000; Reuters, "USA: Update 1—Shuttle Atlantis Lifts Off from Florida," 19 May 2000; Associated Press, "Atlantis Lifts Off on Fourth Try," 18 May 2000; Associated Press, "Six Americans, One Russian Bound for Space Station," 19 May 2000.
[925] *Washington Post*, "Spacewalkers Begin Repairs on Orbiting Station; Wobbly Crane Is Secured; Another Awaits Completion," 22 May 2000; Associated Press, "Outside Repairs Finished, Astronauts Ready To Head Inside Space Station," 22 May 2000.

batteries. The crew hoped that the latter procedure would stop the gradual drop in elevation that the ISS had been experiencing during the months before *Atlantis*'s visit.[926]

*28 May*
NASA controllers began sending signals to the Compton Gamma Ray Observatory to perform a series of maneuvers that would ultimately end in the satellite crashing into the Pacific Ocean. The endgame commands marked the conclusion of a successful nine-year mission, which had recorded data on more than 2,500 gamma-ray bursts. The US$670 million satellite, launched on 5 April 1991 aboard Mission STS-37, had provided researchers with never-before-seen images of explosions throughout the universe. "It will be like losing a member of the family," remarked one NASA scientist who had worked on the program since 1979. NASA officials had decided to end the satellite's space tenure because of the possibility that the aging satellite might become unresponsive and plummet out of control to Earth. By initiating the satellite's demise, NASA could control the landing area of the 17-ton (15,400-kilogram or 15.4-tonne) spacecraft. Some critics charged NASA with being too conservative, because it had based its decision on the 1-in-1,000 odds that an uncontrolled descent of the satellite to Earth would result in injuries.[927]

*29 May*
*Atlantis* touched down at KSC in Cape Canaveral, Florida, only the 14th time that a NASA Shuttle had landed in darkness. NASA controllers used infrared cameras to monitor the Shuttle's final descent. The return of the Shuttle ended the crucial mission to repair the ISS and reset the space station's orbit to an optimal altitude. NASA's Mission Control, congratulating the crew on a "super mission," began almost immediately to anticipate the flurry of missions that would finally complete construction on the ISS. *Atlantis* had sustained minor damage during the trip, mainly dents and scratches on its wings caused by ice. Three members of the returning crew, astronauts Susan J. Helms and James S. Voss and cosmonaut Yury V. Usachev, had a special interest in monitoring the ISS's completion, because they had been selected for the second research crew to the ISS and would make a long-term stay aboard the research station.[928]

## JUNE 2000

*2 June*
NASA announced a partnership with Dreamtime Holdings Inc., a start-up company backed by Lockheed Martin to provide the International Space Station (ISS) with high-definition television and to create digital archives for NASA. The agreement was NASA's first involving a commercial partner that would provide support for the ISS. Administrator Daniel S. Goldin announced the partnership at NASA's Ames Research Center, located in the high-technology hub of California's Silicon Valley, predicting that the partnership would move NASA to the forefront of the information age and terming it "innovative government at its best." Dreamtime planned to spend US$100 million on the project, which would give civilians access to never-

---

[926] *Los Angeles Times*, "Shuttle Gives Space Station a Boost," 24 May 2000; Associated Press, "Atlantis Lifts Space Station to Right Orbit," 24 May 2000; Associated Press, "Space Station Gets Fresh Batteries," 24 May 2000.
[927] James Orberg, "Plan To Scrap Satellite Sacrifices Science," *USA Today*, 24 May 2000; Associated Press, "NASA To Destroy Compton Satellite After Nine Years," 28 May 2000.
[928] William Harwood, "Shuttle Ends Repair Trip to Station," *Washington Post*, 30 May 2000; *USA Today*, "Shuttle Crew Returns from Space Station Repair Mission," 30 May 2000; Associated Press, "Nighttime Landing Ends Shuttle Repair Mission," 30 May 2000.

before-seen images of the Space Shuttles and of NASA missions. The pact resulted from the Commercial Space Act of 1998, enacted to increase commercial involvement in and funding for the ISS.[929]

*4 June*
NASA de-orbited the Compton Gamma Ray Observatory, after nine years in orbit, in a controlled reentry into Earth's atmosphere. Director of NASA's Goddard Space Flight Center Alphonse V. Diaz called the event "a bittersweet day for NASA." The spacecraft, NASA's "gamma-ray equivalent" of the Hubble Space Telescope (HST), launched in 1991 with four instruments designed to observe gamma rays. During the course of its impressive tenure, the Compton Gamma Ray Observatory had yielded valuable data regarding the previously little-known gamma-ray sky. The 17-ton (15,400-kilogram or 15.4-tonne), US$670 million research craft had orbited Earth for long after its predicted lifespan. Some critics had opposed bringing the craft crashing to Earth while it was still providing useful data. In 1999, however, NASA engineers had detected problems with one of the satellite's control gyroscopes. After considering sending maintenance missions to repair the craft or training the Shuttle crew to capture the Compton Gamma Ray Observatory and bring it back to Earth for service, NASA officials had determined that the prudent response was controlled de-orbit and preemptive destruction of the satellite. NASA engineers used a series of engine burns to guide the satellite in its reentry into Earth's atmosphere and to direct the pieces of debris from the craft to crash into the Pacific Ocean. Like NASA's efforts to control the reentry of Skylab in 1979, this maneuver was one of the first times that NASA had intentionally destroyed one of its own craft by guiding its reentry into Earth's atmosphere.[930]

*5 June*
At the American Astronomical Society convention, scientists Karl Gebhardt and Douglas O. Richstone announced that an extensive survey of more than 30 galaxies had revealed that black holes reach their large size through gradual growth. Previously, some research had suggested the opposite—that black holes were simply "born" big. Using the HST, Gebhardt and Richstone had led a team of astronomers in determining the size of black holes in different galaxies. They found that, although small galaxies tend to have smaller black holes, the largest known galaxies have mammoth black holes. The astronomers suspected that the black holes had "grown up" along with the galaxies where they resided, feeding on gas and stars to acquire increasing mass. Although further research must determine the precise correlation between black hole and galaxy size, the scientific community greeted news of the discovery with great interest. The HST had not only made the discovery possible but had provided the opportunity for scientists to continue "black hole hunting," to tabulate more exact answers.[931]

---

[929] NASA, "NASA, Dreamtime Partnership Propels Space Information Age to New Heights," news release 00-87, 2 June 2000; Frank Morring Jr., "Startup To Spend $100 Million Digitizing Station, NASA Archives," *Aerospace Daily*, 2 June 2000.
[930] NASA, "Status Report # 2: Second Deorbit Burn for Observatory Successful; Time Change for Last Two Burns," news release, 1 July 2000; NASA, "Final Status Report: Compton Gamma Ray Observatory Safely Returns to Earth," news release, 4 June 2000; Michael E. Ruane, "A Satellite's Bittersweet Splashdown; Scientists Nostalgic, Angry About Demise of Md.-Based Project," *Washington Post*, 2 June 2000.
[931] NASA, "Black Holes Shed Light on Galaxy Formation," news release 00-88, 5 June 2000.

Scientists at the American Astronomical Society convention also unveiled the first detailed map of the region of the universe inhabited by Earth. Using an Australian robotic telescope, which measured distances between more than 100,000 galaxies, researchers had charted the locations of galaxies, dark regions, and cosmic clusters. Perhaps most significantly, the new map supported a long-standing theory regarding the size limits of cosmic structures. Because of the relatively small scope of previous mapping surveys, scientists had questioned whether the known standards of cosmic size were representative. The newest map, however, did not reveal any cosmic structure outside of previous standards. Therefore, scientists could support with greater conviction the idea that size limits existed as the universe evolved. The researchers termed the limits the "end of greatness."[932]

*6 June*
After two earlier unsuccessful trials, Russia launched, without incident, a modified Proton rocket carrying a large Gorizont communications satellite. Propulsion-system flaws had triggered explosions of the Proton rocket in previous trials. The United States viewed the successful test launch as a positive sign of Russia's commitment to the ISS project, a boost for the strained relationship between the two countries. After the failed tests, a joint NASA-Russian Space Agency research crew had determined that production flaws in the turbomachinery of the rocket's second and third stages, including the presence of contaminants such as asbestos cloth and metal fragments, had caused the problems. Russia planned to test the Proton rocket again before scheduling the much-delayed delivery of its crew module to the ISS.[933]

NASA announced that its Chandra X-ray Observatory had revealed a "luminous spike of x-rays," known as a hot spot, located approximately 800,000 light-years away from its black-hole source. Scientists commented that the images captured x-ray behavior that astronomers had not predicted. Andrew S. Wilson of the University of Maryland explained, "the brightness and the spectrum of the x-rays are very different from what theory projects." A possible explanation offered for the brilliant hot spot of x-rays was that a series of shock waves had catapulted across the galaxy electrons with energies as high as 50 thousand billion times the energy of light.[934]

*13 June*
Having restored *Liberty Bell 7*, which had spent 38 years at the bottom of the Atlantic Ocean, NASA placed the space capsule on display at Kennedy Space Center (KSC). *Liberty Bell 7* had carried astronaut Virgil I. "Gus" Grissom into space on 21 July 1961, in NASA's second piloted spaceflight. After its pathbreaking, 15-minute, suborbital flight, the capsule had descended to the ocean. The hatch of *Liberty Bell 7* had blown off just after landing, causing the capsule to fill with water and almost drowning Grissom. A helicopter present at the landing had tried to pull the *Liberty Bell 7* out of the water, but the waterlogged capsule had become too heavy, and the crew had to cut it loose, leaving it to sink. Regarding the hatch's failure, astronaut Grissom had maintained until his death in the 1967 *Apollo* launchpad fire, that he had followed proper protocol. Nearly 40 years after the ocean landing, the Discovery Channel had funded a recovery mission that had brought the historic spacecraft out of the ocean. Although divers had not found the hatch, previous NASA investigations had supported Grissom's claim that human error had

---

[932] James Glanz, "Robotic Telescope Affirms Assumption on Universe's Birth," *New York Times*, 7 June 2000.
[933] Mark Carreau, "Rocket Gives Space Station a Life," *Houston Chronicle*, 7 June 2000.
[934] NASA, "Spectacular X-ray Jet Points Toward Cosmic Energy Booster," news release 00-89, 6 June 2000.

not caused the hatch failure. NASA had restored the capsule, placing it in a clear plastic display case so that visitors to KSC could peer inside the spacecraft.[935]

*16 June*

After spending more than two months aboard the *Mir* space station, Russian cosmonauts returned to Earth. According to MirCorp, the mission had been the first privately funded crewed mission to *Mir*. Sergei V. Zalyotin and Alexander Y. Kaleri landed safely in Kazakhstan, having spent most of their time aboard the Russian spacecraft conducting crucial repairs. To keep the station running, MirCorp had signed a lease agreement for *Mir* and paid for the mission. However, with the return of Zalyotin and Kaleri, Russia placed on hold the prospect of continuing *Mir*, even if supported with private funds. "If there is no money, *Mir* will not fly," Russian Space Agency Director General Yuri N. Koptev stated at the end of the mission. Russian officials indicated that the Russian Space Agency would leave the space station unpiloted once again, until private investors could fund another mission. The continuing focus on *Mir*, while the building of Russia's ISS module ran far behind schedule, agitated some NASA officials and international space leaders.[936]

NASA and the Boeing Company announced an agreement to use Boeing's Delta Launch Services Inc. for a variety of NASA missions. The contract, estimated to be worth at least US$168 million, included options to extend the agreement. NASA also awarded both Boeing and Lockheed Martin Commercial Launch Services Inc. indefinite delivery–indefinite quantity contracts for 10 years. The agreements solidified NASA's partnerships with the two major aerospace companies. A NASA spokesperson clarified the reasons for the extension of the relationships, saying that Boeing and Lockheed "were selected based on their ability to meet NASA's highly critical future mission requirements and their proven track records for providing a quality product." At the time of the agreement with Boeing, the company had flown 82 missions for NASA with a success rate of 98 percent.[937]

*19 June*

NASA celebrated the 20th anniversary of its Summer High School Apprenticeship Research Program (SHARP), created to encourage promising high school students, especially minorities, to pursue careers in math, science, and engineering. During the first 20 years of the program, nearly 3,000 students participated, and 3,300 NASA employees served as SHARP mentors. Most SHARP student participants had served as interns at one of NASA's research centers during the summer months. NASA had begun the program and had continued to fund it as part of its educational mission.[938]

*21 June*

---

[935] Associated Press, "Grissom's Capsule Back at Space Center," 14 June 2000.
[936] Associated Press, "Mir Cosmonauts Return to Earth," 15 June 2000; Associated Press, "Mir Cosmonauts Prepare To Return to Earth," 15 June 2000; *CNN.com*, "First Privately Funded Manned Space Mission Blasts Off for Mir, 4 April 2000, http://archives.cnn.com/2000/TECH/space/04/04/russia.mir.01 (accessed 15 July 2008).
[937] The Boeing Company, "Boeing Delta Rockets To Launch Next-Generation NASA Space Craft," news release, 16 June 2000; NASA, "NASA Awards Launch Services Contracts," news release c00-e, 16 June 2000; *Seattle Post-Intelligencer* (WA), "Boeing Wins $168 Million NASA Order for 3 More Delta II Launches, Options," 19 June 2000.
[938] NASA, "NASA's SHARP Program Celebrates 20 Years of Excellence," news release 00-96, 19 June 2000.

NASA reported to the public that Mars Global Surveyor had captured images depicting erosion and soil deposits consistent with the presence of flowing water on the planet in the past. Researchers had long postulated that, billions of years ago, Mars's surface held abundant water, which had dried up as the planet's atmosphere thinned. Because scientists had suggested that the one-time presence of water on Mars might indicate the one-time presence of life on the planet, the new evidence created significant excitement in the scientific community. In announcing the find, NASA reported that it "could turn out to be a landmark discovery," but cautioned against drawing conclusions from the images without further analysis. Furthermore, the images suggested that water might have been present on the planet more recently than previously thought. Associate Administrator for Space Science Edward J. Weiler explained, "For two decades scientists have debated whether liquid water might have existed on the surface of Mars just a few billion years ago. With today's discovery, we're no longer talking about a distant time. The debate has moved to present-day Mars."[939]

*22 June*
After a lengthy legislative battle, the U.S. House of Representatives approved a US$60 million increase for NASA in its FY 2001 budget. Although the amount was less than the Clinton administration had requested for NASA, the spending proposal marked the end of several years of cost cutting for NASA. Even during a period of budget surpluses, NASA had not achieved the victory without a fight. Some members of Congress, concerned over the delay in the completion of the ISS, had introduced a floor amendment to delete all funding for the ISS. However, the House had roundly defeated the budget-slashing amendment, as it had in previous years, with a vote of 325 to 98. Noting that the ISS was nearly complete, Representative Robert E. Cramer (D-AL) spoke for NASA proponents when he said, "This is not the time to pull the rug out from under this program."[940]

*30 June*
NASA launched the most advanced communications satellite it had ever developed, the Tracking and Data Relay Satellite-H (TDRS-H) created to provide communication between the Space Shuttles, ISS, HST, and NASA control centers. Intended to replace a communications satellite that had been in place since 1983 and had surpassed its mission-designed lifetime, the new TDRS-H was the first of three communications satellites planned to link NASA's space-bound vehicles. NASA had designated nearly US$500 million for the design, building, and launch of the three new satellites. The TDRS-H launched aboard a Lockheed Martin Atlas IIA rocket, separating from it after about 30 minutes of flight. Once the satellite reached 171° longitude, it became operational.[941]

---

[939] Kathy Sawyer, "Mars Craft Finds Evidence of Water," *Washington Post*, 22 June 2000; William F. Nicholson, "Reports: NASA Finds Evidence of Water on Mars," *USA Today*, 21 June 2000; NASA, "New Images Suggest Present-Day Sources of Liquid Water on Mars," news release 00-99, 22 June 2000.
[940] Tamara Lytle, "House OKs Giving NASA $13.7 Billion in 2001 Fiscal Year," *Orlando Sentinel* (FL), 22 June 2000; Gannett News Service, "House Votes Down Measure To Cut Space Station," 22 June 2000.
[941] Lockheed Martin Inc., "International Launch Services and Lockheed Martin Provide Mission Success for NASA Satellite Launch," news release, 30 June 2000; NASA, "Advanced Communications Satellite Ready To Serve New Millennium Space Projects," news release 00-98, 21 June 2000.

## JULY 2000

*4 July*

Russia launched a new defense and spy satellite aboard a Proton-K rocket from Baikonur Cosmodrome. More concerned with the launch vehicle than with the military satellite, U.S. space officials praised the successful operation of the Proton-K rocket, a rocket similar in type to those that had been involved in several launch accidents during 1999. The United States had a particular interest in the launch, because Russia planned to carry a module built for the International Space Station (ISS) aboard a Proton rocket. Russia had built the *Zvezda* module to provide living space for the ISS crew. A spokesperson for the Russian Space Agency stated that the rocket's successful launch confirmed Russia's ability to transport its *Zvezda* module: "This launch is important in estimating the readiness of the Proton-K booster rocket for a more important mission July 12 when the same kind of booster will carry the Russian-built *Zvezda* service module to the ISS." The Russian Space Agency had overcome significant financial shortfalls to complete production on both the Proton-K rocket and the *Zvezda* module.[942]

*7 July*

Engineers completed a weeklong series of tests on the U.S.-built *Destiny* laboratory, planned to launch to the ISS in 2001, ensuring that the module had no leaks and would provide a safe, vacuum-sealed environment in which the ISS astronauts could conduct research. NASA and the Boeing Company had worked together to test the 32,000-pound (14,500-kilogram) research laboratory. Placing *Destiny* in a pressurized chamber, engineers had monitored gas levels inside the module over the course of the week. Tip Talone, Director of the ISS Payload Process at Kennedy Space Center (KSC), characterized the results as "a large step in meeting the lab's 'Destiny'," stating that the laboratory had "exceeded expectations."[943]

*12 July*

After nearly two years of delays and many questions regarding the Russian Space Agency's commitment to the ISS, Russia successfully launched the *Zvezda* module aboard a Proton rocket. Building the 22-ton (20,000-kilogram or 20-tonne), US$320 million module, to provide living quarters for researchers aboard the ISS, had severely taxed the Russian space program. To bring the project to fruition, Russia had relied on the resourcefulness of its scientists and engineers, as well as resorting to creative financing arrangements, such as selling the U.S.-based Pizza Hut Company an advertisement for display on the side of the Proton launch rocket. Russia had insured *Zvezda*'s launch for more than US$1 billion. Hundreds of spectators, representing aerospace companies from Russia and Western countries, witnessed the launch from Baikonur Cosmodrome in Kazakhstan. Watching the launch with Director General of the Russian Space Agency Yuri N. Koptev, NASA Administrator Daniel S. Goldin praised the successful completion of the long-awaited module: "The Russians have gone through all sorts of difficulty with their economy, the political changes, and a whole variety of other problems, and they came through and did what they said they were going to do." Koptev made clear the high stakes for future Russian space projects, based on the successful launch and linkup of *Zvezda*: "This is 10 years of work and the success of this launch will determine to a large extent whether the Russian space programme continues or not." The international space community greeted news of the

---
[942] Reuters, "Russia: Russia Launches Proton Rocket Ahead of Space Lab," 5 July 2000.
[943] Dow Jones Newswire, "Space State Laboratory Exceeds Expectations in Vacuum Chamber Test," 7 July 2000.

launch as a sign that the much-postponed ISS was nearing completion. Although Russia had plans to contribute further portals to the ISS, none of its subsequent contributions had as much potential to delay the international effort.[944]

*13 July*

Astronaut Eileen M. Collins, speaking for NASA, publicly responded to the report of the Commission for the Advancement of Women and Minorities in Science, Engineering, and Technology Development. The Commission had reported that a severe shortage of high-tech workers threatened economic growth in the United States, highlighting the need to increase funding from both private and governmental sources, to improve education in science and engineering. The report had emphasized that improved education in these fields was necessary if the United States hoped to continue training engineers and researchers qualified to bring NASA's exploratory mission to fruition. Citing the example of her own educational and career path, Collins called for further support for research and education in mathematics, science, engineering, and technology. Both the Commission's report and Collins's response continued an ongoing debate over how to change the trend among students to choose paths of study outside of the fields of science and technology. In 1999 NASA Administrator Daniel S. Goldin had also testified before the U.S. House Committee on Science and Technology regarding the challenge of drawing young people to careers as scientists and engineers.[945]

*14 July*

NASA released to the Internet a database of 1.9 million celestial images, the largest collection of images of stars and other celestial bodies ever made freely available for public use. The rapid improvement of technology had made possible the release, which brought to home computers information that could have filled 6,000 CD-ROMs. The Two-Micron All Sky Survey (2MASS), which used two 51-inch (130-centimeter) telescopes to survey the sky, had collected the images. The University of Massachusetts at Amherst and the Infrared Processing and Analysis Center of NASA's Jet Propulsion Laboratory had collaborated on 2MASS, creating a database of great value to scholars and the public alike.[946]

William G. Fastie, known for helping establish the prestigious Johns Hopkins University space program and for inventing the spectrometer, died in Baltimore at the age of 83. Fastie, often called the father of the Hopkins space program, was one of the United States' preeminent astrophysicists. His spectrometer, which measured the spectrum of light, helping scientists gather data about other planets, was not only innovative but also rugged enough to travel into space.[947]

*15 July*

---

[944] Associated Press, "Russian Launching Clears Way for Space Station," 13 July 2000; *USA Today*, "Russia Sends Module of Space Station into Orbit," 12 July 2000; *Newsday* (Long Island, NY), "Launch of Key Module for Space Station Is Set," 11 July 2000; Reuters, "Russia Space Station Launch Insured for over $1 Bln," 12 July 2000; Reuters, "Kazakhstan: Update 2-Russia Launches Key Module for ISS," 12 July 2000.
[945] NASA, "NASA Supports High-Tech Workforce Reflecting Diversity of America," news release 00-106, 13 July 2000.
[946] NASA, "Twin-Telescope Sky Survey 'Gives You the Stars'," news release 00-108, 14 July 2000.
[947] Alice Lukens, "William G. Fastie, 83, Hopkins Astrophysicist Designed Spectrometer," *Baltimore Sun* (MD), 17 July 2000.

As Russia's *Zvezda* module, its major contribution to the ISS, continued its journey toward the ISS, NASA celebrated the 25th anniversary of one of the first international, collaborative missions. The Apollo-Soyuz Test Project had launched on 15 July 1975, when American astronauts Thomas P. Stafford, Vance D. Brand, and Donald K. Slayton lifted off aboard an *Apollo* spacecraft atop a Saturn 1B rocket. During the course of their nine-day mission, the NASA crew had docked the United States' *Apollo* with Russia's *Soyuz 19* spacecraft, successfully and without incident. The Cold War–era mission, testing the ability of NASA spacecraft to link with Russian ones, had represented a significant step toward the development of international collaboration in future space exploration[948]

The National Aviation Hall of Fame inducted four new members at a black-tie dinner and formal enshrinement ceremony at the Dayton Convention Center in Dayton, Ohio. *Apollo* astronaut Edwin E. "Buzz" Aldrin Jr. joined fellow *Apollo 11* astronaut Neil A. Armstrong in the Hall of Fame. Aviators previously inducted to the Hall of Fame included Eugene A. Cernan, the last *Apollo* astronaut to walk on the Moon; Laurence C. "Bill" Craigie, the first U.S. military aviator to fly a jet aircraft; and Thomas B. McGuire Jr., the second-leading fighter ace in U.S. Air Force history. The National Aviation Hall of Fame's Class of 2000 joined 166 award-winning aviators already chosen for the honor.[949]

*16 July*
The first of two Cluster spacecraft launches took place from Baikonur Cosmodrome in Kazakhstan. NASA and the European Space Agency had formed a partnership to develop and build the exploration craft, designed to travel around the Earth in a tetrahedral formation, collecting data on solar wind. Space officials planned to launch two additional Cluster craft one month after the first two had entered their orbiting pattern. With four satellites launched and flying in formation, scientists expected to harvest data about the "turbulent battle" raging between Earth's magnetic field and the solar winds blowing at an estimated rate of 1–2 million miles (1.6–3.2 million kilometers) per hour. Researchers also hoped the new information would help scientists understand other interactions between the Earth and the Sun. Each of the four Cluster spacecraft carried instruments to measure the patterns of electrons and the presence of protons and helium, as well as instruments to monitor the characteristics of electronic fields, solar wind, and plasma waves.[950]

*18 July*
NASA announced that SPX Services Solutions had obtained a license to use NASA-patented technology to monitor more carefully levels of exhaust produced by the United States' millions of motor vehicles. NASA had developed its atmospheric, remote-sensing technology to track greenhouse gases and to monitor Earth's ozone layer. Combining government and private forces, researchers intended to focus on the pollution output of motor vehicles, particularly that of individual automobiles. SPX Services planned to design a device that would enable a driver to test his or her own vehicle to determine its compliance with U.S. Clean Air Act standards, rather

---

[948] NASA, "Anniversary of Apollo Soyuz Test Project Observed," news release, N00-30, 11 July 2000; Reuters, "US, Russia Mark Space Jubilee as New Era Beckons," 11 July 2000.
[949] *Dayton Daily News* (OH), "Astronaut Meets a Fan; Aviation Hall of Fame Enshrines 4 More," 16 July 2000.
[950] NASA, "New Cluster Mission To Provide Unprecedented Detail about Space Weather," news release 00-105, 13 July 2000.

than requiring the driver to take it to an auto mechanic for an emissions test. The goal of self-monitoring suited the increasingly popular idea of encouraging individuals to accept personal responsibility for protecting Earth from noxious pollutants.[951]

*20 July*

The 20 July 2000 edition of *Astrophysical Journal Letters* published research of Gibor Basri chronicling the first sighting of a flare from a failed star, or brown dwarf. Using the Chandra X-ray Observatory, a team of scientists from the University of California at Berkeley had spotted the brown dwarf and its bright x-ray flare. Scientists believed that examining the x-ray flare would help them better understand dying stars and the explosive activity and magnetic fields of low-mass stars. Early analyses indicated that the brown dwarf's x-ray flare was approximately 1 billion times more powerful than the x-ray flares emitted from the planet Jupiter. Scientists reacted to news of the discovery with excitement and surprise. Robert E. Rutledge of California Institute of Technology stated that he and his colleagues were "shocked," clarifying: "we did not expect to see flaring from such a lightweight object. This is really the mouse that roared." The task of spotting an x-ray flare had been difficult, because the brown dwarf emitted the flares only periodically. However, the intermittent flares made the successful capture of an image all the more significant. The observation of the brown dwarf's occasional release of x-ray flares confirmed the theory that brown dwarfs only release energy when they heat to temperatures above 4,500°F (2,500°C).[952]

*24 July*

Cape Canaveral, Florida, celebrated 50 years as a launch site. Long before NASA had begun to use the site, the U. S. Army had launched test rockets from the swampy land. The "li'l Bumper" was among the first, according to Dick Jones, a retired U.S. Army master sergeant who had helped with the 1950 launches. During its first 50 years, the Army had launched about 3,200 rockets and missiles from the site, which would eventually become KSC.[953]

*26 July*

The *Zvezda* service module successfully docked at the ISS, immediately increasing the size of the space station by 50 percent. Through a series of rocket firings, *Zvezda* had drawn steadily closer to the ISS during its two-week journey to its new home 220 miles (354 kilometers) from Earth. Experts had deemed the successful launch and docking of the Russian component of the ISS vital to the long-term success of the collaborative venture. The Russian-built *Zvezda*, nearly identical to a core section of the Russian *Mir* space station, was to provide living quarters for researchers on the station, as well as vital electronic, computer, and communications systems. Anticipating significant interest in the historic linkup, NASA provided information on its Web site explaining how to view the space station as it traveled over Earth.[954]

---

[951] NASA, "NASA Satellite Technology To Monitor Motor Vehicle Pollution," news release 18 July 2000.
[952] Gibor Basri et al., "An Effective Temperature Scale for Late M and L Dwarfs, from Resonance Absorption Lines of Cs I and Rb I," *Astrophysical Journal Letters* 538, no. 1 (20 July 2000): 363–385; NASA, "Chandra Captures Flare from Brown Dwarf," news release 00-103, 11 July 2000.
[953] Eliot Kleinberg, "50 Years at the Cape: Opening the Door to Space," *Palm Beach Post* (FL), 24 July 2000; Associated Press, "Cape Canaveral: A Half-Century of Launches Begin with Bumper," 19 July 2000.
[954] Warren E. Leary, "Space Station Is Connected to Module from Russia," *New York Times*, 26 July 2000; NASA, "No Telescopes Needed: NASA Web Sites Let Stargazers Track Impending Space Station 'Nuptuals'," news release

*28 July*
Nearly one year after the highly publicized failure of a probe to reach the surface of Mars, NASA announced plans for at least one, possibly two, robotic rovers to explore Mars by 2003. The new design plan, the outcome of an overhaul of NASA's Mars Program in the aftermath of the unsuccessful mission, called for a rover larger than Mars Sojourner, which had successfully navigated the Martian surface in 1997. NASA planned for the new probe to continue the search for water on the surface of the Red Planet.[955]

## AUGUST 2000

*2 August*
Mitsuyuki Ueda of the Aeronautics and Space Development Division of Japan's Science and Technology Agency announced that Japan had decided to freeze the development of an unpiloted space shuttle, which was four years behind schedule, because of problems in Japan's shuttle program, as well as in its H-2 rocket program. Japan had originally conceived its planned 20-ton (18,000-kilogram or 18.1-tonne) shuttle in the 1980s, modeling it after the U.S. Space Shuttle, and designing it to conduct scientific experiments and to carry into space payloads of up to 3 tons (2,700 kilograms or 2.7 tonnes). Technical and financial setbacks within Japan's shuttle program contributed to the freeze. In addition, because the original plan had called for Japan to launch the space shuttle aboard a Japanese-designed H-2 rocket, a launch failure of an H-2 rocket had influenced the decision to halt Japan's shuttle program. In the failed launch of November 1999, officials had to destroy the rocket in midair, leading an advisory panel to suggest that Japan's planned shuttle should launch from a reusable high-speed jet plane, rather than atop a rocket.[956]

Space Media Inc. and S. P. Korolev Rocket and Space Corporation Energia (RSC Energia), prime contractor of the Russian service module of the International Space Station (ISS), announced a new multimedia partnership called Enermedia LLC. Under the new partnership, Space Media Inc. would use the Russian Space Program archives to develop and provide multimedia, which the Russian service module would broadcast on television and the Internet.[957]

*4 August*
New Skies Satellites N.V. selected Lockheed Martin Commercial Space Systems (LMCSS) to build a second geosynchronous satellite to provide high-speed Internet access and other multimedia communications to a large coverage area, extending from the eastern Mediterranean and southern Africa to Australia, Japan, and Korea. New Skies selected the LMCSS A2100

---

00-114, 20 July 2000; NASA, "First Opportunity Tonight for *Zvezda* To Dock with International Space Station," news release N00-36, 25 July 2000.
[955] Kathy Sawyer, "NASA Moves To Send Robot Rovers To Explore Mars Surface," *Washington Post*, 28 July 2000; Peter N. Spotts, "New Rover To Look for Signs of Water on Mars," *Christian Science Monitor*, 31 July 2000; Tribune News Services, "NASA Unveils Plans for 2003 Mission to Mars," 28 July 2000.
[956] Associated Press, "Japan Freezes Plans To Make Unmanned Space Shuttle," 2 August 2000.
[957] SPACEHAB, "Space Media Inc and RSC Energia Join Forces Forming New Multimedia Partnership," news release, 2 August 2000, http://www.spacehab.com/news/2000/00_08_02.htm (accessed 8 May 2008).

satellite because it was able to meet challenging demands for bandwidth, power, and in-orbit flexibility.[958]

*6 August*
A Russian Progress cargo spacecraft launched from Baikonur Cosmodrome in Kazakhstan, carrying fuel to the uninhabited ISS for use in adjusting the station's orbit. The craft also carried scientific instruments, linens, and personal-hygiene supplies in preparation for the October arrival of the station's first permanent crew.[959]

*7 August*
NASA announced that the Hubble Space Telescope (HST) had located a "small armada of 'mini-comets'," the fragmented remains of the nucleus of the comet LINEAR. When it had disappeared behind the Sun on 27 July 2000, astronomers initially thought that LINEAR had disintegrated entirely. Upon losing sight of the comet's core, ground-based observers had suggested that the nucleus had disintegrated into a "pile of dust," and, consequently, astronomers at the Space Telescope Science Institute had reprogrammed the HST to search for the nucleus. The exceptional resolution and sensitivity of the HST had revealed the nuclei of a half-dozen mini-comets at a level of detail never before observed in a disintegrating comet. Some astronomers had suggested that the fragments the HST revealed were the "primordial building blocks of the original nucleus." If so, studying them could help scientists understand how the comet had originally formed.[960]

*8 August*
NASA Administrator Daniel S. Goldin began an official visit to Morocco to discuss space cooperation. He planned to meet with several Moroccan officials, particularly in the fields of scientific research and advanced technology, and to visit facilities in Morocco capable of launching Space Shuttles.[961]

*9 August*
A Starsem Soyuz-Fregat rocket launched from Baikonur Cosmodrome in Kazakhstan, carrying into orbit for the European Space Agency (ESA) the final pair of Cluster II scientific satellites, *Rumba* and *Tango*. The launch was the 10th consecutive success for the French-Russian Starsem consortium. The first pair of Cluster II satellites, named *Salsa* and *Samba*, had launched on 16 July 2000. Designed to determine the physical process of the interaction between the solar wind and Earth's magnetosphere, the mission's main objective was to increase understanding of space weather, viewed as "an increasingly significant obstacle to satellite activity," thereby improving scientists' forecasting abilities. Each of the four spacecraft contained a collection of instruments for detecting plasma fields, waves, and particles. The Cluster II mission, designed to last for two years following an initial three-month period of instrument and system commissioning, was an

---

[958] Lockheed Martin, "Lockheed Martin Awarded Contract To Build First Interactive Broadband Multimedia Spacecraft for New Skies Satellites," news release, 4 August 2000, *http://www.lockheedmartin.com/news/press_releases/2000/LockheedMartinAwardedContractBuildF.html* (accessed 6 May 2008).
[959] *Orlando Sentinel* (FL), "Ship with Supplies for Space Station Launches," 7 August 2000.
[960] NASA, "Hubble Discovers Missing Pieces of Comet Linear," news release 00-122, 7 August 2000, *http://www.nasa.gov/home/hqnews/2000/00-122.txt* (accessed 6 May 2008).
[961] BBC, "Morocco: NASA Official Visits for Talks on Space Cooperation," *RTM TV* (Rabat, Morocco), 9 August 2000.

international effort of more than 200 scientists from Canada, China, the Czech Republic, ESA member states, Hungary, India, Israel, Japan, Russia, and the United States. The original Cluster mission had ended during its inaugural launch in 1996, when a malfunction had caused the explosion of Arianespace Inc.'s Ariane 5 rocket.[962]

*10 August*
NASA's Associate Administrator for Space Science Edward J. Weiler announced that NASA had decided to send two large scientific rovers to Mars in 2003, rather than a single craft. The new plan called for the two craft to launch within weeks of each other, reaching Mars in January 2004 after their respective seven and one-half-month-long journeys. The two 300-pound (136-kilogram) rovers, exact duplicates of each other and similar to the highly successful Sojourner rover of 1997, would head to different locations on Mars. Mars Program Director G. Scott Hubbard explained that NASA had undertaken an extensive study of the two-rover option, weighing the excellent launch opportunity in 2003 against resource requirements and schedule constraints. The study teams concluded that it would be possible to successfully develop and launch identical packages, and that the new plan would not only double NASA's scientific return but also add resiliency and robustness to the Mars exploration program. Although NASA had yet to select the two landing sites, Mars Program Scientist James B. Garvin suggested that possible locations included those with evidence of the existence of water in the past. Steven W. Squyres of Cornell University, Principal Investigator for the rovers' Athena science package, explained that the goal of each rover would be to learn about ancient water and climate conditions on Mars. Each craft would operate as a robotic field geologist, reading the geological record at its landing site to discover what conditions had formed the local rocks and soils.[963]

*14 August*
The Southern African Regional Science Initiative (SAFARI 2000) officially commenced in South Africa's Northern Province, with the gathering of scientists from 14 nations and six South African universities. NASA, alongside the University of Witwatersrand, led SAFARI 2000, their goal to "determine how the region's natural ecosystems and human land use affect air quality and atmospheric conditions." The initiative would use specially equipped planes to make four scientific flights each week, over a six-week period, through September 2000. A modern ER-2 version of the U-2 spy plane would fly directly under the *Terra* satellite as it passed over various regions of southern Africa, to verify the satellite's data. Three other aircraft would also fly with the ER-2 "like a stack of pancakes," with the ER-2 flying at about 65,000 feet (19,800 meters) to conduct remote sensing, a Convair 580 aircraft collecting air and gas samples just below the ER-2, and two Aerocommander 690A aircraft conducting similar tests at between 5,000 and 12,000 feet (1,500 and 3,700 meters).[964]

*16 August*

---

[962] European Space Agency, "Lift Off for Second Pair of Cluster II Spacecraft," news release 52-2000, *http://www.esa.int/esaCP/Pr_52_2000_p_EN.html* (accessed 8 May 2008); *Space Business News*, "Soyuz Launches Second Pair of Cluster Probes Igniting Science Mission; Tenth Successful Launch in 18 Months," 16 August 2000.
[963] NASA, "NASA Plans To Send Rover Twins to Mars in 2003," news release 00-124, 10 August 2000, *http://www.nasa.gov/home/hqnews/2000/00-124.txt* (accessed 7 May 2008).
[964] Justin Arenstein for Africa News Service, "Africa's Biggest Science Experiment Kicks Off," 14 August 2000; Africa News Service, "NASA Project Off to a Flying Start," 14 August 2000.

Emergency rescue personnel from Canada, Russia, and the United States participated in training exercises near St. Petersburg in Russia, where they practiced maneuvers for rescuing astronauts returning to Earth from the ISS. The rescuers practiced searching in water and on land for a module carrying three astronauts, as well as practicing first-aid techniques.[965]

NASA announced that its Submillimeter Wave Astronomy Satellite (SWAS) had detected water vapor throughout interstellar space. However, in the very coldest areas, where temperatures are only 30° above absolute zero, the satellite had detected far less water vapor than most theories had predicted. In those areas, SWAS measurements had indicated water vapor concentrations of only a few parts per billion. Ronald L. Snell of the University of Massachusetts at Amherst remarked that the finding presented a "real puzzle to our understanding of the chemistry of interstellar clouds." In warmer regions, such as those within star-producing gas clouds, SWAS had measured water concentrations as much as 10,000 times greater than in the coldest regions. The new results were the product of 18 months of observations using the compact radio observatory, launched in 1998 on a mission to "study the composition of interstellar gas clouds and their collapse to form new stars."[966]

*17 August*
Robert R. Gilruth, "an aerospace scientist, engineer, and a pioneer of the American space program during the glory days of Mercury, Gemini, and Apollo," died at the age of 86. Gilruth had specialized in flight research, and in 1945 he had organized an engineering team to study experimental, rocket-powered aircraft, leading to the establishment of the Pilotless Aircraft Research Division and the creation of the National Advisory Committee for Aeronautics's Wallops Island launching range. In 1952 Gilruth had become Assistant Director of the Langley Laboratory, investigating high-temperature structures and dynamics loads, and conducting hypersonic aerodynamics research at Wallops Island. His focus shifted to spacecraft in 1957, after the Soviet Union launched *Sputnik*. When NASA began in 1958, Gilruth had become Director of the Space Task Group at Langley, ultimately devising all the basic principles of Project Mercury. In 1961 Gilruth had become Director of the Manned Spacecraft Center, later Johnson Space Center (JSC), where he directed 25 human spaceflights over the course of 10 years. George M. Low, Director of the Apollo Lunar Landing Program, once remarked in an interview that the Mercury, Gemini, and Apollo programs would never have existed without Robert Gilruth. NASA Administrator Daniel S. Goldin commented, "his courage to explore the unknown, his insistence on following strict scientific procedures, and his technical expertise directly contributed to the ultimate success of the American manned space program and the landing of a man on the moon."[967]

*23 August*
Russian space officials announced a decision to alter plans for the *Enterprise* module of the ISS, which the company RSC Energia was building in conjunction with the U.S. company SPACEHAB. In the original plans, the module had been noncommercial, providing docking and

---

[965] Associated Press, "Russian, U.S. Rescuers Practice Space Rescue," 16 August 2000.
[966] NASA, "Cosmic Gas Clouds Yield Puzzling Concentrations of Water," news release 00-126, 16 August 2000, http://www.nasa.gov/home/hqnews/2000/00-126.txt (accessed 6 May 2008).
[967] NASA, "Dr. Robert Gilruth, an Architect of Manned Space Flight, Dies," news release 00-127, 17 August 2000, http://www.nasa.gov/home/hqnews/2000/00-127.txt (accessed 6 May 2008).

cargo space, but in an effort to increase revenue, RSC Energia had revised the plans to allow the module to hold multimedia equipment for various business projects.[968]

Lockheed Martin Systems Integration announced that the U.S. Navy had selected the company to build seven SH-60R Multi-Mission Helicopters, under the first SH-60R low-rate initial production (LRIP) contract, moving Lockheed Martin from development and testing to the production phase of the SH-60R program. The contract required Lockheed Martin to integrate the flight avionics systems, mission avionics systems, and stores and defense systems.[969]

NASA announced the results of the most complete HST census of brown dwarfs, finding that the "odd and elusive objects also tend to be loners." Joan Najita of the National Optical Astronomy Observatory in Arizona explained that the properties of brown dwarfs "reveal new and unique insights into how stars and planets form," because they "bridge the gap between stars and planets." The stellar objects are too low in mass to burn hydrogen, yet are more massive than planets; despite being 15 to 80 times more massive than Jupiter, brown dwarfs are difficult to detect because the light they emit is very faint. The HST census had found more low-mass than high-mass brown dwarfs, as is the case with stars, and the isolated brown dwarfs appeared to represent the low-mass counterparts of the more massive classes of stars. In carrying out the census, scientists had used the HST's infrared vision to measure the brightness and temperature of stars in the cluster IC 348 in the constellation Perseus. Najita and colleagues had used the telescope's NICMOS camera, developing a new technique to distinguish brown dwarfs from "the clutter of background stars." The new procedure had measured the "strength of an infrared water-absorption band in the atmospheres of the stars," a sensitive measure of each star's temperature, solving several problems simultaneously. The procedure had enabled the scientists to distinguish the brown dwarfs from background stars and to measure the masses of the brown dwarfs without needing to assume their ages, thereby greatly improving estimates of mass.[970]

*24 August*
NASA awarded four small businesses 90-day contracts to develop concepts and requirements for providing access to the ISS on emerging launch systems. The contracts' purpose was to identify potential backup capability and to augment the station's primary resupply vehicles: the U.S. Space Shuttle, the Russian Progress, the ESA's Automated Transfer Vehicle, and the Japanese H-II Transfer Vehicle. NASA awarded Andrews Space and Technology US$195,000; Microcosm, US$198,000; HMX Ltd., US$245,000; and Kistler Aerospace Corporation, US$264,000, to develop concepts, determine requirements of launch services, and provide suggestions on specific development risk-reduction activities, which NASA would need to perform. NASA's Marshall Space Flight Center would manage the study contracts under the Alternate Access Project of the Space Launch Initiative.[971]

---

[968] Associated Press, "Russia Seeks Space Station Revenue," 23 August 2000.
[969] Lockheed Martin, "Lockheed Martin Awarded First SH-60R Avionics Production Contract," news release, 23 August 2000, *http://www.lockheedmartin.com/news/press_releases/2000/LockheedMartinAwardedFirstSH60R Avio.html* (accessed 6 May 2008).
[970] NASA, "Hubble Gets Head Count of Elusive Brown Dwarf Stars," news release 00-130, 23 August 2000, *http://www.nasa.gov/home/hqnews/2000/00-130.txt* (accessed 6 May 2008).
[971] NASA, "Small Companies To Study Potential Use of Emerging Launch Services for Alternative Access to Space Station," news release C00-g, 24 August 2000, *http://www.nasa.gov/home/hqnews/contract/2000/c00-g.txt* (accessed 6 May 2008).

*25 August*

Margaret G. Kivelson and four coauthors from the University of California at Los Angeles published in the journal *Science* the strongest evidence to date that Jupiter's moon Europa contains a salty, liquid ocean beneath its surface. The team had used data from the Galileo probe's magnetometer to study Europa's magnetic compass, finding that "the presence of a layer of electrically conducting liquid, such as saltwater" best explained the behavior of the moon's magnetic compass. Kivelson explained that the team had inferred that the conductor must be a liquid ocean, since ice is not a good conductor of electricity. However, Torrence V. Johnson of NASA's Jet Propulsion Laboratory (JPL) remarked that the magnetometer data by itself was insufficient to conclude the presence of a liquid ocean. Johnson explained that scientists needed several further steps of inference, such as precise measurements of gravity and altitude, indicating the effects of tides.[972]

*27 August*

In the second on-orbit failure of Hughes Electronic Corporation's HS601-model satellite, the Spacecraft Control Processor (SCP) of Mexico's *Solidaridad I* communications satellite malfunctioned, disrupting television, radio, and pager services. The first incident had occurred on the *Galaxy IV* satellite, causing similar communications disruptions in the United States during 1998. In the current malfunction, operators in Iztapalapa and Hermosillo received a series of alarms indicating that the only functioning SCP on board the Satmex-owned craft had turned itself off. The satellite had two SCPs on board and needed only one to carry out its communications relay properly, but *Solidaridad I* already was running on its backup SCP, having lost its first processor a year ago. Engineers suspected that the cause of the incident was the growth of a tiny crystalline structure that leads to electrical shorts. Hughes investigators had identified this type of malfunction in their satellites containing tin-plated relay switches and, therefore, had begun to use nickel-plated switches in new satellites to prevent future occurrences of such electrical shorts. However, *Solidaridad I*, designed for 14 years of service in space, had launched in 1993 before the conversion to nickel-plated switches. The *Solidaridad* series, comprising two satellites, provided service to all of Mexico, also extending to the southwestern United States, the Caribbean, and Central and South America. The malfunction led to the loss of educational television programming in 12,000 schools in Mexico, mostly in remote rural areas where some students rely entirely on televised courses. Satmex was transferring *Solidaridad I* users to the three other satellites in its fleet.[973]

*28 August*

A Russian Proton-K rocket launched from Baikonur Cosmodrome in Kazakhstan, carrying into orbit the Russian Ministry of Defense's *Raduga-1-5* spacecraft, a military communications satellite with the alternate names Globus-1 and Cosmos 2372. Although the Globus-1 was a classified program, news sources reported that the craft would use a geostationary orbit to serve

---

[972] NASA, "Galileo Evidence Points to Possible Water World Under Europa's Icy Crust," news release 00-131, 25 August 2000, *http://www.nasa.gov/home/hqnews/2000/00-131.txt* (accessed 6 May 2008); Margaret Kivelson et al., "Galileo Magnetometer Measurements: A Stronger Case for a Subsurface Ocean at Europa," *Science* 289, no. 5483 (25 August 2000): 1340–1343.
[973] Justin Ray, "Satellite Failure Causes Communications Blackout," *Spaceflight Now*, 29 August 2000.

as a relay link for Russian military forces. At the time of the launch, many *Raduga* craft remained in orbit, but only five were operational.[974]

*30 August*
NASA's JPL named Thomas C. Duxbury Project Manager for JPL's Stardust Mission, launched in February 1999 to collect a sample from Comet Wild-2 and return it to Earth. Duxbury, who had served as the Mission's Acting Project Manager for the past year, replaced Kenneth Atkins, who had been heading a program to develop leadership for JPL's projects. Duxbury had joined the Stardust project in 1996 as Mission Manager, responsible for navigation, mission design, the ground-data system, science-data management, and mission operations. Before working on the Stardust project, Duxbury had served on planetary mission teams, including several Mariner missions, the Mars Viking mission, Pioneer 10 and 11 to Jupiter and Saturn, Voyager 1 and 2 to the outer planets, the Soviet Phobos Mission to Mars, the Mars Observer Mission, and the Clementine Mission to study the Moon. Concurrent with his position as Stardust Project Manager, Duxbury planned to continue to serve as a member of the science teams for Mars Global Surveyor's laser altimeter and for the ESA's Mars Express orbiter and lander, as well as continuing as lead scientist for geodesy and cartography in the Mars Exploration Office.[975]

The U.S. General Accounting Office (GAO) released a report discussing workforce and safety issues of NASA's Space Shuttle Program. The report assessed the impact of workforce reductions on the Shuttle program; the challenges that NASA would face as it addressed its workforce issues; and the status of planned safety and supportability upgrades to the Space Shuttle. GAO identified 26 technical skills needed at JSC in Houston, Texas, as well as at other facilities in Florida, Alabama, and Mississippi. In addition, GAO found that the decrease from 3,000 to 1,800 federal employees in the Shuttle program workforce since 1995 had placed NASA at a "critical juncture." The Shuttle launch rate had dropped from eight launches in 1997 to three in 1999, a trend on the cusp of reversing itself in the aftermath of the July 1999 launch of Russia's *Zvezda* module. Moreover, GAO reported signs of overwork and stress among NASA staff. Worker demographics—twice as many workers over age 60 as under age 30—had compounded the problem of a declining workforce. NASA officials welcomed the report, although the results reflected "a much harsher assessment of [S]huttle safety issues [than the assessment] prepared by NASA itself in response to an electrical short that accompanied the July 1999 launching of the [S]huttle *Columbia*." GAO also reported that NASA had terminated downsizing plans in December 1999 and had initiated efforts to begin hiring new staff. Although it was developing safety and supportability upgrades, which it would implement over the next five years, NASA still faced programmatic and technical challenges, such as a demanding schedule and undefined design and workforce requirements.[976]

---

[974] Justin Ray, "Proton Rocket Lofts Russian Military Satellite," *Spaceflight Now*, 28 August 2000; *Spacewarn Bulletin*, no. 562, 1 September 2000, *http://nssdc.gsfc.nasa.gov/spacewarn/spx562.html* (accessed 6 August 2008).
[975] NASA Jet Propulsion Laboratory, "Duxbury Named Project Manager of Stardust Mission," news release 2000-084, 30 August 2000, *http://www.jpl.nasa.gov/releases/2000/duxbury.html* (accessed 6 May 2008).
[976] U. S. General Accounting Office, "Space Shuttle: Human Capital and Safety Upgrade Challenges Require Continued Attention" (report no. GAO/NSIAD/GGD-00-186, Washington, DC, August 2000), *http://www.gao.gov/archive/2000/n200186.pdf* (accessed 9 May 2008); Mark Carreau, "Undersized Work Force Puts NASA at 'Critical Juncture'," *Houston Chronicle*, 31 August 2000.

# SEPTEMBER 2000

*5 September*
A *Sirius-2* communications satellite launched aboard a Russian Proton-K rocket from Baikonur Cosmodrome in Kazakhstan. The launch marked the second success for the Sirius Company, which planned to broadcast digital radio to the United States. Sirius had placed its first satellite in orbit on 1 July 2000, scheduling a third launch for November 2000.[977]

*6 September*
Goddard Space Flight Center (GSFC) selected Space Systems and Applications Inc. of Lanham, Maryland, to provide engineering, science, and technology support for NASA's Space and Earth Science Directorates. The value of the cost-plus-award-fee contract was US$204.5 million over a period of five years. Under the contract, NASA required Space Systems and Applications to provide research and information technology services including scientific data analysis, modeling and simulation of physical processes, development of flight-project data systems and large-scale data management, archival and delivery systems, and systems analysis and programming. These services would support a broad range of scientific disciplines—astronomy and astrophysics, climatology and atmospheric science, geodynamics, land processes, oceanography, planetary studies, solar and space plasma physics, and solid earth geographics.[978]

*7 September*
NASA announced that, on 3 September 2000, the Total Ozone Mapping Spectrometer (TOMS) aboard NASA's Total Ozone Mapping Spectrometer Earth Probe (TOMS-EP) satellite had detected an 11 million-square-mile (28.5 million-square-kilometer) ozone-depletion area—surpassing the ozone hole's record of 10.5 million square miles (27.2 million square kilometers) set on 19 September 1998. Scientists investigating the extent of the ozone-depletion area expressed their surprise, suggesting that early-spring conditions and an "extremely intense Antarctic vortex"—an upper-altitude stratospheric air current that moves around the continent—might partly explain the record-setting size. Jack A. Kaye of the Office of Earth Sciences, Research Director at NASA Headquarters, remarked that, although scientists expect variations in the size of the ozone hole from one year to the next, they plan to observe the evolution of the ozone-depletion area in the coming months, to compare it with previous years. Manager of NASA's Upper Atmosphere Research Program Michael J. Kurylo suggested that the new data reinforced concerns about the fragile nature of the ozone layer, indicating that, although international agreements had curbed the production of ozone-destroying gases, concentrations of those gases in the stratosphere were just reaching their peak levels. Therefore, he warned that it would be decades before the depletion area in the ozone layer would no longer occur annually. Associate Administrator for Earth Science Enterprise Ghassem R. Asrar commented that such discoveries demonstrated the value of long-term observations.[979]

*8 September*

---

[977] Reuters, "Kazakhstan: Russia Launches U.S. Satellite from Kazakhstan," 5 September 2000.
[978] NASA, "Contract for Goddard Space Center Support Awarded," news release C00-j, 6 September 2000, http://www.nasa.gov/home/hqnews/contract/2000/c00-j.txt (accessed 25 April 2008).
[979] NASA, "Largest-Ever Ozone Hole Observed over Antarctica," news release 00-137, 7 September 2000, http://www.nasa.gov/home/hqnews/2000/00-137.txt (accessed 25 April 2008).

Space Shuttle *Atlantis* Mission STS-106 launched from Kennedy Space Center (KSC) in Cape Canaveral, Florida, to prepare the International Space Station (ISS) for the arrival of its first crew. The mission's crew comprised Commander Terrence W. Wilcutt, Pilot Scott D. Altman, and Mission Specialists Daniel C. Burbank, Edward T. Lu, Yuri I. Malenchenko, Richard A. Mastracchio, and Boris V. Morukov. The purpose of STS-106, also known as ISS Flight 2A.2b, was to connect power, data, and communications cables to the *Zvezda* service module, as well as to deliver supplies and to perform maintenance. Although NASA officials had expressed concern that STS-106 might encounter a weather delay, the approaching storm had stalled offshore, so the Shuttle program experienced its first punctual launch since John H. Glenn Jr.'s historic flight in October 1998.[980]

*10 September*
Astronaut Edward T. Lu and cosmonaut Yuri I. Malenchenko conducted a spacewalk outside the ISS, connecting nine power, data, and communications cables between the Russian-built *Zvezda* and *Zarya* modules. Lu and Malenchenko also assembled and installed a 6-foot-long (1.8-meter-long) magnetometer boom on the *Zvezda* module, to serve as a compass showing the relation of the ISS to Earth.[981]

*12 September*
NASA announced that scientists using the Chandra X-ray Observatory had confirmed the existence of "middleweight" black holes. Several groups of scientists used Chandra to focus on a mid-mass black hole located 600 light-years from the center of galaxy M82. Previous x-ray data from the German-U.S. ROSAT (Röntgensatellite) and the Japan-U.S. ASCA (Advanced Satellite for Cosmology and Astrophysics) had suggested the existence of a mid-mass black hole in M82. Scientists compared Chandra's new high-resolution images with optical radio and infrared maps to determine that a single, bright source was emitting most of the x-rays. The scientists continued to observe M82 over an eight-month period, noting that the intensity of the x-rays rose and fell every 600 seconds. Philip Kaaret of the Harvard-Smithsonian Center for Astrophysics explained that this behavior was similar to black holes swallowing gas from a nearby star or cloud. Martin Ward of the University of Leicester in the United Kingdom remarked that the findings opened "a whole new field of research."[982]

*13 September*
Astronaut Edward T. Lu and cosmonaut Yuri I. Malenchenko installed three new batteries in the Russian service module *Zvezda*, which had launched with only five of its eight batteries to reduce its launch weight. Meanwhile, astronaut Daniel C. Burbank and cosmonaut Boris V. Morukov installed one of six batteries on the Russian-built control module *Zarya*. Burbank and Morukov needed a hammer and chisel to remove four small nut plates, which were obstructing a

---

[980] Marcia Dunn for Associated Press, "Atlantis Thunders into Orbit on Space Station Mission," 8 September 2000; NASA, "Mission Archives STS-106, International Space Station Flight 2A.2b," *http://www.nasa.gov/mission_pages/shuttle/shuttlemissions/archives/sts-106.html* (accessed 30 October 2008).
[981] NASA, "Mission Archives: STS-106," *http://www.nasa.gov/mission_pages/shuttle/shuttlemissions/archives/sts-106.html* (accessed 30 April 2008); NASA, "STS-106 Extravehicular Activities," *http://spaceflight.nasa.gov/shuttle/archives/sts-106/eva/index.html* (accessed 2 May 2008).
[982] NASA, "Chandra Clinches Case for Unexpected Black Hole Discovery," news release 00-140, 12 September 2000, *http://www.nasa.gov/home/hqnews/2000/00-140.txt* (accessed 25 April 2008).

pair of bolts; the astronauts needed to loosen the bolts so that they could replace a voltage converter.[983]

*14 September*
NASA Administrator Daniel S. Goldin and H. Fisk Johnson, President of the Wisconsin-based private venture-capital company Fisk Ventures Inc. (FVI), signed an agreement to "explore a new frontier in biotechnology." Johnson had approached NASA regarding its efforts to commercialize space activities, suggesting a partnership. NASA and FVI agreed to use NASA's bioreactor to develop commercial medical products. The bioreactor technology creates a near-weightless environment enabling cells to grow three-dimensionally. When raised in a traditional Petri dish, growing cells are flat. FVI had formed a joint venture with In Vitro Technologies Inc., called StelSys LLC, to focus on commercializing microgravity research in areas related to biological systems. StelSys paid NASA a US$100,000 licensing fee and a royalty of 5 percent of the company's profits, capped at US$2 million, for the rights to 13 patents for the bioreactor technology for five years. NASA planned to use the bioreactor on the ISS, and StelSys intended to use it to research infectious diseases and to develop a liver-assist device for patients in need of transplant surgery. Goldin remarked that the agreement was "a symbol of the success that can be achieved when government, private industry, and academia work together on the exploration of new frontiers for scientific, technological, and economic growth."[984]

A GE-7 satellite, built by Lockheed Martin Commercial Space Systems for GE American Communications (GE Americom), launched from Korou, French Guiana, aboard an Ariane 5 rocket. The C-band satellite supplemented GE Americom's fleet of 12 satellites servicing the Americas, replacing the GE SATCOM C1 craft. A member of the A2100 family of satellites, the GE-7 joined four others—the GE-1, GE-2, GE-3, and GE-4—to provide distribution of cable, broadcast television and radio, business television, and broadband data distribution across the contiguous United States, Mexico, the Caribbean, and South America. The GE-7 would provide "service to regional and national customers, in-orbit protection for many current cable and radio customers, plus critical telecommunications services delivered by AT&T Alascom."[985]

NASA Associate Administrator for Space Science Edward J. Weiler appeared before the U.S. House Science Subcommittee on Space and Aeronautics to address the future of NASA's space science programs. Weiler discussed NASA's problems in the previous year, such as the loss of Mars Climate Orbiter and Mars Polar Lander, explaining that, although NASA was developing a response to those losses, it would need more funding if it were to prevent similar failures in the future. House Subcommittee Chairperson Dana Rohrabacher (R-CA) indicated that President William J. Clinton had proposed increasing the space science budget by US$200 million over

---

[983] NASA, "STS-106"; C. Bryson Hull for Associated Press, "Astronauts Installing Fresh Batteries in Space Station," 13 September 2000.
[984] NASA, "Landmark Commercial Agreement Gives Biotechnology Research a New Dimension," news release 00-143, 14 September 2000, *http://www.nasa.gov/home/hqnews/2000/00-143.txt* (accessed 25 April 2008); Laura Heinauer, "NASA Signs First Major Contract for Private Biotech Development," *Wall Street Journal*, 14 September 2000.
[985] Lockheed Martin, "Lockheed Martin-Built GE-7 Satellite Successfully Launched from Kourou, French Guiana Launch Site," news release, 14 September 2000, *http://www.lockheedmartin.com/news/press_releases/2000/LockheedMartinBuiltGE7SatelliteSucc.html* (accessed 25 April 2008).

each of the following four years. NASA's budget reauthorization bill for FY 2001 added an additional US$19 million in funding for space science programs.[986]

*20 September*
Space Shuttle *Atlantis* landed at KSC in Cape Canaveral, Florida before dawn, the 15th nighttime landing in the Shuttle program's history. Among the tasks that the crew had completed during the mission were delivering 3 tons (2,700 kilograms or 2.7 tonnes) of supplies for the first permanent crew—including toiletries, Russian and American meals, medical kits, and camera equipment—and installing the station's toilet, an oxygen generator, a treadmill, and power and television cables. NASA officials had decided to extend the mission an extra day to allow the crew more time to accomplish its work inside the space station. N. Wayne Hale, STS-106 Flight Director, explained that the mission had begun with 52 items on its list of tasks to accomplish, but the number had increased to 74 different tasks, large and small, on board the ISS. The only malfunction during the mission had been that of a newly installed battery, which Russian Mission Control had ordered disconnected when it failed to charge properly.[987]

*21 September*
NASA selected 28 of 119 grant proposals, totaling US$10 million over four years, to conduct microgravity combustion research. NASA's Office of Life and Microgravity Science and Applications, which was sponsoring the grants would provide the researchers with access to its microgravity research facilities, including drop tubes, drop towers, aircraft-flying parabolic trajectories, and sounding rockets. Twenty-six of the grants were for ground-based research, and two were for flight-definition efforts; four grants continued NASA-funded work, but the remaining 24 entailed new research efforts.[988]

Russian police announced that cosmonaut German S. Titov, the Soviet Union's second man in space and the first person to spend more than one day in orbit, had been found dead at home at the age of 65. Although the police had not reported an official cause of death, the media speculated that the cause was either carbon monoxide poisoning or a heart attack. Titov had spent 25 hours and 18 minutes aboard the tiny Soviet *Vostok-2* spacecraft on 6 and 7 August 1961. Fellow Soviet Yuri Gagarin had made the first flight on *Vostok-1* on 12 April, flying for less than 2 hours, and U.S. astronaut Alan B. Shepard Jr. had followed Gagarin's mission with a 15-minute, suborbital flight on 5 May 1961. Titov worked on the Buran program during the 1980s, attempting to create a Russian space shuttle, but the Soviet Union abandoned the project after the Russian shuttle had made one unpiloted flight. After the fall of the Soviet Union, Titov had entered politics, representing the Communist Party on the defense committee of the State Duma lower house.[989]

Following a 24-hour delay caused by a computer anomaly, NASA and the National Oceanic and Atmospheric Administration (NOAA) successfully launched the Lockheed Martin–built, NOAA-

---
[986] *Space Business News*, "Money, Management Are Key To Avoiding NASA Failures, House Panel Hears," 27 September 2000.
[987] Dow Jones Newswire, "Space Shuttle Atlantis Touches Down Successfully," 20 September 2000.
[988] NASA, "NASA Announces Research Grants in Microgravity Combustion Science," news release 00-144, 21 September 2000, http://www.nasa.gov/home/hqnews/2000/00-144.txt (accessed 25 April 2008).
[989] Reuters, "Russia's 2nd Cosmonaut, German Titov, Dies in Sauna," 21 September 2000.

L spacecraft from Vandenberg Air Force Base, aboard a Titan II rocket. Ground controllers confirmed that the craft had deployed its solar array and verified a power-positive condition. NASA and NOAA had designed the satellite, the second in a series of five polar-orbiting spacecraft, to collect meteorological data and to transmit the information to users worldwide. NOAA's National Weather Service planned to use the craft's data for long-range weather and climate forecasts. Although NASA had managed the project of building and launching the satellite, it intended to transfer operational control to NOAA 10 days following the launch.[990]

NASA released time-lapse movies created from images produced by the Hubble Space Telescope, showing "spectacular outbursts from young stars," changing "dramatically over a period of just weeks or months." NASA had combined individual images of young star systems XZ Tauri and HH 30 in the Taurus-Auriga molecular cloud, captured over several years, to create the time-lapse movies. Documenting the activity of the early stages of stars' lives, the movies demonstrated that images "taken of the universe today won't necessarily look the same as those snapped a few months from now."[991]

*26 September*
Administrator Daniel S. Goldin presented awards to three minority contractors at NASA's annual Minority Business and Advocates Awards Ceremony. RS Information Services (RSIS), based in McLean, Virginia, won Minority Contractor of the Year; Rigging and Welding Specialists Inc., a Native American–owned business, won Minority Subcontractor of the Year; and Pace and Waite Inc. won Women-Owned Business of the Year. In addition, NASA recognized advocates for innovative approaches to using minority- and women-owned businesses, awarding NASA's Exceptional Achievement Medal to Kenneth Martindale and Rodney J. Etchberger of Johnson Space Center and to Shantaram S. Pai of Glenn Research Center. Furthermore, NASA recognized five individuals for outstanding achievements and three NASA field centers for meeting or exceeding all socioeconomic business goals for FY 1999.[992]

NASA named award-winning broadcast journalist Bob Jacobs as Chief of News and Information at NASA Headquarters. Before joining NASA, Jacobs had served for four years as Projects Manager for the Washington-based Broadcast Technology Division of the Associated Press, where he helped to develop and implement newsroom management technology. Clients had included the British Broadcasting Corporation, ESPN, National Public Radio, and CBS News. Jacobs had won an Emmy award and other regional honors for excellence in journalism.[993]

Sally K. Ride, the first female astronaut from the United States to travel into space, resigned her position as President of Space.com to concentrate on her education career. She had worked with Space.com founder Lou Dobbs, former host of CNN's *Moneyline*, since the company's start-up

---

[990] NOAA, "NOAA-L Weather Satellite Successfully Launched," news release 2000-069, 21 September 2000, *http://www.publicaffairs.noaa.gov/releases2000/sep00/noaa00069.html* (accessed 29 April 2008); Associated Press, "Weather Satellite Launched After Delay," 21 September 2000.
[991] NASA, "Hubble Movies Show the Changing Faces of Young Stars," news release N00-044, 21 September 2000, *http://www.nasa.gov/home/hqnews/note2edt/2000/n00-044.txt* (accessed 25 April 2008).
[992] NASA, "NASA To Honor Minority Businesses," news release 00-148, 25 September 2000, *http://www.nasa.gov/home/hqnews/2000/00-148.txt* (accessed 25 April 2008).
[993] NASA, "News Chief Named at NASA Headquarters," news release 00-150, 26 September 2000, *http://www.nasa.gov/home/hqnews/2000/00-150.txt* (accessed 25 April 2008).

in June 1999 and had been the company's president for the past year. Ride would continue her leave of absence from the University of California at San Diego where she is a physics professor throughout the remainder of the fall academic term.[994]

NASA announced that a team of Lockheed Martin scientists using NASA's Transition Region and Coronal Explorer (TRACE) spacecraft to observe coronal loops—coils of hot, electrified gas—believed they had located the source of the heating mechanism that makes the Sun's corona 300 times hotter than its visible surface. A thirty-year-old theory had assumed that the coronal loops heated evenly, but the TRACE observations had indicated that most of the heating occurs at the base of the loops, near the point from which they emerge and return to the solar surface. The team had observed 41 loops extending from 2,500 miles (4,000 kilometers) to more than 180,000 miles (290,000 kilometers) from the solar surface and had found that, although threads within shorter loops heat more evenly, longer threads cool noticeably as they attain height. The team had calculated the loops' energy levels and estimated that "heating typically occurs in the first 6,000 miles of a loop's length." Richard Fisher, head of the Laboratory for Astronomy and Solar Physics at NASA's GSFC, remarked that understanding how the coronal loops function could shed light on coronal-mass ejections (CMEs). CMEs can disrupt or destroy satellite components orbiting Earth and prompt surges in electrical transmission lines, causing blackouts on Earth.[995]

*27 September*
Arianespace, the France-based European satellite launch company, announced that it had won a contract to launch three more satellites for INTELSAT, the international satellite corporation based in Washington, DC. Arianespace already held a contract to launch three other satellites in the INTELSAT IX series. INTELSAT had scheduled six of the series' seven spacecraft for launch from mid-2001 through the end of 2002, on Ariane 4 or Ariane 5 launch vehicles, from Europe's spaceport in French Guiana. California-based Space Systems/Loral was building the 4.7-tonne (4,700-kilogram or 5.2-ton) satellites. In 1965 INTELSAT had launched the world's first commercial communications satellite, *Early Bird*.[996]

Lockheed Martin announced that for the third time in four years, the Air Traffic Control Association had awarded the company its Industry Award, recognizing Lockheed's outstanding achievement in and contribution to the science of air traffic control. The award specifically recognized Lockheed Martin Air Traffic Management for its worldwide support of air traffic–control systems. The Federal Aviation Administration (FAA), the German Civil Aviation Authority, and the United Kingdom's National Air Traffic Services Ltd. Had all nominated the company for the award.[997]

---

[994] Bloomberg News, "Ride Leaves Space.com Top Post—For 1st U.S. Female Astronaut in Space, It's Back to Education," 27 September 2000.
[995] NASA, "Fountains of Fire Illuminate Solar Mystery," news release 00-146, 26 September 2000, *http://www.nasa.gov/home/hqnews/2000/00-146.txt* (accessed 25 April 2008); Peter N. Spotts, "Scientists Begin To Unravel a Stubborn Solar Mystery," *Christian Science Monitor*, 27 September 2000.
[996] Reuters, "France: Arianespace in Deal To Launch 3 INTELSAT Satellites," 27 September 2000.
[997] Lockheed Martin, "Lockheed Martin Receives Prestigious Air Traffic Control Association Award," news release, 27 September 2000, *http://www.lockheedmartin.com/news/press_releases/2000/LockheedMartinReceivesPrestigiousAi.html* (accessed 25 April 2008).

*28 September*
NASA's Langley Research Center awarded contracts valued at US$1.8 million to two companies for operating airport-surface surveillance systems intended to prevent runway accidents, a top safety priority of the FAA and the National Transportation Safety Board. The contracts required Sensis Corporation of Dewitt, New York, and Rannoch Corporation of Alexandria, Virginia, to operate the Airport Traffic Identification System (ATIDS) and the Dynamic Runway Occupancy Measurement Systems (DROMS) at Detroit's Metropolitan Wayne International Airport, for six months, establishing and validating the systems in a "live" environment. The contracts also required Sensis to deploy and evaluate DROMS at Memphis International Airport and to interface the system with the existing Sensis-operated ATIDS at that airport.[998]

*29 September*
NASA announced its agreement with Lockheed Martin on a plan for the X-33 space plane program. The agreed plan included aluminum fuel tanks for hydrogen fuel, a revised payment schedule, and a target launch date in 2003, contingent on Lockheed Martin's winning additional funding under the Space Launch Initiative for completing the project. NASA had funded the project through March 2001. The restructured plan focused on "providing milestone payments" to Lockheed Martin's team, following the completion of tests and the delivery of hardware and software systems this year. The plan also gave greater emphasis to mission safety, quality, and mission success. Arthur G. Stephenson, Director of NASA's Marshall Space Flight Center, pointed out that, despite setbacks, the program's successes had included the creation of technology applicable to the space program and to the commercial aircraft industry, such as a revolutionary new rocket engine; a robust, reusable, metallic thermal-protection system; and software and sensors that automatically determine and predict failures and errors before they affect the flight.[999]

NASA announced the restructuring of the Office of Life and Microgravity Sciences and Applications (OLSMA), part of the Human Exploration and Development of Space Enterprise (HEDS). OLSMA, renamed the Office of Biological and Physical Research (OBPR), would become a separate enterprise focusing on scientific research, working closely with HEDS to facilitate long-term exploration of space. The OBPR would comprise five divisions: the Physical Sciences Division, the Fundamental Space Biology Division, the Biomedical and Human Support Research Division, the Division of Research Integration, and the Division of Policy and Program Integration. NASA named Kathie L. Olsen as Acting Associate Administrator, to return to her position as Chief Scientist after NASA had filled the position permanently.[1000]

## OCTOBER 2000

*1 October*

---

[998] NASA, "Contracts Awarded for Airport Surface Surveillance Systems," news release C00-m, 28 September 2000, *http://www.nasa.gov/home/hqnews/contract/2000/c00-m.txt* (accessed 25 April 2008).
[999] NASA, "NASA, Lockheed Martin Agree on X-33 Plan," news release 00-157, 29 September 2000, *http://www.nasa.gov/home/hqnews/2000/00-157.txt* (accessed 25 April 2008).
[1000] NASA, "NASA Creates New Enterprise Focusing on Biology," news release 00-158, 29 September 2000, *http://www.nasa.gov/home/hqnews/2000/00-158.txt* (accessed 25 April 2008).

New NASA research published in the 1 October issue of the American Meteorological Society's *Journal of Climate* warned that clouds might not help counteract climate-warming trends. Anthony D. Del Genio, of NASA's Goddard Institute for Space Studies in New York City, had analyzed observations of low clouds over land, collected between 1994 and 1997 as part of the U.S. Department of Energy's Atmospheric Radiation Measurement program. Del Genio had found that clouds are thinner when temperatures were higher, a phenomenon occurring in any weather condition or season and at any time of day. Some climate theories had predicted that high temperatures would cause thicker clouds because of an increase in water vapor, but Del Genio explained that during warmer temperatures, the bottoms of clouds rise and become thinner. This occurs because clouds that form over a warm, dry air mass must rise higher before becoming sufficiently saturated with water to form a cloud base. Thinner clouds are less capable of reflecting sunlight back into space and are, therefore, unable to act as a "natural sun shield." Del Genio's findings corroborated long-term, worldwide satellite observations, published using the NASA-funded International Satellite Cloud Climatology Project (ISCCP) database. The ISCCP, a global composite of cloud observations from international weather satellites, began showing a link between cloud thinning and temperature in 1992.[1001]

International Launch Services (ILS), a joint venture of Lockheed Martin, Khrunichev State Research, and S. P. Korolev Rocket and Space Corporation Energia, launched a Russian Proton rocket from Baikonur Cosmodrome in Kazakhstan. The Proton launch, the fourth in the year for ILS, successfully placed into orbit a United States–built GE-1A communications satellite serving customers in parts of Asia, including China, the Philippines, and India.[1002]

*3 October*
Russian space experts agreed that the *Mir* space station was in fine condition to continue orbiting Earth, and the Russian government extended the station's time in space indefinitely, after the Netherlands-based MirCorp agreed to fund a new mission. Russia's council of space engineers, known as the Council of Chief Designers, authorized plans to launch a Progress cargo spacecraft, in the middle of October, to deliver fuel and equipment to *Mir*.[1003]

*5 October*
European Space Agency (ESA) engineers announced the discovery of a flaw in the European receiver on NASA's Cassini spacecraft, prompting an inquiry into why engineers had not identified the problem before NASA had launched the craft, and how the issue could be resolved before the craft reached its destination. Cassini's probe data relay subsystem (PDRS) lacked sufficient bandwidth to navigate the Doppler shift with the Huygens probe as the probe parachuted toward the surface of Saturn's moon Titan. Therefore, the system would be unable to recover all of the data that Huygens's six instruments would generate. Cassini and Huygens launched on 15 October 1997, with Huygens scheduled to separate from Cassini on 6 November 2004, break through Titan's atmosphere on 27 November 2004, and descend to the moon's

---

[1001] NASA, "Revised—NASA Scientist Predicts Less Climate Cooling from Clouds," news release 00-151, 3 October 2000, http://www.nasa.gov/home/hqnews/2000/00-151.txt (accessed 1 April 2008).
[1002] Jim Banke, "Proton Rocket Lifts American Satellite into Earth Orbit," *Space.com*, 2 October 2000, http://www.space.com/missionlaunches/launches/proton_launch_001001.html (accessed 14 April 2008).
[1003] MirCorp, "MirCorp Statement on the Mir Space Station's Future," news release, 3 October 2000; Associated Press, "Russian Space Experts Determine Mir Fit To Continue Orbital Flight," 3 October 2000.

surface using a parachute system. The mission plan called for Huygens to measure the composition and winds of Titan's atmosphere and to collect images. The probe would send the data via the S-band PDRS to Cassini, which would then transmit the data to Earth. However, the ESA stated that the inadequate bandwidth of the PDRS meant that link margins would degrade, because the Doppler shift on the data subcarrier would be outside the bandwidth of the receiver-phase lock loop, leading to a potential loss in the link of 10dB over what engineers had assumed for the mission.[1004]

*6 October*
NASA announced the selection of Sverdrup Technology Inc. to fulfill a contract providing engineering, science, and technical services at Marshall Space Flight Center (MSFC). The two-year contract, with three one-year priced options valued at US$300 million over five years, would go into effect on 15 October.[1005]

*9 October*
The High Energy Transient Explorer Mission (HETE-2) launched aboard a Hybrid Pegasus expendable launch vehicle, from Kwajalein missile-range facility on the Kwajalein Atoll in the Pacific Ocean. The Center for Space Research at Massachusetts Institute of Technology headed the HETE-2 mission, designed to detect and localize gamma-ray bursts (GRBs) using a set of instruments that would allow simultaneous observations of the GRBs. The craft would compute locations of GRBs and immediately transmit the coordinates to ground-based observers. HETE-2, an international collaboration among the United States, Japan, France, and Italy, replaced the original HETE spacecraft, which was lost following a rocket malfunction in November 1996.[1006]

*11 October*
Space Shuttle *Discovery* launched from Kennedy Space Center (KSC) in Cape Canaveral, Florida, on Mission STS-92, also known as ISS Assembly Flight 3.3A, to carry out an 11-day construction mission requiring four scheduled spacewalks at the International Space Station (ISS). *Discovery*'s crew—Commander Brian Duffy, Pilot Pamela A. Melroy, and Mission Specialists Koichi Wakata, Leroy Chiao, Peter J. K. Wisoff, Michael Lopez-Alegria, and William S. McArthur—was transporting a truss and a docking port to the station, planning to install them before the arrival of the first scheduled ISS crew at the end of October.

*13 October*
Responding to a 25 January 2000 article in the *New York Times*, alleging that grants in the mid-1990s had benefited the Russian civilian agency Biopreparat, NASA's Office of Inspector General published a report stating that NASA had not followed the U.S. Department of State's guidelines for administering biotechnology grants to Russia, thereby inadvertently funding germ-warfare laboratories. The Russian government owned 51 percent of Biopreparat, the organization

---

[1004] Frank Morring Jr., "ESA Sets Inquiry After Tests Reveal Flaw in Titan Probe Link," *Aerospace Daily*, 6 October 2000.
[1005] NASA, "Tennessee Technology Firm Selected for $300 Million Contract," news release c00-n, 6 October 2000, http://www.nasa.gov/home/hqnews/contract/2000/c00-n.txt (accessed 1 April 2008).
[1006] NASA Goddard Space Flight Center, "The High Energy Transient Explorer Mission (HETE-2)," http://imagine.gsfc.nasa.gov/docs/sats_n_data/missions/hete2.html (accessed 14 April 2008); NASA, "HETE-2 Flies in Search of Gamma-Ray Bursts," news release 00-160, 2 October 2000, http://www.nasa.gov/home/hqnews/2000/00-160.txt (accessed 1 April 2008).

that had conducted most of the Soviet Union's biological warfare research since the 1970s. At the time of the article's publication, Biopreparat was continuing to research pathogens that had been included in the Soviet biological warfare program. The Inspector General found that "almost three-fourths of US$1.68 million in NASA grants intended for space biotechnology work found its way to 'institutions that had been affiliated with Russia's biological warfare program'." Therefore, the Inspector General asserted that NASA had not followed the Department of State's guidelines requiring NASA to "regularly visit and participate in the research it was funding at Russian institutes that had been part of the Soviet biological warfare program." NASA agreed with the Inspector General's sole recommendation: to practice "invasive collaboration" in the future, when funding biotechnology research in countries with known or suspected biological weapons programs.[1007]

Space Shuttle *Discovery* arrived at the ISS. Shuttle Commander Brian Duffy successfully executed the rendezvous without the Shuttle's radar, relying on a star-tracking system and on handheld lasers operated by his crew. The Ku-band antenna had failed after launch, disrupting the radar and television link. The Shuttle docking was NASA's first conducted without radar.[1008]

The ESA signed a four-year contract with two industry groups—Sarcom, led by Spot Image of Toulouse, France, and the Emma consortium, led by Eurimage of Rome—granting the two groups "priority access to the Envisat radar remote-sensing satellite" scheduled for launch in 2001, as well as imagery from the ERS-2 spacecraft, already in orbit. About the new business relationship, a major policy shift for the ESA, Acting Envisat Mission Manager Günther Kohlhammer remarked that the ESA had realized that it needed to change so that it could develop new markets. The contract gave commercial orders for the satellite images priority over scientific requests, setting new limits on the type and number of users ESA would grant access to data without paying commercial prices.[1009]

*14 October*
Despite unexpected problems, astronauts aboard Space Shuttle *Discovery* installed a 9-ton (8,200-kilogram or 8.2-tonne) structural truss on the *Unity* module of the ISS, completing one of the major objectives of the mission. The Z1 truss held four mass gyroscopes and the station's main Ku-band and S-band communications antennas. In addition, the truss provided a temporary attachment point for a set of solar arrays scheduled to arrive during a December Shuttle mission. The astronauts faced several obstacles during the installation process. A short in the Shuttle's payload circuitry disabled the computerized camera system needed to orient the truss for attachment, causing a 2½-hour delay. However, the astronauts were able to wire a backup computer into an alternate power source, restoring power to the Space Vision System, which NASA had programmed to "determine the truss's precise location and orientation with respect to the station by measuring the locations of the targets on its surface as viewed from different

---

[1007] NASA Office of Inspector General, "NASA Oversight of Russian Biotechnology Research 1994–1997" (NASA report no. G-00-07, Washington, DC, 13 October 2000), *http://oig.nasa.gov/old/inspections_assessments/g-00-007.pdf* (accessed 8 April 2008); Frank Morring Jr., "NASA Seen Funding Russian Germ-Warfare Labs Through Lax Oversight," *Aerospace Daily*, 18 October 2000.
[1008] Associated Press, "Bad Antenna Blacks Out Shuttle's Television," *Charleston Gazette* (SC), 13 October 2000; *New York Times*, "Astronauts Rendezvous with Space Station," 14 October 2000; *Los Angeles Times*, "Shuttle Discovery Docks Safely with Space Station," 14 October 2000.
[1009] Peter B. de Selding, "ESA Hands Radar Satellite Responsibility to Industry," *Space News*, 30 October 2000.

payload bay cameras." The astronauts had to delay their work again when the Space Vision System failed to generate reliable data during certain orbital lighting conditions, requiring astronaut William S. McArthur to recalibrate the system.[1010]

*15 October*
Astronauts Leroy Chiao and William S. McArthur made a spacewalk outside the ISS to link electrical cables to the station's new truss and to deploy the main antenna. The pair routed six power- and data-cables between the truss and the station, as well as four backup cables. Tests indicated that all connections worked. The astronauts moved one of the truss's two antennas to a temporary location out of the way of future construction work; mounted the main antenna, a 78-inch-wide (2-meter-wide) dish, to a 12-foot (3.7-meter) boom; and extended the main antenna. According to NASA's schedule, the ISS crew would turn on the antenna during an April 2001 Shuttle mission.[1011]

*16 October*
NASA Administrator Daniel S. Goldin announced the appointment of Baruch S. Blumberg, 1976 Nobel Prize Laureate in Physiology and Medicine, as Senior Advisor to the Administrator, effective immediately. While continuing in his position as Director of NASA's Astrobiology Institute, Blumberg would also provide guidance to NASA's newly created Office of Biological and Physical Research (OBPR), an interdisciplinary effort combining research in biology, physics, chemistry, and engineering. Goldin also announced that NASA would begin a search, headed by Blumberg and NASA Chief Scientist Kathie L. Olsen, to fill the position of Associate Administrator for the new OBPR and other key positions. Olsen would serve as Acting Associate Administrator until NASA had filled the position.[1012]

Three astronauts attached a new docking port to the ISS, with two of the crew spacewalking and one operating *Discovery*'s robotic arm from inside the Shuttle. Astronauts Peter J. K. "Jeff" Wisoff and Michael Lopez-Alegria, assigned to work outside the ISS, had difficulty loosening the latches securing the docking port inside *Discovery*'s cargo bay. However, once they had freed it, Japanese astronaut Koichi Wakata lifted the 2,700-pound (1,200-kilogram) port using the Shuttle's robotic arm and positioned it on the space station. When the port was within 1 foot (0.3 meters) of its position, the astronauts outside gave Wakata "invaluable" instructions, necessary because the crew was working without the assistance of the camera at the end of the robotic arm. The camera had been out of commission since the short circuit on 14 October.[1013]

*17 October*
Russia launched a Progress supply spacecraft to deliver fuel to the uninhabited *Mir* space station, which was slipping from orbit. Increased solar flare activity had expanded the atmosphere, creating friction between *Mir* and the thin gases above Earth and causing the station to descend

---
[1010] William Harwood, "Structural Truss Added to Spine of Space Station: Shuttle Astronauts Overcome Camera System's Short Circuit," *Washington Post*, 16 October 2000.
[1011] Michael Cabbage, "Discovery Spacewalk Goes Off Without a Hitch: Crew Members Finished a Six-Hour Task that Included Hooking Up a Truss and an Antenna," *Orlando Sentinel* (FL), 16 October 2000.
[1012] NASA, "Nobel Prize Laureate Appointed as Senior NASA Advisor," news release 00-164, 16 October 2000, http://www.nasa.gov/home/hqnews/2000/00-164.txt (accessed 1 April 2008).
[1013] Marcia Dunn for the Associated Press, "Docking Port Nudged Carefully into Place," *Sun-Sentinel* (Ft. Lauderdale, FL), 17 October 2000.

steadily, ever since its last crew had returned to Earth in June after a 73-day mission. The fuel would enable the station to achieve a higher orbit, preventing it from falling into thicker layers of the atmosphere and burning up. Some Russian politicians had called for disposal of the station, and U.S. space officials had urged the Russian government to dedicate its "scarce space funds" to the ISS. However, MirCorp, a private company based in the Netherlands, had leased time on *Mir* and agreed to fund the Progress launch, thereby saving the station.[1014]

*18 October*
NASA Administrator Daniel S. Goldin recognized members of the Space Shuttle's ice and debris inspection team for their work on 10 October, leading to the discovery of a stray 4-inch (10-centimeter) pin near the Shuttle's external fuel tank, hours before the scheduled launch of STS-92. In response to the find, NASA had delayed the launch for 24 hours to allow the team to retrieve the pin, thereby preventing the damage that would have ensued if the pin had been sucked into the Shuttle's thermal protection system or into a main engine. Goldin awarded employees Gregory N. Katnik and Jorge E. Rivera NASA's Exceptional Achievement Medal, an honor recognizing significant contributions to the mission of NASA through substantial and significant improvements in operations, efficiency, service, financial savings, science, or technology. United Space Alliance employees Michael Barber, John B. Blue, and Thomas F. Ford received NASA's Public Service Medal, an award given to individuals who are not government employees but make exceptional contributions to NASA's mission.[1015]

In preparation for the December arrival of a pair of solar panels, astronauts Michael Lopez-Alegria and Peter J. K. "Jeff" Wisoff cleaned up the top of the newly installed truss on the ISS. The two men also deployed a tray on the truss, which would hold the connections between the station's central structure and the *Destiny* laboratory module built by the United States and scheduled for installation in January. Next, the spacewalkers tested "mini-jetpacks," known as Simplified Aid for EVA Rescue or SAFER. NASA intended the nitrogen-powered jetpacks for use only in emergencies, such as rescuing an injured or incapacitated spacewalker. With only limited fuel supplies in the jetpacks, each of the astronauts' moves had to be deliberate; each test maneuver lasted approximately 3 minutes, while the astronaut traveled a distance of about 50 feet (15 meters). The astronauts remained tethered to the Shuttle's robotic arm throughout the tests.[1016]

*19 October*
During their only full day inside the ISS, Shuttle crew members spent their last day before heading back to Earth transferring supplies, wiping down the station's walls with fungicide to prevent mold and mildew, and testing four gyroscopes installed earlier in the mission. The crew spun the gyroscopes briefly at 100 revolutions per minute, and the gyroscopes functioned as expected.[1017]

---

[1014] Associated Press, "Russia Launches Progress Cargo Space Ship," 17 October 2000.
[1015] NASA, "Space Shuttle Inspection Team Rewarded for Its 'Eagle Eyes'," news release 00-165, 18 October 2000, http://www.nasa.gov/home/hqnews/2000/00-165.txt (accessed 1 April 2008).
[1016] C. Bryson Hull for Associated Press, "Discovery Astronauts Head Out for Final Day of Spacewalks," 18 October 2000; Associated Press, "Discovery's Crew Zips Around Bay on New Jetpacks," *USA Today*, 19 October 2000.
[1017] C. Bryson Hull for Associated Press, "Shuttle, Station To Part Ways After Weeklong Construction Mission," 20 October 2000.

*20 October*
SkyCorp signed a commercial Space Act agreement with NASA providing that SkyCorp deploy into orbit, either from the Space Shuttle or from the ISS, a satellite containing an Internet server. NASA and SkyCorp planned to test the idea as a "proof of concept" for a Low Earth Orbit 544 satellite constellation, offering Internet access on a global scale. SkyCorp chose to launch an Apache server on a small satellite containing an Apple 500 MHz Power Mac G4. The agreement with NASA outlined SkyCorp's novel idea of launching a partially assembled satellite aboard a Space Shuttle. Astronauts would assemble the craft in space before tossing it from a Shuttle or ISS airlock into an independent orbit. SkyCorp named its concept Gossamer. Typically, engineers assemble a satellite fully on Earth, and it must withstand the stress of launch. A satellite built according to the Gossamer concept would only need to retain structural integrity in microgravity, greatly simplifying guidance, navigation, and control requirements.[1018]

*23 October*
Chairman and founder of Beal Aerospace Technologies Inc. Andrew Beal announced his decision to cease all business operations effective immediately. The company had sought to develop low-cost, reliable, commercial space launch systems and had "made significant advances in low cost hydrogen peroxide propulsion systems." Beal expressed confidence that, despite the company's cost overruns and schedule delays, Beal Aerospace Technologies could have successfully developed its BA-2C rocket-launch system if it had remained in operation. Beal outlined the risk factors that had been beyond the company's control and had led to the decision to cease operations: 1) NASA's and the U.S. government's commitment to the subsidization of competing launch systems; 2) federal laws mandating the company's "potential liability for pre-existing environmental contamination at the only available Cape Canaveral launchpads"; and 3) uncertainty about receiving approval from the U.S. Department of State to launch from the company's launch facilities in Guyana. Beal remarked that government subsidization of launch systems was the primary reason that the company had to choose, either to become a government contractor, like the Boeing Company and Lockheed Martin, or to cease operations.[1019]

*24 October*
NASA's Langley Research Center selected Swales and Associates Inc. to provide research and development, engineering, and support services, under a contract valued at up to US$240 million over five years. Specific work areas covered under the contract included full-spectrum aerodynamics, gas dynamics, fluid dynamics, aerothermodynamics, acoustics and aeroacoustics, metallic and nonmetallic structures and materials, and spaceborne and airborne systems.[1020]

Space Shuttle *Discovery* landed safely at Edwards Air Force Base in California, ending the 100th mission of the Shuttle program. Poor weather conditions had prevented the Shuttle's scheduled

---

[1018] SkyCorp, "SkyCorp Announces Space Act Agreement with NASA To Fly First Internet Webserver in Space," news release, 20 October 2000; Keith Cowing, "SkyCorp Signs Agreement with NASA To Fly the First Webserver in Space—and It Will Be a Mac G4," *SpaceRef.com*, 23 October 2000.

[1019] Beal Aerospace Technologies Inc., "Statement from Andrew Beal Regarding Cease of Operations by Beal Aerospace," news release, 23 October 2000.

[1020] NASA, "NASA Awards Support Services Contract to Maryland Firm," news release C00-o, 24 October 2000, *http://www.nasa.gov/home/hqnews/contract/2000/c00-o.txt* (accessed 1 April 2008).

landing in Florida on 22 October. Edwards had opened as a backup site on 23 October, but wind and rainy weather had prevented a landing at either site until 24 October. The landing was the first Shuttle landing at Edwards since poor weather had prevented a Shuttle landing in Florida in March 1996. In the early years of the program, Shuttles had typically landed at Edwards Air Force Base, but later, both launches and landings became the functions of KSC in Cape Canaveral, Florida.[1021]

*26 October*
NASA announced the findings of an investigation into the 16 June incident involving the Space Shuttle's main engine. The team, headed by Robert L. Sackheim, Assistant Director for Space Propulsion at NASA's MSFC, had determined that 24 square inches (155 square centimeters) of tape had fallen into the Shuttle engine's fuel system and that nobody had noticed the tape before the engine's test firing. The tape's location had caused the engine's temperature to increase rapidly, beyond normal operating limits, damaging some components of the engine's fuel pump. According to Sackheim, the engine controller had performed according to its design, shutting down the engine 5 seconds into the planned 200-second test when it sensed a temperature exceeding safe limits. Sackheim's team had concluded, "the handling of, accounting for, and inspecting for loose materials, used to process and rebuild engines during normal operations, were inadequate." The test had been a "temperature margin" demonstration, undertaken as part of the developmental phase of a more robust Pratt & Whitney Advanced Technology High Pressure Fuel Turbopump. The engine had not been in flight configuration but had been a unit in a testing process aimed at validating the prototype engine's ability to operate at higher-than-normal temperatures.[1022]

The Russian government earmarked funds to send two Progress cargo spacecraft to the *Mir* space station, announcing that it would wait until February 2001 to make a final decision about the station's fate. Earlier, Deputy Prime Minister Ilya I. Klebanov had explained that the decision the Russian government had been wavering about for more than a year rested on the availability of private funds to keep the station in orbit. Executives from MirCorp had promised the Russian government that the company would raise between US$100 million and US$170 million by next year, but Russian space officials were skeptical of MirCorp's ability to meet this commitment. MirCorp had yet to pay the US$10 million it owed Russia for the 17 October launch of a Progress cargo spacecraft delivering fuel to *Mir*.[1023]

NASA announced that it planned to implement a new Mars Exploration Program over the following two decades. NASA planned six major missions during the next 10 years, launching the Mars Odyssey orbiter mission in 2001, twin Mars Exploration Rovers in 2003, and Mars Reconnaissance Orbiter—a powerful scientific orbiter—in 2005. Additional plans included the development and launch, possibly as early as 2007, of a "long-range, long-duration mobile science laboratory," which would "pave the way for a future sample return mission." NASA also

---

[1021] *Washington Times*, "Shuttle Lands at Edwards for First Time since 1996: Marks 100th Mission of U.S. Program," 25 October 2000; Matthew Fordahl for Associated Press, "Discovery Lands in California," 25 October 2000.
[1022] NASA, "Shuttle Main Engine Test Investigation Points to Fuel System Contamination," news release 00-170, 26 October 2000, *http://www.nasa.gov/home/hqnews/2000/00-170.txt* (accessed 1 April 2008).
[1023] Vladimir Isachenkov for Associated Press, "Russia Earmarks Funds for Mir," 26 October 2000.

proposed to create a new line of small missions called Scout missions, involving airborne vehicles or small landers. NASA would select designs for the Scout spacecraft from proposals submitted by the scientific community. Besides sending additional scientific orbiters, rovers, and landers to Mars during the second decade of the program, NASA announced its plans to launch its first sample-return mission in 2014, with a second mission in 2016. The new program incorporated "lessons learned from previous mission successes and failures" and built on recent scientific discoveries. Although NASA led the revamped program, the Mars missions also included international participants, particularly France and Italy, whose space agencies had agreed to conduct collaborative scientific orbital and surface investigations, as well as to contribute sample collection-and-return systems, telecommunications assets, and launch services.[1024]

*30 October*
NASA awarded Lockheed Martin Space Systems a US$1.15 billion six-year contract to produce 35 additional super-lightweight external tanks for the Space Shuttle Program, the sixth production of tanks and the first composed entirely of super-lightweight tanks. The contract covered the manufacture, assembly, test, and delivery of the tanks, as well as the operations and maintenance of NASA's Michoud Assembly Facility in New Orleans and activities at MSFC and KSC. Jerry W. Smelser, Manager of MSFC's External Tank Project Office, described the contract as designed to "assure the delivery of a quality product and to reward the contractor and employees on the basis of performance."[1025]

President William J. Clinton signed into law H. R. 1654, the National Aeronautics and Space Administration Authorization Act of 2000, authorizing funding for "a robust space and aeronautics program for the Nation." Clinton commended bipartisan efforts to pass an authorization bill funding NASA's priorities, such as building the ISS, improving Space Shuttle safety, and using the Space Launch Initiative to reduce the cost of access to space. However, Clinton criticized the bill for limiting "NASA's flexibility to pursue a promising commercial habitation module" for the ISS, saying that the bill included several "objectionable provisions" and only one of the 14 legislative provisions that his administration had proposed.[1026]

*31 October*
The first crew to inhabit the ISS launched from Baikonur Cosmodrome in Kazakhstan on its four-month mission. A veteran of three U.S. Space Shuttle missions, William M. Shepherd commanded Expedition I. Crew members, with whom Shepherd had trained since 1996, were two cosmonauts, Pilot Yuri P. Gidzenko of the Russian air force and Flight Engineer Sergei K. Krikalev, a veteran of four Russian and U.S. space missions. With 484 days logged in orbit, Krikalev was one of the world's most experienced space travelers.[1027]

---

[1024] NASA, "NASA Outlines Mars Exploration Program for Next Two Decades," news release 00-171, 26 October 2000, *http://www.nasa.gov/home/hqnews/2000/00-171.txt* (accessed 1 April 2008).
[1025] NASA, "NASA Awards $1.15 Billion Contract for Shuttle External Tanks," news release C00-p, 30 October 2000, *http://www.nasa.gov/home/hqnews/contract/2000/c00-p.txt* (accessed 1 April 2008).
[1026] U.S. Newswire, "Clinton Statement on Signing of the 'National Aeronautics and Space Administration Authorization Act of 2000'," 30 October 2000.

## NOVEMBER 2000

*2 November*
The first International Space Station (ISS) crew, Expedition I, arrived at the new station where they planned to live for the next four and one-half months. The crew's *Soyuz* craft automatically docked at a port on the *Zvezda* habitation module. The Russian-built, NASA-financed *Zarya* propulsion-and-storage module, attached to the other end of *Zvezda*, was loaded with supplies and equipment delivered by recent Shuttle missions. Unless an emergency or breakdown occurred, the multi-hatch *Unity* module would remain off limits. A Shuttle crew had installed a structural truss on the *Unity* module in October, and during a mission scheduled for early December, another crew would attach a set of solar arrays to the truss. Without the solar arrays, the station did not have enough power to heat the module properly or to control the dew point, to prevent dangerous condensation. The ISS crew and their U.S. and Russian managers temporarily named the station *Alpha*, at least for the duration of the Expedition I mission.[1028]

*3 November*
NASA announced that the Chandra X-ray Observatory had detected iron-emission lines in the afterglow of a gamma-ray burst (GRB), marking the first time such lines had been "unambiguously detected" in association with GRBs, and the first time scientists had been able to measure their properties precisely in x-ray wavelengths. Luigi Piro, lead author of a paper published in the 3 November issue of the journal *Science*, explained that the discovery provided an important clue to understanding the origins of the bursts. The new data enabled scientists to rule out the theory that two neutron stars or black holes collide to cause a GRB, one of various theories about how the bursts originate. Instead, Piro suggested that the bursts were more likely the result of "something similar to a supernova explosion, but much more powerful."[1029]

*6 November*
Jason Allen Diekman of Mission Viejo, California, pled guilty to federal charges of computer crimes after negotiating a deal with prosecutors. Prosecutors had charged Diekman in September with illegal hacking and with using stolen credit card numbers to purchase US$6,000 worth of computer equipment and other items. He had confessed to investigators that he had hacked into "hundreds, maybe thousands" of computers over two years, including systems at the University of California at Los Angeles, Harvard University, Cornell University, the University of California at San Diego, and California State University at Fullerton. Diekman's most serious confessed invasions had occurred at NASA's Jet Propulsion Laboratory (JPL), where he had gained root-level access to two computer systems, and at Stanford University, where he had

---

[1027] John Daniszewski, "Crew of 3 Blasts Off To Make a Home of Space Station," *Los Angeles Times*, 31 October 2000.
[1028] William Harwood, "Crew of 'Alpha' Males Moves into Space Station: Docking Complete, NASA Chief Allows Name's Use for Now," *Washington Post*, 3 November 2000.
[1029] NASA, "NASA's Chandra Captures Telling Gamma-Ray Afterglow," news release 00-173, 3 November 2000, http://www.nasa.gov/home/hqnews/2000/00-173.txt (accessed 7 April 2008); L. Piro et al., "Observation of X-ray Lines from a Gamma-Ray Burst (GRB991216): Evidence of Moving Ejecta from the Progenitor," *Science* 290, no. 5493 (3 November 2000): 955–958.

gained control over 24 systems, two of them owned by NASA and containing flight-control software for NASA's satellites. The U.S. District Judge set sentencing for 5 February 2001.[1030]

*8 November*
An eruption of solar-flare activity prompted NASA to order the ISS crew to take shelter in the Russian-built *Zvezda* module for 12 hours. The federal Space Environment Center in Boulder, Colorado, rated the solar-flare event an S-4, making it the fourth-largest solar-radiation storm since 1976. NASA calculated that the ISS would travel through danger zones during the seven or eight orbits of a 12-hour period on 9 November. Officials advised the crew to activate radiation-detection monitors in *Zvezda*'s living quarters and to remain in the aft section of the module during working hours, except for two periods of 15 to 20 minutes during each 90-minute orbit, when they could safely move around the cabin. Michael J. Golightly, Chief of Space Science at Johnson Space Center (JSC), explained that the *Zvezda* module had a shield heavy enough to reduce the crew's radiation exposure by 60 percent.[1031]

*9 November*
The Chandra X-ray Observatory team won a Current Achievement award from the Smithsonian Institution's National Air and Space Museum for "its efforts in building, placing in orbit, and operating the most sophisticated astronomical observatory ever built." NASA had first proposed the Chandra in 1976 and had placed the telescope in orbit during a Space Shuttle mission in July 1999. Since its deployment, Chandra had refined scientists' knowledge of the nature of galactic nuclei, confirmed the existence of supermassive black holes at the centers of galaxies like Andromeda and the Milky Way, captured images of an active starburst galaxy, analyzed the compositions of supernovas, and examined x-ray stars in the Orion Nebula. The museum also honored engineer and aviator A. Scott Crossfield with a Lifetime Achievement award. Crossfield had begun his career in the 1950s as a research pilot for NASA's predecessor agency, the National Advisory Committee for Aeronautics, and had become the first pilot to exceed Mach 2, accomplishing this feat while flying the rocket-powered Douglas D-558-2. Crossfield had joined North American Aviation in 1955 and had worked on all aspects of the X-15 program. In addition to his work on the X-15, Crossfield had developed the ground-control test methodology that became standard in the Mercury, Gemini, and Apollo space programs. He had served as a division vice president at Eastern Airlines between 1967 and 1975 and, after that, as a technical advisor to the U.S. House Science Subcommittee for Transportation, Aviation, and Weather until his retirement in 1993.[1032]

NASA announced that the Hubble Space Telescope (HST) had captured an image of a lone neutron star traveling close to Earth, one of several hundred million neutron stars that scientists believed to exist in the Milky Way. Hubble scientists believed that, at 200 light-years away, this neutron star was the nearest to Earth. No companion star affected its appearance, allowing

---
[1030] David Rosenzweig, "Hacker Pleads Guilty to Invading JPL, Stanford Computers Crime: Mission Viego Man, 20, Accepts Federal Plea Bargain. He Had Gained Control of NASA Computers at Both Sites," *Los Angeles Times*, 7 November 2000.
[1031] Frank D. Roylance, "Space Station Astronauts Take Shelter from Solar Radiation: Crew Retreats to Module with more Shields During Unusually Severe Flares," *Baltimore Sun* (MD), 11 November 2000.
[1032] Smithsonian National Air and Space Museum, "Chandra X-ray Observatory Team and Scott Crossfield Are Smithsonian's National Air and Space Museum Trophy 2000 Winners," news release, 13 November 2000, *http://www.nasm.si.edu/events/pressroom/releaseDetail.cfm?releaseID=88* (accessed 15 April 2008).

astronomers to more easily test and confirm stellar theories against its physical properties, such as size, brightness, and true age. Frederick M. Walter of State University of New York at Stony Brook remarked that, as the closest and brightest of the few known isolated neutron stars, this star would be "the easiest to study and is an excellent test bed for nuclear astrophysical theories."[1033]

United Space Alliance finalized its investigation into the loose pin that had delayed the October launch of Space Shuttle *Discovery*, reporting that the pin had come from the Vehicle Assembly Building. Technicians had incorrectly installed the pin, intended to secure a platform above part of the Shuttle's fuel tank. During the Shuttle's transport to the launchpad on 8 September, the 4-inch (10-centimeter) metal pin had fallen 73 feet (22 meters), landing on a fuel line near one of the tank's connections to the Shuttle. After inspectors had found the pin on 10 October, NASA officials had delayed *Discovery*'s launch 24 hours, so that technicians could remove the pin, preventing damage during liftoff.[1034]

*10 November*
NASA's Stardust spacecraft entered safe mode after a stream of high-energy protons emanating from a solar flare hit the craft. Stardust, on a mission to return samples of a comet, had been 130 million miles (210 million kilometers) from the Sun, traveling at about 12,000 miles (19,000 kilometers) per hour, when the solar flare erupted on 9 November. The solar wind had brought the stream of protons into contact with the craft hours later, in the middle of the night, confirming that NASA engineers and scientists monitoring the situation had reason to feel concerned.[1035]

*13 November*
The NASA flight team for the Stardust spacecraft commanded it to leave safe mode after having successfully inspected the craft's cameras. Once back in normal operating mode, the craft controlled its orientation in space perfectly.[1036]

*15 November*
Iridium Satellite LLC announced that the U.S. Bankruptcy Court for the Southern District of New York had approved its bid to purchase the operating assets, property, and intellectual property of Iridium LLC. Iridium Satellite LLC would continue providing commercial satellite communication services to the U.S. government.[1037]

*16 November*
The Russian government decided that it could no longer afford to maintain the *Mir* space station and would destroy it in February 2001 during a controlled descent. The Russian government had previously decided to abandon *Mir*, but had extended the space station's time in orbit after the private MirCorp, based in the Netherlands, had leased time on the station, financing its operation.

---

[1033] NASA, "Hubble Sees Lone Neutron Star Streaking Across Galaxy," news release 00-176, 9 November 2000, http://www.nasa.gov/home/hqnews/2000/00-176.txt (accessed 7 April 2008).
[1034] Kelly Young, "Pin Found on Shuttle from VAB," *Florida Today* (Brevard, FL), 11 November 2000.
[1035] Associated Press, "NASA's Stardust Spacecraft Survives Encounter with Solar Flare," 21 November 2000.
[1036] Associated Press, "NASA's Stardust Spacecraft."
[1037] Iridium LLC, "Statement for Iridium Satellite LLC," news release, 15 November 2000.

According to the Russian government, a primary factor in its decision to end *Mir*'s orbit had been increasing skepticism over MirCorp's continued ability to finance *Mir*.[1038]

*17 November*
When the automatic docking system failed, cosmonaut Yuri P. Gidzenko used remote control to dock the first Progress cargo spacecraft arriving at the ISS. The craft, loaded with food, boots, and supplies, had traveled smoothly to the ISS until within 300 feet (91.4 meters) of the station. When the spacecraft's automatic docking system failed to lock onto the space station, Gidzenko took control, guiding the craft to within 15 feet (4.6 meters) of the station. Glaring sunlight and a fogged camera lens forced him to wait 40 minutes, before he was able to complete the maneuver safely.[1039]

*20 November*
EarthWatch Inc., a Colorado-based company developing a network of satellites to capture images of various regions of the Earth for commercial applications, launched its *QuickBird 1* craft from Plesetsk Cosmodrome, aboard a Russian Cosmos-3M rocket. However, ground stations did not pick up its signals as planned. U.S. tracking data listed the craft's orbit as "decayed," and a Russian Aerospace Agency spokesperson described the craft as "effectively lost." EarthWatch had lost its first satellite, *Early Bird 1*, four days after its 24 December 1997 launch because of a power-system problem.[1040]

Arizona State University and JPL, a division of California Institute of Technology, announced the creation of the Arizona State University Planetary Imaging Facility and Advanced Training Institute (PIF-ATI) for the study of Mars. Arizona State University and JPL would jointly fund the new facility, which the two institutions intended for scientists and students to use. PIF-ATI expanded on a facility originally planned in support of the Thermal Emission Imaging System (THEMIS), scheduled to fly on the 2001 Mars Odyssey spacecraft. PIF-ATI would offer scientists and students outside the project greater access to instruments and data, providing access for university students, for students of the fifth through twelfth grades, and for their teachers.[1041]

*21 November*
Two Earth-monitoring satellites launched from Vandenberg Air Force Base aboard a Boeing Delta II rocket, successfully reaching orbit. NASA launched its Earth Observing I (EO-I) satellite, part of its New Millenium Program, to test several advanced technologies for possible use in future missions. NASA planned to fly the US$193 million craft within 2 miles (3 kilometers) of the *Landsat 7* satellite to collect the same images, with the *Landsat 7* images serving as the benchmark for the new imager aboard the EO-I. The second satellite—the multinational SAC-C satellite, a joint venture of NASA, Argentina, Brazil, Denmark, France,

---

[1038] Vladimir Isachenkov for Associated Press, "Russia To Dump Mir Space Station," 16 November 2000.
[1039] *Chicago Tribune*, "Cosmonaut Forced To Dock Cargo Ship by Remote Control," 19 November 2000.
[1040] Stephen Clark, "Commercial Eye-in-the-Sky Appears Lost in Launch Failure," *Spaceflight Now*, 21 November 2000.
[1041] NASA Jet Propulsion Laboratory, "New Mars Research Facility To Involve Scientists, Kids," news release 2000-115, 20 November 2000, *http://www.jpl.nasa.gov/releases/2000/marsasu.html* (accessed 17 April 2008).

and Italy—carried 11 instruments for studying the Earth's surface, atmosphere, and magnetic field, including an instrument to observe the migration of the Franca whale.[1042]

NASA selected six teams of scientists to participate in the first new mission of its Origins Program. The teams would use the new Space Infrared Telescope Facility (SIRTF), scheduled for launch in July 2002, to study the formation of galaxies, stars, and dust discs, under the following projects: "Galaxy Birth and Evolution," "Black Holes and Galaxies," "Unveiling Hidden Stars," "Inside the Milky Way," "From Gas to Stars," and "Planet Formation: When the Dust Settles." The teams, selected from 28 proposals submitted by astronomers located around the world, comprised the SIRTF Legacy Science Program.[1043]

NASA selected 41 proposals of the 109 scientists had submitted, to conduct research on Earth and in space using NASA's microgravity research facilities. NASA intended the researchers to use the facilities to "enhance understanding of physical, biological, and chemical processes associated with fundamental physics." NASA was funding the research with more than US$15 million in grants over a four-year period, under the sponsorship of its Office of Biological and Physical Research. Sixteen of the grants continued work already funded by NASA, and twenty-four were for new research efforts. Thirty-six grants funded ground-based research, and the remaining five funded flight-definition projects.[1044]

The U.S. Department of Energy announced its selection of the Idaho Engineering and Environmental Laboratory to produce plutonium-238, the most radioactive form of plutonium. NASA planned to use the isotope, which generates electricity to "keep things from freezing up," to fuel a spacecraft on a mission to Pluto planned for after 2020.[1045]

*22 November*
Gerald A. Soffen, Director of University Programs at NASA's Goddard Space Flight Center (GSFC), died of a heart ailment. Soffen had begun his career at NASA at JPL, working on biological instrumentation development, moving in the mid-1970s to Langley Research Center (LARC), where he worked on the Viking Mars Project as a project scientist. He became LARC's chief environmental scientist before transferring to GSFC in 1983. At Goddard, Soffen had helped develop to NASA's astrobiology program and to establish NASA's Astrobiology Institute. In 1978 NASA had appointed him Life Sciences Director at NASA Headquarters and, since its establishment in 1990, he had served as Director of the University Programs Office. GSFC Director Alphonse V. Diaz remarked that science and students were Soffen's two loves, and that "the Agency and the nation will continue to benefit enormously from the talented young people he has brought into the scientific community."[1046]

---

[1042] Associated Press, "Two Earth-Monitoring Satellites," 21 November 2000.
[1043] NASA, "The Sky's the Limit: Science Teams Chose for Space Observatory," news release 00-184, 21 November 2000, http://www.nasa.gov/home/hqnews/2000/00-184.txt (accessed 7 April 2008).
[1044] NASA, "NASA Announces Research Grants in Fundamental Physics," news release 00-183, 21 November 2000, http://www.nasa.gov/home/hqnews/2000/00-183.txt (accessed 7 April 2008).
[1045] Associated Press, "A Type of Plutonium," 22 November 2000.
[1046] NASA, "NASA Astrobiology Architect, Dr. Gerald Soffen, Dies," news release 00-186, 24 November 2000, http://www.nasa.gov/home/hqnews/2000/00-186.txt (accessed 7 April 2008); *Washington Post*, "Gerald Alan Soffen, NASA Official," 27 November 2000.

The government of the People's Republic of China released a document outlining its policy regarding its future human spaceflight program and satellite-launch industries. The document, which provided the outside world with information about China's top-secret space program, described the Chinese space industry as "an integral part of the state's comprehensive development strategy" and stated, "exploration and utilization of outer space should be for peaceful purposes and benefit the whole of mankind." The paper provided details of China's satellite program, stating that as of October 2000, China had developed and launched 47 satellites with a success rate of over 90 percent. China had four satellite series in operation: the Dongfanghong telecommunications series; the Fengyun meteorological series; the Shijian research and technology series; and the Ziyuan Earth-observation series, which had recently launched its first spacecraft. The paper also outlined China's efforts toward international cooperation, notably China's signing of an agreement with a dozen countries, including the United States, Russia, Japan, and several European countries, allowing for scientific exchange and joint development of spacecraft components, as well as for commercial launch services. To date, China had launched 27 foreign-made satellites using its Long March series of rockets.[1047]

*24 November*
NASA Director of Media Services Brian D. Welch died after suffering a heart attack. Welch, who had led many of NASA's public outreach efforts, was responsible for its news operations, for NASA Television, and for NASA's Internet efforts. Welch had begun his career as a cooperative-education student of public affairs at LARC in 1979. In 1981 he had become editor of *Space News Roundup*, the newspaper of NASA's JSC. Throughout the 1980s and 1990s, Welch had worked as a public affairs mission commentator, Deputy News Chief at Mission Control Center, Manager of JSC's Mission Commentary Team, and Newsroom Manager during Shuttle flights. NASA had appointed Welch as Chief of News and Information in 1994, and he had become Director of Media Services in 1998. Administrator Daniel S. Goldin spoke of how Welch's sudden death had stunned NASA, saying that Welch's "love and enthusiasm for spaceflight and exploration" was "infectious," and describing Welch's passionate approach to his work as the true embodiment of "the spirit of this agency."[1048]

*28 November*
NASA announced the selection of a science team for the Space Interferometry Mission (SIM), scheduled for launch in 2009. The team consisted of 10 principal investigators leading key science teams and 5 mission specialists. NASA intended SIM, part of its Origins Program, to search for Earth-sized planets around other stars; to measure precisely the locations and distances of stars throughout the Milky Way Galaxy; and to study other celestial objects, helping to answer fundamental questions about the origin and evolution of the galaxy. NASA planned to place the SIM spacecraft in an Earth-trailing orbit around the Sun, enabling the craft's multiple telescopes to gather the Sun's light and producing information normally only obtained with much larger telescopes. NASA considered identification of potential observing targets for the Terrestrial Planet Finder a critical part of SIM's mission. NASA planned the Terrestrial Planet Finder to

---

[1047] Stephen Clark, "China Temporarily Lifts Its Veil on Its Secret Space Program," *Spaceflight Now*, 24 November 2000.
[1048] NASA, "Brian Welch, NASA Director of Media Services, Dies," news release 00-187, 27 November 2000, *http://www.nasa.gov/home/hqnews/2000/00-187.txt* (accessed 7 April 2008).

capture images of planetary systems around other stars and to search for chemical signatures suggesting the possibility of life.[1049]

The National Polar-Orbiting Operational Environmental Satellite System (NPOESS) Integrated Program Office (IPO) selected Raytheon Company to provide imaging-sensor instruments for the new NPOESS craft, which would replace the U.S. Department of Commerce's Polar-Orbiting Operational Environmental Satellites (POES) and the Defense Meteorological Satellite Program (DMSP) satellites. Under a contract valued at US$152.8 million, Raytheon would design, develop, and test the Visible Infrared Imaging Radiometer Suite (VIIRS) instrument and develop the algorithms to produce environmental data records from VIIRS data.[1050]

Lockheed Martin and NASA entered into a consolidated space operations contract (CSOC), allowing NASA to use the services of two commercial, satellite ground-tracking stations. Under CSOC, NASA customers would have access on a per-pass basis to additional stations on Svalbard Island in Norway, owned by Kongsberg Spacetec-Lockheed Martin Space Data Services, and in Poker Flat, Alaska, owned by DataLynx. Both tracking stations had recently completed CSOC operational-readiness reviews in preparation for inclusion in Lockheed Martin's catalog of services available to NASA programs.[1051]

*30 November*
NASA announced that new technology recently tested at Ames Research Center (ARC) had converted sounds emanating from landing-gear wind noise into color images on computer screens, "enabling engineers to pinpoint loud and preventable aircraft flight noise more easily than in the past." The imposition of nighttime curfews on noisy takeoffs and landings had prompted aircraft manufacturers to develop quieter planes. Paul T. Soderman, leader of the ARC aeroacoustics group, explained that the ability to visualize the cause of wind noise coming from landing gear enables engineers to analyze the problem so that they can find ways to reduce significantly noise. The engineers had created the sound images using a computer linked to a quarter-scale landing-gear model and an array of 70 microphones inside the wall of a wind tunnel. The researchers had then removed various combinations of landing-gear parts in the wind tunnel, resulting in significant noise reduction.[1052]

Space Shuttle *Endeavour* launched from Kennedy Space Center in Cape Canaveral, Florida, on Mission STS-97, carrying to the ISS Commander Brent W. Jett Jr., Pilot Michael J. Bloomfield, and Mission Specialists Joseph R. Tanner, Marc Garneau, and Carlos I. Noriega. The purpose of STS-97, also known as ISS Assembly Flight 4A, was to connect solar arrays, to prepare a docking port for the U.S.-made *Destiny* module, to install Floating Potential Probes, to install

---

[1049] NASA, "Science Team Chosen for Space Interferometry Mission," news release 00-178, 28 November 2000, *http://www.nasa.gov/home/hqnews/2000/00-178.txt* (accessed 7 April 2008).
[1050] Associated Press, "Raytheon Awarded $152.8 Million Contract for Meteorological Satellite Imaging Sensor," 28 November 2000.
[1051] "Lockheed Martin Adds Two Commercial Stations to CSOC Capability," *Aerospace Daily*, 29 November 2000.
[1052] NASA, "NASA Technology Allows Engineers to See Airframe Noise," news release 00-189, 30 November 2000, *http://www.nasa.gov/home/hqnews/2000/00-189.txt* (accessed 7 April 2008).

camera cable outside the *Unity* module, and to transfer supplies, refuse, and equipment between the ISS and the Shuttle.[1053]

## DECEMBER 2000

*1 December*
Raymond Toricelli of New Rochelle, New York, pled guilty to breaking into two NASA computers in 1998, to steal user names and passwords to gain access to other systems. The two computers, which helped launch robotic spacecraft, were at NASA's Jet Propulsion Laboratory (JPL) in Pasadena, California. Toricelli had also set up a chat room to direct people to a pornography site, which paid him 18 cents per referral; broken into a San Jose University computer system to gain access to other networks; and stolen more than 15 credit card numbers online. He stated that he had never intended to damage the computers he illegally entered. Prosecutors planned to seek a prison term of 8 to 14 months at Toricelli's sentencing on 7 March 2001. The maximum penalty for his crime was 27 years in prison and a US$950,000 fine.[1054]

Space Shuttle *Endeavour* docked successfully with the International Space Station (ISS), the first Shuttle docking at an inhabited ISS. William M. Shepherd, Commander of the ISS, noted the festive mood at the space station, even though the two crews would not meet in person until later in the mission. Soon after *Endeavour* docked, Canadian astronaut Marc Garneau used the Shuttle's robotic arm to lift the 49-foot-long (15-meter-long) tower containing the solar array out of the cargo bay, where it would remain suspended 10 feet (3 meters) above the Shuttle while the crew slept.[1055]

*3 December*
*Endeavour* astronauts attempted to deploy the new solar array at the ISS, but only one of the two panels unfurled as planned. Joseph R. Tanner and Carlos I. Noriega conducted a spacewalk to assist in attaching a 17-ton (15,400-kilogram or 15.4-tonne) power assembly, while Marc Garneau maneuvered the Shuttle's robotic arm from inside *Endeavour*. The spacewalkers were unable to open the latches on the storage boxes containing the folded solar panels, delaying their attempts to unfurl the array. They began to unfurl the first panel 1 hour behind schedule, but although it reached its full length in 14 minutes, the panel did not appear to have the correct tension. NASA officials decided to delay expanding the second panel for a full day, to give them time to determine the cause of the inadequate tension of the first panel. The ISS crew did not witness the deployment of the solar array, because the Shuttle crew carried out their work during the ISS crew's sleep period.[1056]

*4 December*
Members of the astronaut crew who had flown aboard NASA's historic 100th Space Shuttle flight, Mission STS-92, visited NASA's Michoud Assembly Facility to present Silver Snoopy

---

[1053] NASA, "NASA Mission Archives: STS-97, International Space Station Assembly Flight 4A," http://www.nasa.gov/mission_pages/shuttle/shuttlemissions/ archives/sts-97.html (accessed 30 October 2008).
[1054] *Chicago Tribune*, "20-Year-Old Pleads Guilty to Hacking into NASA Computers," 3 December 2000.
[1055] Steven Siceloff, "Endeavour Docks at Station: Astronauts To Attach Arrays Today," *Florida Today* (Brevard, FL), 3 December 2000.
[1056] Warren E. Leary, "Space Station Gets Its Wings, but Only One Is Deployed," *New York Times*, 4 December 2000.

awards, "the highest accolade that the astronauts corps gives to people who build flight hardware," to the Lockheed Martin employees who had built the Shuttle's super-lightweight tank and liquid-oxygen tank. NASA had named the Silver Snoopy award after the Peanuts character, the dog that imagined he was a fighter pilot.[1057]

Space Shuttle *Endeavour* Commander Brent W. Jett Jr. successfully unfurled the second panel of the solar array (colloquially referred to as a solar wing) that the crew had deployed on the ISS on 3 December. Jett used computer commands to unfold the panel a few feet at a time, a process that took nearly 2 hours. No problems occurred until the crew had extended the panel almost fully, when it became clear that a panel on each blanket was stuck to a neighboring section. *Endeavour*'s crew turned the Shuttle to allow the Sun to warm the blankets and fired thrusters to shake the array, snapping the wing into place. The panel began generating electricity before it reached its full 115 feet (35 meters). To NASA's relief, the panel was stretched tight. Once the crew had completed the process, Commander Jett radioed to Mission Control that they had two tensioned blankets, and Mission Control replied, "Great work, gentlemen. We think you've earned your solar wings."[1058]

*5 December*
NASA made data from its Advanced Spaceborne Thermal Emission and Reflection Radiometer (ASTER) available to the public. ASTER, a general-purpose imaging instrument featuring 14 spectral bands, extremely high spatial resolution, and stereo-imaging capabilities, had launched aboard NASA's *Terra* satellite in December 1999. One of ASTER's primary goals was "to acquire a one-time cloud-free image of the entire land surface of Earth," intended as a baseline image for monitoring environmental change.[1059]

*Endeavour* astronauts Joseph R. Tanner and Carlos I. Noriega worked outside the ISS to connect power lines, correcting an electrical shortage that had restricted the station's first crew to two of three habitable modules. The two astronauts also inspected the first solar panel that they had unfurled, to determine whether the crew could correct its tension. Wearing a small TV camera on his helmet, Noriega beamed down to Mission Control images of the loose tension cables on the array's right wing. The cables had slipped off their pulleys during the solar panel's deployment on 3 December. Tanner remarked that an astronaut could place one tension cable back on its pulley without too much trouble, but that the other cable would require more work. Mission Control instructed the spacewalkers to try to fix the cables on their third and final spacewalk the following day.[1060]

*7 December*
NASA and the U.S. Federal Emergency Management Agency (FEMA) announced the signing of a memorandum of understanding to form a partnership to use science and space technology to

---

[1057] *New Orleans Time-Picayune*, "Astronauts Thank Lockheed Workers: Company Makes Shuttle Components," 17 December 2000.
[1058] Associated Press, "Astronaut Manages To Spread Station's 2nd Wing," *USA Today*, 5 April 2000; Warren E. Leary, "Astronauts Finnish Adding Solar Wing to Station," *New York Times*, 5 December 2000.
[1059] NASA, "Terra Satellite's ASTER Data Now Available to the Public," news release 2000-124, 5 December 2000, http://www.jpl.nasa.gov/releases/2000/terra.html (accessed 7 April 2008).
[1060] *Chicago Tribune*, "Space Station Gets an Electrical Boost from Solar Wings," 6 December 2000; Associated Press, "Live from Space, 'Carlos Cam'," *Newsday* (Long Island, NY), 6 December 2000.

prevent natural disasters. The agreement called upon NASA and FEMA to apply remote-sensing research images to emergency-management issues, creating accurate and informative maps of flood plains and wildfires, as well as other types of maps, intended to help state and local communities respond to and mitigate natural disasters. The partnership, affiliated with Project Impact: Building Disaster Resistant Communities, would use NASA's Earth Science Enterprise, as "part of an aggressive new strategy devoted to significantly increasing the application of NASA remote sensing data, information, science and technologies to societal needs."[1061]

Astronauts Joseph R. Tanner and Carlos I. Noriega were able to tighten the slack solar panel on the ISS within minutes, using only a hook to place two tension cables back on their pulleys. Although the wing functioned well even though it was slack, if the astronauts had not adjusted its tension, the solar cells might have torn or the support rods might have bent or broken. After repairing the wing, the two astronauts installed a camera cable and a static-electricity monitor on the outside of the ISS. The monitor would help NASA learn how to protect spacewalking astronauts from electrical shocks.[1062]

*8 December*
Michael C. Malin and Kenneth S. Edgett of Malin Space Science Systems published research in the journal *Science* based on images from NASA's Mars Global Surveyor spacecraft. The two researchers had found that the images showed massive sedimentary deposits on Mars, suggesting that the planet "was once a water-rich land of lakes." Malin described the images as revealing "hundreds and hundreds of identically thick layers," features that were "almost impossible to have without water." J. William Schopf, head of the Center for the Study of Evolution and the Origin of Life at the University of California at Los Angeles, commented that the images provided the "strongest evidence yet for what appear to be sedimentary units on Mars," supporting the theory that, billions of years ago, Mars had been wetter, warmer, and possibly hospitable to life.[1063]

Hatches between the ISS and Space Shuttle *Endeavour* opened for the first time, and the three members of the space station crew, who had been living aboard the ISS for five weeks, received their first visitors. ISS Commander William M. Shepherd commended *Endeavour*'s crew for their installation of the giant solar array on the space station. The two crews spent their day together transferring supplies, such as food, water, mail, and gifts, to the space station and removing trash, damaged items, and unneeded items for return to Earth.[1064]

*11 December*

---

[1061] NASA, "NASA, FEMA Partner To Use Science and Space Technology for Disaster Prevention," news release 00-192, 7 December 2000, *http://www.nasa.gov/home/hqnews/2000/00-192.txt* (accessed 7 April 2008).
[1062] Marcia Dunn for Associated Press, "Astronauts Breeze Through Wing Repair: 'All Finished'," 7 December 2000.
[1063] NASA Jet Propulsion Laboratory, "Evidence of Martian Land of Lakes Discovered," news release 2000-123, 4 December 2000, *http://www.jpl.nasa.gov/releases/2000/marslakes.html* (accessed 7 April 2008); Associated Press, "Satellite Photos Indicate Mars Was Water Plentiful," *Washington Times*, 5 December 2000; Michael C. Malin and Kenneth S. Edget, "Sedimentary Rocks of Early Mars," *Science* 290, no. 5498 (8 December 2000): 1927–1937.
[1064] Reuters, "Hatch Is Opened; Endeavour, Space Station Crews Become One," *Chicago Tribune*, 9 December 2000; *Los Angeles Times*, "Endeavour's Crew Visits Space Station," 9 December 2000; Warren E. Leary, "Space Station Crew Greets First Visitors," *New York Times*, 9 December 2000.

Space Shuttle *Endeavour* landed at KSC in Cape Canaveral, Florida, after completing the 11-day Mission STS-97 at the ISS, where the crew had installed and deployed the station's new solar array. Shortly before touchdown, the ISS sailed above KSC, a "reminder of the pressure" on NASA to support more flights to the ISS as the station's assembly accelerated. Shuttle Program Manager Ronald D. Dittemore remarked that the successful mission of *Endeavour* to the ISS was "a good way to end this year—a very successful five missions."[1065]

A team of three high school students from the North Carolina School for Science and Mathematics in Durham won first place in the Siemens-Westinghouse Science and Technology Competition. The team won for discovering the first evidence of a neutron star in the nearby supernova remnant IC443, using data from NASA's Chandra X-ray Observatory and the National Science Foundation's Very Large Array (VLA) Observatory. The students had located a point-like source of x-rays embedded in a supernova and had determined that the central object was most likely a pulsar, a young and rapidly rotating neutron star. Bryan M. Gaensler, a pulsar expert at the Massachusetts Institute of Technology who had reviewed the team's paper, remarked that the students had produced "a really solid scientific finding."[1066]

*12 December*
The Netherlands-based MirCorp, the company originally established to keep the *Mir* space station in orbit for commercial purposes, announced that, following a board meeting during the previous week, the company had decided to "retool the company to work with other Russian space equipment." The company planned to develop a new orbiter capable of docking with the ISS, to market existing space technologies, such as cargo spacecraft, and to explore commercial projects involving access to a future Russian module on the ISS.[1067]

*16 December*
Ganymede, the largest moon in Jupiter's solar system, joined the planet Mars and Jovian moons Europa and Callisto as the only known bodies showing strong evidence of the presence of liquid water beneath their surfaces. During the annual meeting of the American Geophysical Union, Margaret G. Kivelson, a planetary scientist at the University of California at Los Angeles, announced that NASA's Galileo spacecraft had found evidence of a vast sea of liquid beneath the surface of Jupiter's moon Ganymede. Using data collected from Galileo's magnetometer during May 2000 and earlier, Kivelson's team had measured the moon's magnetic field. The magnetometer had registered readings "best explained by a thick layer of water—about as salty as Earth's oceans—hidden about 120 miles [190 kilometers] beneath" the moon's surface. Thomas B. McCord, a geophysicist at the University of Hawaii who also presented research at the conference, had used Galileo's Near Infrared Mapping Spectrometer to discover that portions of Ganymede appeared "to have types of salt minerals that would have been left behind by exposure of salty water near or onto the surface." Gene D. McDonald, an astrobiologist at JPL, which manages the Galileo Mission, remarked that Ganymede might be a more promising destination than Europa for robotic spacecraft searching for life. Ganymede, farther from Jupiter

---

[1065] William Harwood, "Endeavour Returns from Station Work," *Washington Post*, 12 December 2000.
[1066] NASA, "Students Using NASA and NSF Data Make Stellar Discovery; Win Science Team Competition," news release 00-195, 11 December 2000, *http://www.nasa.gov/home/hqnews/2000/00-195.txt* (accessed 7 April 2008).
[1067] Andrew Kramer for Associated Press, "Russian Company To Build Spacecraft To Carry Customers," 13 December 2000.

than Europa, has better protection than Europa from Jupiter's intense and deadly radiation. Therefore, biomarker compounds emerging on the surface of Ganymede would survive longer there than on the other Jovian moons, as would any spacecraft orbiting or landing on Ganymede.[1068]

*19 December*
NASA suspended all of Cassini's observations of Jupiter that required the spacecraft to point—to take magnetic-field measurements—in order to capture images, after the craft developed a problem with a maneuvering system. NASA permitted Cassini to continue making observations that did not requiring pointing. On 17 December, Cassini's No. 2 reaction wheel had begun to require extra force to turn, prompting the craft to switch from electrical power to a hydrazine-thrusting system, which Cassini needed to conserve for use in its primary Saturn mission. Cassini Program Manager Robert T. Mitchell remarked that, if tests of the maneuvering system were favorable, the craft could begin its observations again in a week to 10 days.[1069]

*20 December*
NASA's Office of Earth Sciences selected Gencorp Aerojet to build the Advanced Technology Microwave Sounder (ATMS), a spaceborne instrument for measuring microwave energy emitted and scattered by the atmosphere. NASA intended the ATMS to replace instruments currently used on polar-orbiting weather satellites. NASA planned for the ATMS, working alongside an infrared sounder instrument, to produce daily global atmospheric temperature, humidity, and pressure profiles, essential for accurate weather forecasting and long-term climate research. Associate Administrator for NASA's Office of Earth Science Enterprise Ghassem R. Asrar remarked that the advanced technology would improve the accuracy of weather forecasting, from the current three-to-five day forecast span to seven-to-ten day predictions. NASA planned to fly the first unit as part of the National Polar-Orbiting Operational Environmental Satellite System (NPOESS) Preparatory Project Bridge mission, intended to ensure continuity of research-quality data through the bridging of sounding data between NASA's Earth Observing System research missions and future NPOESS operational missions. Goddard Space Flight Center would provide oversight of the contract, worth US$206.6 million.[1070]

NASA announced it had begun searching worldwide for proposals from principal investigators and institutions to develop its first mission to Pluto. In the announcement of opportunity, NASA solicited proposals for a complete mission to the Pluto-Charon system and the Kuiper Belt, requiring that submitted proposals include an expendable launch vehicle and spacecraft, its bus and systems, and the scientific instrumentation package. Although NASA placed no restrictions on the mission's launch date, its stated goal was to reach Pluto by 2015, at a cost of no more than US$500 million in FY 2000 dollars. This was the first time that the Office of Space Science had solicited proposals for a mission to an outer planet on a competitive basis. NASA had chosen this

---

[1068] NASA Jet Propulsion Laboratory, "Solar System's Largest Moon Likely Has a Hidden Ocean," news release 2000-130, 16 December 2000, *http://www.jpl.nasa.gov/releases/2000/aguganymederoundup.html* (accessed 7 April 2008); Kathy Sawyer, "Evidence of Liquid Found on Jupiter's Ganymede; Largest Moon May Be a Place To Look for Life," *Washington Post*, 17 December 2000.
[1069] *Los Angeles Times*, "NASA Halts Spacecraft's Observation of Jupiter," 21 December 2000; Associated Press, "NASA Cassini Spacecraft Trouble Free Approaching Jupiter," 28 December 2000.
[1070] NASA, "NASA Selects Firm To Build Next Generation Weather Instrument," news release C00-q, 20 December 2000, *http://www.nasa.gov/home/hqnews/contract/2000/c00-q.txt* (accessed 7 April 2008).

approach to elicit "creative ideas from innovative thinkers" capable of developing highly focused missions rapidly and at a relatively low cost. NASA made the announcement after it had issued a stop order on the Pluto/Kuiper Express (PKE) mission on 12 September because of the PKE mission's unacceptably high increases in cost. NASA also appointed Colleen Hartman, the Deputy Director of the Research Division of the Office of Space Science, as the new Outer Planets Program Director, to act as the point of contact at NASA Headquarters for budget, content, and policy direction.[1071]

The Huygens Communications Link Enquiry Board released its findings, recommendations, and conclusions, after conducting interviews and hearings with the European Space Agency (ESA), NASA, and industry project staff and experts. The Board had reviewed the unexplained anomalies in the communication subsystem of the Huygens Probe, discovered during a Probe Relay Link Test in February 2000. The ESA Director General had formed an independent inquiry board to assess the current status of the Huygens communication link; to recommend means of safeguarding the mission objectives and guaranteeing full scientific data return; and to recommend ways to prevent similar problems in future projects. The Board had found that the entire project structure had led to the anomaly, outlining the specific problems in its report. The Board also made six recommendations and provided 10 recovery options. The Board's key recommendation was that, in the future, when a mission carries new hardware, the mission plan should be sufficiently flexible to allow for changes from ground command.[1072]

*21 December*
NASA's JPL announced that the Cassini spacecraft, after successfully undergoing a series of tests, was resuming the use of its electrically powered reaction wheels to control its orientation. Although mission scientists were uncertain of the cause of the spacecraft's switch to a different maneuvering system, they suggested that a small piece of material, perhaps from the motor's magnets, might have lodged in an area where it caused friction. Because the scientists were unable actually to detect any such material, they speculated further that either centrifugal force had subsequently forced the material out, or the motor had ground it up. Scientists also suggested that reduced lubrication in the bearings, during prolonged operation at reduced speeds, might have caused the problem, and that, perhaps, the increased motor speeds used during the tests had restored the lubrication.[1073]

*26 December*
Ground controllers in Russia restored communication with *Mir*, after they had been unable to contact the uninhabited space station for 24 hours. Spokesperson for Russian Mission Control Valery Lyndin allayed fears that *Mir* would crash to Earth shortly, announcing that Russia had received no indication that the space station was losing pressure. Mission Control Chief Vladimir A. Solovyov explained that the station's batteries had lost power and no longer had sufficient energy to communicate with ground controllers. Once ground controllers had identified the

---

[1071] NASA, "NASA Seeks Proposals for Pluto Mission; Plans To Restructure Outer Planet Program," news release 00-201, 20 December 2000, *http://www.nasa.gov/home/hqnews/2000/00-201.txt* (accessed 7 April 2008).
[1072] Huygens Communications Link Enquiry Board, "Findings, Recommendations, and Conclusions," 20 December 2000, *http://klabs.org/richcontent/Reports/Failure_Reports/ESA_Cassini/huygens_enquiry_board.PDF* (accessed 21 April 2008).
[1073] NASA Jet Propulsion Laboratory, "Resuming Use of Reaction Wheels," news release, 21 December 2000, *http://saturn.jpl.nasa.gov/news/press-release-details.cfm?newsID=6* (accessed 21 April 2008).

problem, they had switched off specific systems to direct more energy to ground communications. Although *Mir*'s solar panels were recharging the batteries, the cause of the power shortage remained unclear.[1074]

*27 December*
Russian Space Agency Director General Yuri N. Koptev reiterated Russia's commitment to the international community, stating that Russia definitely would bring *Mir* safely down in February 2001. Koptev "angrily dismissed Communist demands to keep the nearly 15-year-old station in orbit," referring to the 20-hour loss of contact earlier in the week as a "final warning that time was up" and an indicator of the level of the station's wear and tear. Communist lawmakers, including cosmonaut Svetlana Y. Savitskaya, who called the move a concession to NASA, had criticized the government's decision to destroy the space station. However, Koptev had countered the criticism, saying that, if *Mir* spun out of control, the Russian president and government would have to explain to the world where the station would fall and how much damage it would cause.[1075]

*28 December*
Cassini Program Manager Robert T. Mitchell, at NASA's JPL, announced the resumption of the spacecraft's observations. Cassini had operated without any problems, since scientists had reactivated its reaction-wheel system on 21 December.[1076]

---

[1074] *Washington Times*, "Contact with Mir Restored After Loss of Battery Power: Mission Control Allays Fears of Crash," 27 December 2000.
[1075] Vladimir Iachenkov for Associated Press, "Space Officials Say Mir Back Under Control, Will Be Dumped in February," 27 December 2000.
[1076] Associated Press, "NASA Cassini Spacecraft Trouble Free Approaching Jupiter," 28 December 2000; NASA Jet Propulsion Laboratory, "Cassini Passes Through Asteroid Belt," news release, 28 December 2000, *http://saturn.jpl.nasa.gov/news/press-release-details.cfm?newsID=4* (accessed 21 April 2008).

# APPENDIX A

# TABLE OF ABBREVIATIONS

| | |
|---|---|
| 2MASS | Two-Micron All Sky Survey |
| | |
| ACE | Advanced Composition Explorer |
| ACRIM | Active Cavity Radiometer Irradiance Monitor |
| ACRP | Advanced Concepts Research Projects |
| ADEOS | Advanced Earth Observing Satellite |
| AIRSAR | Airborne Synthetic Aperture Radar |
| AMM | Auroral Multiscale MIDEX Mission |
| AMS | Alpha Magnetic Spectrometer |
| ARC | Ames Research Center |
| ARGOS | Advanced Research and Global Observation Satellite |
| ASCA | Advanced Satellite for Cosmology and Astrophysics |
| ASCE | Advanced Solar Coronal Explorer |
| ASTER | Advanced Spaceborne Thermal Emission and Reflection Radiometer |
| ATIDS | Airport Traffic Identification System |
| ATMS | Advanced Technology Microwave Sounder |
| AXAF | Advanced X-ray Astrophysics Facility |
| | |
| BA | Beal Aerospace Technologies Inc. |
| Baracuda | Boldly Advanced and Refined Aircraft Concept Under Development (for BESS) |
| BATSE | Burst and Transient Source Experiment |
| BeppoSAX | Italian-Dutch satellite, named in honor of physicist Giuseppe Occhialini, with acronym SAX, for "Satellite per Astronomia X," Italian for "X-ray Astronomy Satellite" |
| BESS | Balloon Borne Experiment with a Superconducting Solenoidal Magnet |
| BOOMERANG | Balloon Observations of Millimetric Extragalactic Radiation and Geophysics |
| BRSP | Brown and Root Services and Pioneer Contract Services |
| | |
| CAMEX-3 | Third Convection and Moisture Experiment |
| CASST | Commercial Aviation Safety Strategy Team |
| CBERS | China-Brazil Earth Resources Satellite |
| CERES | Clouds and the Earth's Radiant Energy System |
| CIA | Central Intelligence Agency |
| CME | coronal mass ejection |
| COBE | Cosmic Background Explorer |
| CRV | crew return vehicle |
| CSA | Canadian Space Agency |
| CSOC | consolidated space operations contract |

## TABLE OF ABBREVIATIONS

| | |
|---|---|
| DARTS | Dynamics Algorithms for Real-Time Simulation |
| DFRC | Dryden Flight Research Center |
| DOD | U.S. Department of Defense |
| DF-31 | Dong Feng-31 |
| DMSP | Defense Meteorological Satellite Program |
| DROMS | Dynamic Runway Occupancy Measurement Systems |
| DTAM | Digital Tectonic Activity Map |
| | |
| EDT | Eastern Daylight Time |
| EO | Earth Observing |
| EOS | Earth Observing Satellite |
| ESA | European Space Agency |
| ESO | European Southern Observatory |
| ESR | emergency Sun reacquisition mode |
| EST | Eastern Standard Time |
| ETR | Experimental Test Range |
| Eutelsat | European Telecommunications Satellite Organization |
| EXPRESS | EXpedite the PRocessing of Experiments to the Space Station |
| | |
| FAA | Federal Aviation Administration |
| FAME | Full-Sky Astrometric Mapping Explorer |
| FAST | Fast Auroral Snapshot |
| FBI | Federal Bureau of Investigation |
| FFC | FutureFlight Central |
| FGB | Functional Cargo Block (known by its Russian abbreviation) |
| FEMA | Federal Emergency Management Agency |
| FUSE | Far Ultraviolet Spectroscopic Explore |
| FVI | Fisk Ventures Inc. |
| | |
| GAO | U.S. General Accounting Office |
| GCN | Gamma Ray Burst Coordinates Network |
| GOES | Geostationary Operational Environmental Satellite |
| GPS | Global Positioning Satellite |
| GRB | gamma-ray bursts |
| GRC | Glenn Research Center |
| GSFC | Goddard Space Flight Center |
| | |
| HAL5 | Huntsville Alabama L5 Society |
| HEDS | Human Exploration and Development of Space Enterprise |
| HESSI | High Energy Solar Spectroscopic Imager |
| HDF-S | Hubble Deep Field South |
| HETE | High Energy Transient Explorer Mission |
| HST | Hubble Space Telescope |

## TABLE OF ABBREVIATIONS

| | |
|---|---|
| ICBM | intercontinental ballistic missile |
| ILS | International Launch Services |
| ISCCP | International Satellite Cloud Climatology Project |
| ISS | International Space Station |
| | |
| JPL | Jet Propulsion Laboratory |
| JSC | Johnson Space Center |
| | |
| KITSAT | Korea Institute of Technology satellite |
| KSC | Kennedy Space Center |
| | |
| LAGEOS | Laser Geodynamics Satellite |
| LARC | Langley Research Center |
| LASCO | Large Angle Spectrometric Coronagraph |
| LED | light-emitting diodes |
| LERC | Lewis Research Center |
| LFBB | Liquid Fly Back Booster |
| LMCSS | Lockheed Martin Commercial Space Systems |
| LoFLYTE | Low Observable Flight Test Experiment |
| LRIP | low-rate initial production |
| | |
| maglev | magnetic levitation |
| MIDEX | Medium-Class Explorer |
| MIRACL | Mid-Infrared Advanced Chemical Laser |
| MISR | Multi-Angle Imaging Spectroradiometer |
| MOD | Masters of Downloading |
| MODIS | Moderate-Resolution Imaging Spectroradiometer |
| MOLA | Mars Orbiter Laser Altimeter |
| MOPITT | Measurement of Pollution in the Troposphere |
| MSFC | Marshall Space Flight Center |
| MSL | Microgravity Science Laboratory |
| | |
| NACA | National Advisory Committee on Aeronautics |
| NASA | National Aeronautics and Space Administration |
| NASDA | National Space Development Agency of Japan |
| NEAR | Near Earth Asteroid Rendezvous |
| NGI | Next Generation Internet |
| NGSS | Next Generation Sky Survey |
| NGST | Next Generation Space Telescope |
| NICMOS | Near Infrared Camera and Multi-Object Spectrometer |
| NIMS | Near Infrared Mapping Spectrometer |
| NOAA | National Oceanic and Atmospheric Administration |
| NPOESS | National Polar-Orbiting Operational Environment Satellite System |
| NPP | NPOESS Preparatory Project |

## TABLE OF ABBREVIATIONS

| | |
|---|---|
| NPR | National Partnership for Reinventing Government |
| NRC | National Research Council |
| NRO | National Reconnaissance Office |
| NSTAR | NASA Solar Electric Propulsion Technology Application Readiness |
| | |
| OAST | Office of Aeronautics and Space Technology |
| OBPR | Office of Biological and Physical Research |
| OLSMA | Office of Life and Microgravity Sciences and Applications |
| OMB | Office of Management and Budget |
| | |
| PDRS | probe data relay subsystem |
| PETI-5 | Phenylethynyl Terminated Imide Oligomers, fifth composition |
| PIF-ATI | Planetary Imaging Facility and Advanced Training Institute |
| PKE | Pluto/Kuiper Express |
| POES | Polar-Orbiting Environmental Satellite |
| POLARIS | Photochemistry of Ozone Loss in the Arctic Region in Summer |
| PSA | Personal Satellite Assistant |
| PSLV | Polar Satellite Launch Vehicle |
| PST | Pacific Standard Time |
| | |
| QuikSCAT | Quick Scatterometer |
| | |
| RESTORE | Reconfigurable Control for Tailless Fighter Aircraft |
| RLV | reusable launch vehicle |
| ROSAT | Röntgensatellit |
| ROTSE | Robotic Optical Transient Search Experiment |
| RSIS | RS Information Services |
| RTG | Radioisotope Thermoelectric Generator |
| RXTE | Rossi X-ray Timing Explorer |
| | |
| SAC | Scientific Applications Satellite |
| SAFARI | Southern African Regional Science Initiative |
| SAFER | Simplified Aid for EVA Rescue |
| SCP | Spacecraft Control Processor |
| SELVS | Small Expendable Launch Vehicle Services |
| SHARP | Summer High School Apprenticeship Research Program |
| SIM | Space Interferometry Mission |
| SIR-C | Spaceborne Imaging Radar-C |
| SIR-C/X-SAR | Spaceborne Imaging Radar-C/X-band Synthetic Aperture Radar |
| SIRTF | Space Infrared Telescope Facility |
| SMEX | Small Explorer |
| SOHO | Solar and Heliospheric Observatory |
| SRB | solid rocket booster |
| STS | Space Transportation System |

## TABLE OF ABBREVIATIONS

| | |
|---|---|
| SUNSAT | Stellenbosch University, South Africa, satellite |
| SWAS | Submillimeter Wave Astronomy Satellite |
| | |
| TDRS | Tracking and Data Relay Satellite |
| THEMIS | Thermal Emission Imaging System |
| THUNDER | Thin-Layer Composite-Unimorph Piezoelectric Driver and Sensor |
| TIROS | Advanced Television Infrared Observation Satellite |
| TOMS | Total Ozone Mapping Spectrometer |
| TOMS-EP | Total Ozone Mapping Spectrometer Earth Probe |
| TRACE | Transition Region and Coronal Explorer |
| TRACE-A | Transport and Chemistry near the Equator of the Atlantic |
| TRMM | Tropical Rainfall Measuring Mission |
| TSS | tethered satellite system |
| | |
| USA | United Space Alliance |
| UV | ultraviolet |
| VIIRS | Visible Infrared Imaging Radiometer Suite |
| VLA | Very Large Array |
| VLBI | Very Long Baseline Interferometry |
| VLT | Very Large Telescope |
| | |
| WIRE | Wide-Field Infrared Explorer |
| | |
| XMM | X-ray Multi-Mirror scientific satellite |
| XTE | X-ray Timing Explorer |
| | |
| Y2K | Year 2000 |

# APPENDIX B

## BIBLIOGRAPHY

Anderson, J. D., W. L. Sjogren, and G. Schubert. "Galileo Gravity Results and the Internal Structure of Lo." *Science* 272, no. 5262 (3 May 1996): 709–712.

Anselmo, Joseph C. "NASA Funds Research for Shuttle Successor." *Aviation Week and Space Technology* 152, no. 7 (14 February 2000): 11–12.

Basri, Gibor, Subjanjoy Mohanty, France Allard, Peter H. Hauschildt, Xavier Delfosse, Eduardo L. Martin, Theirry Forveille, and Bertrand Goldman. "An Effective Temperature Scale for Late-M and L Dwarfs, From Resonance Absorption Lines of Cs I and Rb I." *Astrophysical Journal Letters* 538, no. 1 (20 July 2000): 363–385.

Berenji, H. R., Ping-Wei Chang, and S. R. Swanson. "Refining the Shuttle Training Aircraft Controller." *Fuzzy Systems, 1997: Proceedings of the Sixth IEEE International Conference* 2 (1–5 July 1997): 677–682.

Carlson, R. W., M. S. Anderson, R. E. Johnson, W. D. Smythe, A. R. Hendrix, C. A. Barth, L. A. Soderblom, G. B. Hansen, T. B. McCord, J. B. Dalton, R. N. Clark, J. H. Shirley, A. C. Ocampo, and D. L. Matson. "Hydrogen Peroxide on the Surface of Europa." *Science* 283, no. 5410 (26 March 1999): 2062–2064.

Carlson, R. W., R. E. Johnson, and M. S. Anderson. "Sulfuric Acid on Europa and the Radiolytic Sulfur Cycle." *Science* 286, no. 5437 (1 October 1999): 97–99.

Chakrabarty, Deepto, and Edward H. Morgan. "The Two-Hour Orbit of a Binary Millisecond X-ray Pulsar." *Nature* 394, no. 6691 (23 July 1998): 346–348.

Corliss, William R. "Development of the First Sounding Rockets." Chapter 4 in *NASA Sounding Rockets, 1958–1968: A Historical Summary*. Washington, DC: NASA Scientific and Technical Information Office, 1971.

Covault, Craig. "Shuttle/Spartan Verdict Sparks Station Concerns." *Aviation Week and Space Technology* 148, no. 16 (20 April 1998): 26.

Covault, Craig, and Pierre Sparaco. "French Astronaut Joins Russian/U.S. Mir Crew." *Aviation Week and Space Technology* 145, no. 9 (26 August 1996): 69–70.

Dornheim, Michael A. "Pathfinder Aircraft Hits 71,500 Ft." *Aviation Week and Space Technology* 147, no. 3 (21 July 1997): 40–42.

Dornheim, Michael A. "Pathfinder Surpasses Propeller Altitude Record." *Aviation Week and Space Technology* 146, no. 25 (16 June 1997): 53.

Dornheim, Michael A. "Unexpected Jovian Radiation Hits Galileo." *Aviation Week and Space Technology* 151, no. 8 (23 August 1999): 42.

Farquhar, James, Mark H. Thiemens, and Teresa Jackson. "Atmosphere-Surface Interactions on Mars: $\Delta^{17}O$ Measurements of Carbonate from ALH84001." *Science* 280, no. 5369 (5 June 1998): 1580.

Furness, Tim. "New Evidence Reveals Fire on Doomed Challenger's Booster." *Flight International*, 5–11 February 1997.

Glanz, James. "Chain of Errors Hurled Probe into Spin." *Science* 281, no. 5376 (24 July 1998): 499.

Gloeckler, G., J. Giess, N. A. Schwadron, L. A. Fisk, T. H. Zurbuchen, F. M. Ipavich, R. von Steiger, H. Balsiger, and B. Wilken. "Interception of Comet Hyakutake's Ion Tail at a Distance of 500 Million Kilometers." *Nature* 404, no. 6778 (6 April 2000): 576–579.

Gurnett, D. A., W. S. Kurth, A. Roux, S. J. Bolton, and C. F. Kennel. "Evidence for Magnetosphere at Ganymede from Plasmawave Observations by the Galileo Spacecraft." *Nature* 384, no. 6609 (12 December 1996): 535–538.

Hartmann, William, Michael Malin, Alfred McEwen, Michael Carr, Larry Soderblom, Peter Thomas, Ed Danielson, Phillip James, and Joseph Veverka. "Evidence for Recent Volcanism on Mars from Crater Counts." *Nature* 397, no. 6720 (18 February 1999): 586.

Hassler, Donald, Ingolf E. Dammasch, Philippe Lemaire, Pal Brekke, Werner Curdt, Helen E. Mason, Jean-Claude Vial, and Klaus Wilhelm. "Solar Wind Outflow and the Chromospheric Magnetic Network." *Science* 283, no. 5403 (5 February 1999): 810.

Head, James W., III, Harald Hiesinger, Mikhail A. Ivanov, Mikhail A. Kreslavsky, Stephen Pratt, and Bradley J. Thomson. "Possible Ancient Oceans on Mars: Evidence from Mars Orbiter Laser Altimeter Data." *Science* 286, no. 5447 (10 December 1999): 2134–2137.

Herman, J. R., P. Bhartia, J. Ziemke, Z. Ahmad, and D. Larko. "UV-B Increases (1979–1992) from Decreases in Total Ozone." *Geophysical Research Letters* 23, no. 16 (1996): 2117–2120.

Huygens Communications Link Enquiry Board. "Findings, Recommendations, and Conclusions." Report, 20 December 2000, 1–13. *http://klabs.org/richcontent/Reports/ Failure_Reports/ESA_Cassini/huygens_enquiry_board.PDF* (accessed 21 April 2008).

*Jet Magazine*. "Navy Captain Winston Scott Becomes Second Black To Walk in Space." 5 February 1996, 24.

Jones, Geraint H., Andre Balogh, and Timothy S. Horbury. "Identification of Comet Hyakutake's Extremely Long Ion Tail from Magnetic Field Signatures." *Nature* 404, no. 6778 (6 April 2000): 574–577.

Khurana, K. K., D. J. Stevenson, G. Schubert, C. T. Russell, R. J. Walker, and C. Polanskey. "Induced Magnetic Fields as Evidence for Subsurface Oceans in Europa and Callisto." *Nature* 395, no. 6704 (22 October 1998): 777–780.

Kirschvink, Joseph L., Altair T. Maine, and Hojatollah Vali. "Paleomagnetic Evidence of a Low-Temperature Origin of Carbonate in the Martian Meteorite ALH84001." *Science* 275, no. 5306 (14 March 1997): 1629–1634.

Kivelson, Margaret, Christopher T. Russell, Martin Volwerk, Raymond J. Walker, and Christophe Zimmer. "Galileo Magnetometer Measurements: A Stronger Case for a Subsurface Ocean at Europa." *Science* 289, no. 5483 (25 August 2000): 1340–1343.

Kosovichev, A. G., and V. V. Zharkova. "X-ray Flare Sparks Quake Inside Sun." *Nature* 383, no. 6683 (28 May 1998): 317.

Kouveliotou, C., S. Dieters, T. Strohmayer, J. van Paradijs, G. J. Fishman, C. A. Meegan, K. Hurley, J. Kommers, I. Smith, D. Frail, and T. Murakami. "An X-ray Pulsar with a Superstrong Magnetic Field in the Soft Gamma-Ray Repeater SGR1806-20." *Nature* 393, no. 6682 (21 May 1998): 235–237.

Lawrence, D. J., W. C. Feldman, B. L. Barraclough, A. B. Binder, R. C. Elphic, S. Maurice, and D. R. Thomsen. "Global Elemental Maps and the Moon: The Lunar Prospector Gamma-Ray Spectrometer." *Science* 281, no. 5382 (4 September 1998): 1484–1489.

Luu, Jane, Brian G. Marsden, David Jewitt, Chadwick A. Trujillo, Carl W. Hergenrother, Jun Chen, and Warren B. Offutt. "A New Dynamic Class of Object in the Outer Solar System." *Nature* 387, no. 6633 (5 June 1997): 573–575.

Malin, Michael C., and Kenneth S. Edget. "Sedimentary Rocks of Early Mars." *Science* 290, no. 5498 (8 December 2000): 1927–1937.

Maloney, Lawrence D. "Galileo Probe's Guardian Angel." *Design News* 52 no. 5 (3 March 1997): 74–83.

McCord, T. B., G. B. Hansen, F. P. Fanale, R. W. Carlson, D. L. Matson, T.V. Johnson, W. D. Smythe, J. K. Crowley, P. D. Martin, A. Ocampo, C. A. Hibbitts, J. C. Granahan, and the NIMS Team. "Salts on Europa's Surface Detected by Galileo Near Infrared Mapping Spectrometer." *Science* 280, no. 5367 (22 May 1998): 1242.

McEwen, Alfred S., Michael C. Malin, Michael H. Carr, and William K. Hartmann. "Voluminous Volcanism on Early Mars Revealed in Valles Marineris." *Nature* 397, no. 6720 (18 February 1999): 584.

McKay, David S., Everett K. Gibson Jr., Kathie L. Thomas-Keptra, Hojatollah Vali, Christopher S. Romanek, Simon J. Clemett, Xavier D. F. Chillier, Claude R. Maechling, and Richard N. Zare. "Search for Past Life on Mars: Possible Relic Biogenic Activity in Martian Meteorite ALH84001." *Science* 273, no. 5277 (16 August 1996): 924–930.

Meltzer, Michael. *Mission to Jupiter: A History of the Galileo Project.* Washington, DC: NASA, 2007.

Mojzsis, S. J., G. Arrhenius, K. D. McKeegan, T. M. Harrison, A. P. Nutman, and C. R. L. Frien. "Evidence for Life on Earth Before 3,800 Million Years Ago." *Nature* 384, no. 6604 (7 November 1996): 55–60.

NASA. *Aeronautics and Space Report of the President: Fiscal Year 1996 Activities.* Washington, DC, 1997.

NASA. *Aeronautics and Space Report of the President: Fiscal Year 2000 Activities.* Washington, DC, 2001.

NASA. "U.S. and Russian Human Space Flights 1961–September 30, 2000," in *Aeronautics and Space Report of the President: Fiscal Year 2000 Activities.* Washington, DC, 2001.

NASA Advisory Council. "Report on the Cost Assessment and Validation Task Force on the International Space Station." Washington, DC, 21 April 1998. *http://history.nasa.gov/32999.pdf* (accessed 24 September 2007).

NASA Office of Inspector General. "Assessment of the Triana Mission." Final report G-99-013, Washington, DC, 10 September 1999. *http://oig.nasa.gov/old/inspections_assessments/g-99-013.pdf* (accessed 3 March 2008).

NASA Office of Inspector General. "NASA Oversight of Russian Biotechnology Research 1994–1997." NASA report G-00-07, Washington, DC, 13 October 2000. *http://oig.nasa.gov/old/inspections_assessments/g-00-007.pdf* (accessed 8 April 2008).

Owen, Tobias, Paul Mahaffy, H. B. Niemann, Sushil Atreya, Thomas Donahue, Akiva Bar-Nun, and Imke de Pater. "A Low-Temperature Origin for the Planetesimals that Formed Jupiter." *Nature* 402, no. 6759 (18 November 1999): 269–270.

Phillips, Edward H. "Overmyer Dies in Crash." *Aviation Week and Space Technology* 144, no. 14 (1 April 1996): 29.

Piro, L., G. Garmire, M. Garcia, G. Stratta, E. Costa, M. Feroci, P. Meszaros, M. Vietri, H. Bradt, D. Frail, F. Frontera, J. Halpern, J. Heise, K. Hurley, N. Kawai, R. M. Kippen, F. Marshall, T. Murakami, V. V. Sokolov, T. Takeshima, and A. Yoshida. "Observation of X-ray Lines from a Gamma-Ray Burst (GRB991216): Evidence of Moving Ejecta from the Progenitor." *Science* 290, no. 5493 (3 November 2000): 955–958.

Sawyer, Kathy. *The Rock from Mars: A Detective Story on Two Planets*. New York: Random House, 2006.

Smith, Bruce. "Fast Launched Successfully." *Aviation Week and Space Technology* 145, no. 9 (26 August 1996): 71.

Smith, David E., Maria T. Zuber, Sean C. Solomon, Roger J. Phillips, James W. Head, James B. Garvin, W. Bruce Banerdt, Duane O. Muhleman, Gordon H. Pettengill, Gregory A. Neumann, Frank G. Lemoine, James A. Abshire, Oded Aharonson, C. David Brown, Steven A. Hauck, Anton B. Ivanov, Patrick J. McGovern, H. Jay Zwally, and Thomas C. Duxbury. "The Global Topography of Mars and Implications for Surface Evolution." *Science* 284, no. 5419 (28 May 1999): 1495–1503.

Sonett, C. P., E. P. Kvale, A. Zakharian, Marjarie A. Chan, and T. M. Demko. "Late Proterozoic and Paleozoic Tides, Retreat of the Moon, and Rotation of the Earth." *Science* 273, no. 5271 (5 July 1996): 100–104.

Stern, Robert J., and Mohamed Gamal. "The Origin of the Great Bend of the Nile from SIR-C/X-SAR Imagery." *Science* 274, no. 5293 (6 December 1996): 1696–1698.

Thomas, Peter C., Richard P. Binzel, Michael J. Gaffey, and Alex D. Storrs. "Impact Excavation on Asteroid 4 Vesta: Hubble Space Telescope Results." *Science* 277, no. 5331 (5 September 1997): 1492–1495.

U. S. Air Force Headquarters. *The Roswell Report: Case Closed*. Washington, DC: GPO, 1997.

U.S. General Accounting Office. "Export Controls: International Space Station Technology Transfers." Report no. NSIAD-00-14, Washington, DC, 3 November 1999. *http://www.gao.gov/archive/2000/ns00014.pdf* (accessed 21 March 2008).

U.S. General Accounting Office. "Information Security: Many NASA Mission-Critical Systems Face Serious Risks." Report no. GAO/AIMD-99-47, Washington, DC, May 1999. *http://www.gao.gov/archive/1999/ai99047.pdf* (accessed 7 February 2008).

U.S. General Accounting Office. "Space Shuttle: Human Capital and Safety Upgrade Challenges Require Continued Attention." Report no. GAO/NSIAD/GGD-00-186, Washington, DC, August 2000. *http://www.gao.gov/archive/2000/n200186.pdf* (accessed 9 May 2008).

U.S. General Accounting Office. "Space Station: Russian Commitment and Cost Control Problems." Report no. GAO/NSIAD-99-175, Washington, DC, August 1999. *http://www.gao.gov/archive/1999/ns99175.pdf* (accessed 25 February 2008).

U.S. General Accounting Office. "Space Station U.S. Life-Cycle Funding Requirements: Statement of Allen Li, Associate Director, Defense Acquisitions Issues, National Securit and International Affairs Division." Testimony before the House Committee on Science and Technology. Document no. GAO/T-NSIAD-98-212, 24 June 1998.

U.S. General Accounting Office. "Space Surveillance: DOD and NASA Need Consolidated Requirements and a Coordinated Plan." Report no. NSIAD 98-42, Washington, DC, 1 December 1997.

U.S. General Accounting Office. "Space Transportation: Status of the X-33 Reusable Launch Vehicle Program." Report no. NIASD-99-176, Washington, DC, 11 August 1999. *http://www.gao.gov/archive/1999/ns99176.pdf* (accessed 4 March 2008).

Wijnands, Rudy, and Michiel van der Klis. "A Millisecond Pulsar in an X-ray Binary System." *Nature* 394, no. 6691 (23 July 1998): 344–346.

Zuber, Maria T., David E. Smith, Sean C. Solomon, James B. Abshire, Robert S. Afzal, Oded Aharonson, Kathryn Fishbaugh, Peter G. Ford, Herbert V. Frey, James B. Garvin, James W. Head, Anton B. Ivanov, Catherine L. Johnson, Duane O. Muhleman, Gregory A. Neumann, Gordon H. Pettengill, Roger J. Phillips, Xiaoli Sun, H. Jay Zwally, W. Bruce Banerdt, and Thomas C. Duxbury. "Observations of the North Polar Region of Mars from the Mars Orbiter Laser Altimeter." *Science* 282, no. 5396 (11 December 1998): 2053.

Zuber, Marie T., Sean C. Solomon, Roger J. Phillips, David E. Smith, G. Leonard Tyler, Oded Aharonson, Georges Balmino, W. Bruce Banerdt, James W. Head, Catherine L. Johnson, Frank G. Lemoine, Patrick J. McGovern, Gregory. A. Neumann, David D. Rowlands, and Shijie Zhong. "Internal Structure and Early Thermal Evolution of Mars from Mars Global Surveyor Topography and Gravity." *Science* 287, no. 5459 (10 March 2000): 1788–1893.

The writers used NASA news releases, contract announcements, policy directives, NASA's *HQ Bulletin*, *Spacewarn Bulletin*, and *NASA Daily News Summary*; news releases of aeronautics industry companies and universities, such as Ball Aerospace & Technologies Corporation, The Boeing Company, Cordant Technologies, Lockheed Martin, Medialink Worldwide, SkyCorp, SPACEHAB, California Institute of Technology, and Cornell University; news articles of wire services, including Africa News Service, Agence France-Presse, Armed Forces Newswire Service, Associated Press, Bloomberg News, Business Wire, Dow Jones Newswire, Federal Document Clearing House, Gannett News Service, Knight-Ridder News Service, PR Newswire, Reuters, Tribune News Services, United Press International, and U.S. Newswire; newspaper articles from *Aerospace Daily, Aerospace News, Albuquerque Journal, Antelope Valley Press, Arizona Daily Star, Atlanta Journal, Baltimore Sun, Birmingham News, Boston Globe, Boston Herald, Buffalo News, Charleston Gazette, Chicago Tribune, Christian Science Monitor, Cleveland Plain Dealer, Dayton Daily News, Defense Daily, Denver Post, Des Moines Register, Deseret News, Federal Computer Week, Financial Times, Flight International, Florida Today, Houston Chronicle, Huntsville Times, Kansas City Star, Los Angeles Daily News, Los Angeles Times, Miami Herald, New Orleans Times-Picayune, New York Times, Newsday, Oregonian, Orlando Sentinel, Palm Beach Post, Press Democrat, Roanoke Times, Rocky Mountain News, Russia Today, Sacramento Bee, Salt Lake Tribune, San Diego Union-Tribune, San Francisco Chronicle, San Jose Mercury News, Seattle Post-Intelligencer, Seattle Times, Space Business*

*News*, *Space Flight Now*, *Space News*, *St. Petersburg Times*, *Star-Ledger*, *Sun-Sentinel*, *Sunday Times*, *Times-Dispatch*, *UniSci Science and Research News*, *USA Today*, *Virginian-Pilot*, *Virginian-Pilot and the Ledger-Star*, *Wall Street Journal*, *Washington Post*, and *Washington Times*; and television news reports from CNN.com and BBC News.